Agrarian Policies in Communist Europe

Studies in East European and Soviet Russian
Agrarian Policy: *Volume I*

Studies in East European and Soviet Russian Agrarian Policy
  Alec Nove, general editor

VOLUME I
Agrarian Policies in Communist Europe:
A Critical Introduction
by Karl-Eugen Wädekin

VOLUME II
The Organization of Agriculture in the Soviet Union and
Eastern Europe
edited by Everett M. Jacobs

VOLUME III
Agriculture in Eastern Europe and the Soviet Union:
Comparative Studies
edited by Karl-Eugen Wädekin

Agrarian Policies in Communist Europe

A Critical Introduction

by
KARL-EUGEN WÄDEKIN

Edited by Everett M. Jacobs

Allanheld, Osmun *Publishers*
Martinus Nijhoff *Publishers*
The Hague/London

ALLANHELD, OSMUN & CO. PUBLISHERS, INC.

Published in the United States of America in 1982
by Allanheld, Osmun & Co. Publishers, Inc.
(A Division of Littlefield, Adams & Company)
81 Adams Drive, Totowa, New Jersey 07512

Copyright © 1982 by Karl-Eugen Wädekin

Published in The Netherlands in 1982
by Martinus Nijhoff Publishers
c/o Distribution Centre
P.O. Box 322
3300AH Dordrecht, The Netherlands
ISBN: 90 247 2572 0

Library of Congress Cataloging in Publication Data

Wädekin, Karl Eugen.
  Agrarian policies in communist Europe.

  (Studies in East European and Soviet Russian
agrarian policy: v. 1)
  Bibliography: p.
  Includes index.
  1. Agriculture and state—Europe, Eastern.
2. Agriculture—Economic aspects—Europe, Eastern.
I. Title. II. Series.
HD1918 1981.W33   338.1'847   79-55000
ISBN 0-916672-40-9   AACR2

82 83 84 / 10 9 8 7 6 5 4 3 2 1

Printed in the United States of America

Contents

List of Tables

Table

Foreword

The perennial crisis in Communist agriculture and the distinctive policies and forms of agricultural organization in the Soviet Union and Eastern Europe have drawn considerable attention in the West over the years. Up to now, most books on Soviet and East European agriculture have been either studies of individual countries, or else edited collections of usually unrelated contributions on particular aspects of agriculture in the individual countries. What comparative works have appeared have tended to be articles focusing on selected areas of agricultural policy. As valuable as these efforts have been, it has been difficult, if not impossible, for the nonspecialist or even the specialist to gain an overall picture of developments. In recent years, there has been an increasing trend among most East European Communist countries to modify the Soviet model of agricultural policy and organization to suit their own conditions more closely, but until now, no study has existed to pinpoint the areas of diversity and similarity or to put the changes into an area-wide or East-West perspective.

This remarkable book by Karl-Eugen Wädekin is the first attempt to treat agrarian policies in Communist Europe in a comprehensive, systematic, cross-national way. At long last, we have a full-length study combining a deep comparative knowledge of agriculture and agricultural policies with a thorough understanding of the political, social, and economic systems of the countries involved. Going far beyond mere description, the author presents a cogent analysis and interpretation of developments. At the same time, his clear and concise exposition makes his subject accessible and understandable to the nonspecialist, for whom the book is basically designed. But the specialist, too, will find much of interest in this work. Apart from the careful scholarship and perceptive use of a wealth of source material (only to be expected by those who have read his monumental work, *The Private Sector in Soviet Agriculture* and his other books and many articles on Soviet and East European agriculture), this study covers a very wide range of topics and countries, and presents many new insights.

In the text itself, the author's main contribution relates to the period since the 1960s, especially his analysis of performance and the factors of agricultural production, labor incomes, agricultural producer prices, the reforms of the 1960s, and the progress of agro-industrial integration. However, this is not to minimize the importance of the extensive bibliography of English-language publications, which should facilitate further research into comparative, and individual countries', agricultural policies.

In its approach and scope, this book opens new horizons in the study of Soviet and East European agriculture. I am confident that it will quickly establish itself as the standard reference for those interested in comparative agricultural policy in Communist Europe, and I am most pleased and honored to have been associated with it as editor.

Everett M. Jacobs

March 1981
University of Sheffield, England.

Author's Acknowledgments

Work on the present book was started as part of the comparative project "East European and Soviet Agrarian Policies", as conceived by Alec Nove (University of Glasgow) and financed by the Ford Foundation. However, contrary to the original intention, the author wrote two volumes in German, which did not fit into the project. Only much later he continued writing the initial version, which now is based on the German book, but shortened, updated, and changed in many ways. This work was greatly helped by a grant of the German Fulbright Commission, enabling the author to complete a first draft during the winter of 1978/79 in the stimulating intellectual atmosphere of the Kennan Institute of the Wilson International Center for Scholars, Washington, D. C.

The final book could not have been completed without the advice and critical judgment of a number of colleagues who specialize in the various countries covered. For this reason, the writer wishes to express his gratitude towards Endre Antal, Vladislav Bajaja, the late Eberhard von Dalwig-Nolda, Stanislawa Hegenbarth, Edeltrude Hudea, Günter Jaehne, Ivan Loncarević, and Eberhard Schinke (who all are, or for a time were, working at the Zentrum für kontinentale Agrar- und Wirtschaftsforschung at the Justus Liebig University, Giessen, West Germany), and also towards Dionisie Ghermani (Munich), Christian Krebs (West Berlin), Konrad Merkel (West Berlin), Constantin Sporea (Munich), Bernhard Tönnes (Munich), and Paul Wiedemann (Vienna). The comprehensive archives of Radio Free Europe and Radio Liberty (Munich) and the assistance rendered by their staff in searching for figures, facts, and sources were of invaluable help. It is appropriate specially to thank Everett M. Jacobs (Sheffield) who contributed advice on a very early version of the first chapters and, above all, did an admirable job in not only correcting and polishing the author's English but also by offering criticism, comments and suggestions for the final whole. Apart from his being the editor, he acted as a competent colleague and friend.

The various helpful minds made the author come to grips with his wide and many-sided subject, but any remaining errors or misjudgments are, of course, entirely his own.

March 1981   K.-E. W.
University of Giessen

Introduction

To STUDY Communist agrarian policies mainly from the view of the socialization of agricultural production was clearly justified in the past. After all, collectivization of agriculture, together with the establishment of a sizable state farm sector, has more or less become a synonym for Communist agrarian policy. Yet such an identification has not always been the unanimous theoretical opinion among nineteenth-century Socialists or twentieth-century Communists. Nor is collectivization presently being emphasized by Communists with regard to non-Communist countries, where Communists have considerable influence or may hope to gain it. Nevertheless, where Communists have obtained supreme power, in practice they have soon proceeded to socialize, that is, to collectivize and/or nationalize (turn into government property), the farm sector. Where they did not persevere with such a policy, this was due to specific economic and political circumstances, as in Poland and Yugoslavia. However, decollectivization in both cases was considered a reluctant temporary retreat, not forsaking the ultimate goal of "the socialist transformation of agriculture."

Another anomaly in the political practice of implementing collective farming has been the continued existence of a private sector of mini-plot and animal farming (including a few remaining small farmers) in all Communist countries after collectivization. This, too, was a rather reluctant concession to peasant psychology and, more important, resulted from the shortcomings of the large socialized farms, especially in labor-intensive branches such as animal, fruit, potato, and other vegetable production. This concession is explicitly part of the official policy today, although it was not always so in the past. Nowadays it is held that the private sector still has an important economic function and therefore will not have to disappear in the immediate future. In the long term, however, Communist theory has doomed this sector for dissolution, to be brought about either by economic forces or with the help of the policy makers, except that its quality as an economically unimportant sphere for spending leisure time remains undisputed.

Where the socialization of agriculture has already been achieved, it can no longer be a policy goal, although to persevere with the achieved ownership relations and the corresponding organization of agriculture remains a policy object of high priority. It has a parametrical function for other measures on which the attention of political and economical

1

decision makers now is increasingly focused. The new focus clearly is on productive performance in the first place, and on improving the living standards in the countryside in the second. The fast pace of industrialization, combined with repressed and now increasingly more acute consumer demand due to rising incomes, has resulted in a situation where the supply of high-quality food does not meet that demand at given prices. Output growth of such food is therefore of paramount importance.

In the Communist countries, politicians and specialists increasingly realize that past agrarian policies and the resulting unsatisfactory productive performance have become the Achilles heel of their socioeconomic systems, making them inferior in an important aspect of "peaceful competition" and even dependent on Western food supplies. A Soviet author recently admitted:

> The well-known lag of the level of development of the productive forces of agriculture of the socialist countries, compared with the ranks of the developed capitalist states, is explained by *objective* reasons. First of all, there were the economic and political conditions in which the socialist countries found themselves. The low general initial level of economic development of the majority of them demanded in the first instance the creation of a powerful industry. The construction of socialism in [conditions of] encirclement by aggressive imperialist states caused the necessity of diverting significant resources for defense. The insufficiently rapid development of the technical base of agriculture in the socialist countries is explained also by *subjective* factors: *by mistakes and errors* in the leadership of the branch, which were often connected with *non-observance of the demands of objective economic laws* [emphasis added].[1]

The change of agrarian policy goals has made itself felt since the early 1960s, when the socialization of agriculture was achieved in all European Communist countries except Poland, Yugoslavia, and—until 1967—Albania, and when economic reforms were soon inaugurated in the nonagricultural sectors. Among other things, the reforms provided for more material labor incentives, and thereby for rapidly rising expectations of wage earners. The new period also brought, on the common basis of achieved socialization along Soviet lines, a greater differentiation of the organization of agricultural production. This was so because it was necessary not only to raise gross output, but also to make the utilization of the limited land, labor, and capital resources more efficient.

Thus, to achieve a current understanding of agrarian policies in Communist Europe, it is necessary to go beyond the study of the socialization of agricultural production to examine the comparative productive performance of Communist agriculture. It is equally necessary to investigate generally the means by which these Communist governments in Eastern Europe and the Soviet Union operated in the past and are operating at present within the framework of their political-ideological system. In the present book, a short introductory outline will be given of the historical roots of Marxist-Leninist agrarian policies and of the leading role the Soviet Union assumed in their Russian and Stalinist application after the October Revolution. Chapters 3 and 5 then deal with the expansion of Russian domination over Eastern Europe after World War II and the application of the Soviet model in that area, as well as with the general uniformity and partial diversity developing during the 1950s. The main emphasis of this study, though, will be on the agrarian policies and their results under the conditions of the various European Communist countries after 1962 and up to the end of the 1970s.

Notes

1. M. E. Bukh, *Problemy effektivnosti sel'skogo khozyaistva v evropeiskikh stranakh SEV* (Moscow, 1978), p. 17, n. 16.

1. The Agrarian Question in Marxist and Leninist Thought

KARL MARX's and Friedrich Engel's perspective of the future was concerned almost exclusively with the development of industrial society and its revolutionary transition to Communist society. They had little to say on the agrarian question. Famous is their view on the "idiocy of rural life" in the Manifesto of the Communist Party (1848). Also, the following excerpts from a characteristic passage of that document show how they assessed the role of the peasant in the revolution of the proletariat.

> Of all the classes that stand face to face with the bourgeoisie today, the proletariat alone is a really revolutionary class. . . . The lower middle class, the small manufacturer, the shop-keeper, the artisan, *the peasant,* all these fight against the bourgeoisie, to save from extinction their existence as fractions of the middle class. They are therefore not revolutionary, but conservative. . . . If by chance they are revolutionary, they are so only in view of their impending transfer into the proletariat; they thus defend not the present but their future interests [emphasis added]. . . .

Yet the founders of Marxism did at times spell out their conviction that the proletarian revolution would bring about large-scale production in agriculture after the industrial model. They did not, however, consider the "concentration of the means of production" in agriculture under capitalism to be an indispensable precondition for establishing the Communist order of society. This concentration was to take place in industry and was to lead to revolution in the urban and industrial centers. Differing from Proudhon, Marx and Engels had no perspective on the inherently "conservative" peasant way of life and production between the (at the time) progressive capitalist way and the more progressive way of a future socialized economy. To them, the peasants played a secondary role at best. If capitalist concentration of production did not already apply to agriculture, such concentration in some way or other, preferably by cooperation, was to spread out to the countryside after the proletarian revolution. The question as to which size of enterprise might be optimal for agriculture was not empirically explored.

Engels showed interest in winning over the rural proletariat (landless agricultural workers) to the cause of Marxism. Certain successes of Communist propaganda on the German and French countryside and the fact that the agrarian question was becoming a much disputed issue

4

among Socialists, prompted him in 1894, one year before his death, to make the following statements in his "The Peasant Question in France and Germany." They concerned the correct attitude towards the estate-owners and the segments of the peasantry after a successful Socialist revolution.

> I flatly deny that the Socialist workers' party of any country is charged with the task of taking into its fold, in addition of the rural proletarians and small peasants, also the middle and big peasants and perhaps even the tenants of big estates, the capitalist cattle breeders and other capitalist exploiters of the national soil. To all of them land-owning feudalism may appear to be a common foe. On certain questions we may make common cause with them and be able to fight side by side with them for definite aims. We . . . have no use whatever for any groups representing capitalist, middle-bourgeois or middle-peasant interests. . . .
>
> Secondly, it is just as evident that when we are in possession of state power we shall not even think of forcibly expropriating the small peasants (regardless of whether with or without compensation), as we shall have done in the case of the big landowners. Our task relative to the small peasant consists, in the first place, in effecting a transition of his private enterprise and private possession to cooperative ones, not forcibly but by dint of example and the profer of social assistance for this purpose. And then of course we shall have ample means of showing the small peasant prospective advantages that must be obvious to him even today. . . .
>
> As soon as our party is in possession of political power it has simply to expropriate the big landed proprietors just like the manufacturers in industry. . . . The big estates thus restored to the community are to be turned over by us to the rural workers who are already cultivating them and are to be organised into cooperatives. They are to be assigned to them for their use and benefit under the control of the community. Nothing as yet be stated as to the terms of their tenure. . . . And the example of these agricultural cooperatives would convince also the last of the still resistant smallholding peasants and surely also many big peasants, of the advantages of cooperative, large-scale production.

Two things should be stated here, in view of what was later to be Lenin's policy toward the peasants. First, Marx and Engels, as well as the German Social Democrats of 1891, rejected any promise to peasants, in order to revolutionize them, that the revolution would hand over to them as individual producers the land of the estates. Indeed, such a promise would have been of minor importance in a fully industrialized country, where the proletariat formed the majority of the population, such as they foresaw for the revolution to succeed. Secondly, and no less important for the interpretation of later events, the founders of Marxism did not give thought to the question of what should be done by the victorious proletariat if the peasants refused to be allured by the advantages of cooperative production. Yet, as the peasants were not the bearers of the proletarian revolution and its ideology, it is hard to see why the socialist way of life of the proletariat should seem alluring to them, not to speak of the questionable material advantages.

Neither early Marxist writers nor Party resolutions drew a consistent distinction between the form of ownership in agriculture (private or social) and the organization of production (small or large farms). In fact, one may well imagine large-scale production based on private ownership not only of big capitalists or feudal landlords, but also of small peasants remaining as individual owners but cooperating to form large units of production, either with their whole farms or with specialized branches of their production. Similarly, with social or state ownership of the land, even small-scale peasant production does not seem an impossible thing for a socialist state. Such small-scale production was dominant in Soviet Russia in 1918–28. However, it might elso be imagined on the basis of a stricter form of dependency, that is tenant enterprise.

In 1868, the majority (against the Proudhonists) of the Third Congress of the First International passed a resolution, that was reemphasized one year later by the Fourth Congress. It demanded socialization of the land and its renting out to agricultural cooperatives. This meant large-scale, not small-scale, production. In the German Social Democratic Party, the attitude toward the peasantry was extensively discussed at Party congresses during the early 1890s, and in the end, the line of the Erfurt Program of 1891 remained valid. This Program regarded the small peasant farm as already doomed for ruin under capitalism, and foresaw a future for the peasant only if he accepted his transition into the proletariat, that is, into a class of workers on large farms.

There was also the question of the peasantry (as distinct from the landless agricultural workers) being a potential ally, although not part of the proletariat. Marx and Engels made a number of passing remarks on this subject, but it was only in his "The Peasant Question in France and Germany" that Engels did so specifically and in some detail. Karl Kautsky similarly considered the peasants as potential *temporary* allies, although he remained convinced that small-scale peasant farming was doomed. The question of a closer alliance was discussed, but in the end, the line advocating such an alliance was rejected.

The great debate of the early 1890s concerning the agrarian question also strongly influenced the emergence of Marxist revisionism. The impulse came from the exigencies of practical party policy. Eduard Bernstein, theorist and the "father of revisionism," joined in only somewhat later. Yet after Kautsky's victory over the Revisionists at the Party Congress of the German Social Democrats in 1895, the agrarian question was no longer an issue, or rather, it was decided along dogmatist lines. Most other social-democratic parties sooner or later followed the lead. The issue of an agrarian program was taken up again only after World

War I, when Lenin's Communism was no longer part of Social-Democracy. Thus, the later stage of the social-democratic discussion is not relevant for an account of Communist agrarian policy. For geographical reasons, the more recent stands of Communist Parties in Western and Third World countries are also beyond the scope of the present book.

In Eastern Europe and Russia, the Marxist movement did not gain importance until World War I, but Russia constituted a special case. This was so not only because of Russia's sheer size but also because of special inherited features of her rural sector. There existed, side by side with feudal land ownership, the *mir* and the *obshchina*, traditional village community institutions with communal land ownership through periodic repartition in much of the country. These were not as old, though, and in fact not a remnant of "primitive communism," as some thought. After the liberation of the Russian peasants from serfdom in 1861 and following years, there were still many remnants of feudalism, including latifundia and the administrative system, though capitalism and individual ownership of land increasingly penetrated the rural system. The level of agricultural methods remained low, the yields were correspondingly meager. The small farms were impoverished, and the rural population increased rapidly. All this made large segments of the Russian peasantry long ever more for the land of the estates to be divided among them. Moreover, part of the nonagricultural working class shared these aspirations, as many workers in the expanding industries were of rural origin and preserved their ties with the village community, often returning to it seasonally or after a number of years. Under these circumstances, it was the Russian peasantry who, to a certain extent, backed the Bolsheviks during the Civil War that ensued after October 1917 and thus saved the Revolution from extinction. The peasantry gained the land of the estates from the Revolution, only to lose it by force a decade later in Stalin's "communist course" of collectivization.

Lenin combined two elements of the nineteenth-century Marxist debates on the agrarian question into a theory and a practice well adapted to the underdeveloped, predominantly peasant countries of Eastern Europe and especially Russia: radical revolutionism, and an "alliance" of the proletariat with the peasants by granting them for a time what they most wanted—individual land ownership. Engels, in his old age, had to a certain degree anticipated Lenin's alliance theory by declaring (in "The Peasant Question in France and Germany") that the agrarian policy of the socialists should be a means for seizing power with the help of the small peasants, either by elections or by revolutionary force. For Lenin, the actual achievement of revolution in backward Russia was the primary goal to which all other considerations were subordinate. As he believed that this revolution would, in Marx's and Engels's words (foreword to Plekhanov's Russian edition of the Communist Manifesto, 1882), "sound the signal for a proletarian revolution in the West," his trend of thought was not far from what they had deemed possible. But Lenin did not connect this with a possible preservation of the old *mir* community, as Marx and Engels had done. His ideas, and especially the "alliance" concept, were later to become the guiding principle of Communism (Marxism-Leninism) with regard to the agrarian question. In political practice, however, Communist agrarian policy has·deviated from Lenin, as will be shown.

The young Lenin had visited Western Europe for the first time and stayed in Berlin in 1895, just as the debate on an agrarian program for Marxism reached its climax. Already then he stated that for Russia he did not agree with Kautsky and the dogmatic outcome of the debate. His "Draft Program," written in 1895/96, for the not yet existing Social Democratic Party of Russia, contained demands in Section E endorsing the main peasant aspirations of the time: land and the reduction of payments and taxes. In his second Draft Program (1899) he stated:

> The peasant question in Russia differs substantially from the peasant question in the West, the *sole* difference being that in the West the question is almost exclusively one of a peasant in a capitalist, bourgeois society, whereas in Russia it is one of a peasant who suffers no less (if not more) from *pre-capitalist* institutions and relations, from the *survivals of serfdom.*

Lenin was against any protection or support of the small peasant and landowner, but he nevertheless insisted

> that the working-class party should inscribe on its banner *support* for the peasantry (*not by any means* as a class of small proprietors or small farmers), *in so far as the peasantry is capable of a revolutionary struggle against the survivals of serfdom in general and against the autocracy in particular* . . . it would be senseless to make the peasantry the *vehicle* of the revolutionary movement, that a party would be insane to *condition* the revolutionary character of its movement upon the revolutionary mood of the peasantry. . . . We say only that a working-class party cannot, without violating the basic tenets of Marxism and without committing a tremendous political mistake, *overlook* the revolutionary elements that exist among the peasantry and not afford these elements support [emphasis in original].

Such an "alliance" was meant to be temporary, it is true. However, under Russian conditions of the time—a backward economy and a great part of the peasantry ready for revolt—it could be more effective and last longer than in capitalist Western Europe, without violating Marxist principles. Lenin at the time held the opinion that the revolution should and would usher in a capitalist development of the economy and society and that only then a truly proletarian and socialist revolution would follow, effected by a proletariat greatly strengthened by the preceeding events. When, in 1902, he drafted a special agrarian program along the same lines, he explained, "We, Russian Social-Democrats, will also try to make some use of the experience of Europe, and begin to attract the 'country-folk' to the socialist working-class movement at a much earlier stage and much more zealously than was done by our Western comrades. . . . [We] shall take much that is ready-made 'from the Germans', but in the agrarian sphere we may perhaps evolve something new." Lenin remained in the framework of truly Marxist thinking, but took up and emphasized some ideas of Marx and Engels that had been neglected by Western Social Democrats, and at the same time somewhat simplified matters.

It was essential for Lenin's way of reasoning—in accordance with Marx and Engels—to differentiate between the various social classes and layers within the peasantry. This he did explicitly in his book "The Development of Capitalism in Russia," written 1896–99, and also in his extensive article "The Agrarian Program of Social-Democracy in the First Russian Revolution," written in late 1907. Apart from the landless workers and the owners of feudal latifundia, he saw three categories: the overwhelming majority of "small ruined peasants," the "middle peasants," and the "wealthy peasants and bourgeoisie." He held that the middle group was decreasing, most of them joining the small peasants, while a minority of them became wealthy peasants and rural bourgeois. Thus, in his view, class fronts would become clearer and the small peasant, as a future rural proletarian, had interests in common with the proletariat.

In his "Agrarian Program" article, he dwelt at great length on what should be done with the agricultural and forest land after a victorious revolution. Among the possible roads to be taken—repartitioning among peasants, municipalization, and nationalization—he advocated the latter. Socialization was not considered, as the expected revolution was a "bourgois-democratic" one, with a socialist and proletarian one to follow at a later stage. Nationalization to Lenin meant that the state—not yet a socialist one—should be the owner of all land, but that this land should be rented out to any "direct tiller." He called it a "grave error" to think that "nationalization has anything to do with socialism."

For the sake of the alliance of proletariat and peasantry, the mood of the more radical peasants was decisive in Lenin's eyes. In his words, "the beginnings of capitalism in landlord economy [which was a characteristic of Russian agriculture at that time] can and must be sacrificed to the wide and free development of capitalism on the basis of renovated small farming." When, afterwards, these capitalist peasants will demand "that the holdings they have rented from the state be converted into their property . . . the proletariat will uphold the revolutionary tradition against all such strivings and will not assist them." The heart and the motivation of Lenin's reasoning was not hidden: "It is undoubtedly in the interests of the working class to give the most vigorous support to the peasant revolution. More than that: it must play the leading part in that revolution." He left no doubt, though, as to the temporary nature of the "alliance" and the change of socio-economic policies after the first (the "bourgeois-democratic") revolution, and spelled that out very distinctly in 1918 in "The Proletarian Revolution and the Renegade Kautsky." In a second stage the Communists together with the poorest peasants and half-proletariat would fight the rural rich, the *kulaks,* and the speculators, and only then the revolution would become socialist.

Lenin's attitude remained consistent after he formed his opinion on the agrarian question in the mid- and late 1890s, and it was not kept secret. But if it is true, as Stalin later reported (in the foreword to the first volume of his own collected works), that at the very beginning of the twentieth century Lenin envisaged a very short lapse between the "bourgeois-democratic" and the "socialist" revolution, he seems to have deliberately concealed this intended timing from those concerned. This may have been part of his overall revolutionary tactics and must, in retrospect, change the true meaning of the earlier statements. That Lenin's agrarian politics in practice sometimes varied from those implied in his earlier writings was perhaps a consequence of the fact that only half a year elapsed between the "bourgeois-democratic" (February 1917) and the "socialist" (October 1917) revolutions in Russia. Thus, what had been promised to the peasant was bound to be taken away soon afterwards.

Just how this was to be effected, whether by persuasion, economic incentives, or force, was now the paramount question. In the pamphlet against Kautsky, Lenin seems to have envisaged a certain voluntary nature of the process and a gradualistic approach, but in very indefinite terms. Moreover, the issue became intertwined with the "grain problem," that is, with decreasing grain production and marketings. Thus, creating state and collective farms at that early time was also intended as a means of increasing food deliveries to the state.

One of the very first steps of the new government after the October Revolution was the famous decree of 8 November (26 October, old calender) 1917, "Concerning the Land," which in practice reintroduced the allotment system with periodical reapportionment of the land among peasants and abolished other private ownership rights. In this, Lenin essentially adopted the agrarian program of the Socialist-Revolutionaries, the numerically strong radical party representing mainly the peasants. The land was declared the property of all the people, to be used by those who actually tilled it: selling, renting, and mortgaging of land was prohibited. But in a speech on the day of the decree, Lenin also said, "Of course, it will not be possible to make a transition everywhere at once to collective farming. . . . But the more the peasantry is persuaded by example, by its own experience, of the superiority of communes, the more successful will the movement become."

A few months later, on 19 February 1918, with only the left wing of the Socialist-Revolutionaries still participating in the government, another decree followed "On the Socialization of the Land." In one of its articles (11e), "the development of collective farming" was envisaged without detailed instruction. Article 13 provided for the organization of state farms on the basis of former estates, a wish Lenin had always held, but which in practice had been frustrated by peasant action in 1917. Legislation favoring the formation of collective (of varying forms) and state farms followed, beginning in early summer 1918, including model charters for agricultural communes (21 July and 3 August 1918). The positive official attitude toward collective farming was reemphasized in the law of 14 February 1919 "On Socialist Land Organization and Measures for the Transition to Socialist Agriculture," which explicitly incorporated the principle of state ownership of land; individual farming was declared to be "withering away." Also, the organization of state farms was begun in summer 1918 and was spelled out in greater detail in the law of February 1919. By 1920, a tendency to enforce such forms of farming made itself felt, in part under the influence of insufficient food deliveries by individual peasants, from whom the state extracted produce in the name of the proletariat without adequate recompense. But on the whole, the collective and state farm sectors did not gain much economic importance, and the productive performance of these new farms was anything but a success. Nor did overall agricultural output recover.

Before the revolution, Lenin had not given much thought to exactly what forms socialized farming should have, except in the most general terms. He personally must not necessarily be identified with the early Bolshevik measures taken in this respect; he may have only considered them as an approach of trial and error. Yet these measures could not have been enacted against his will. Possibly, he sincerely believed that there was a movement toward collective farming originating from the peasants themselves and that only its course had to be influenced by Communist authorities. If so, this was a grave misjudgment. In a way, Lenin and his heirs admitted that themselves, later calling the "War Communism" of 1918–21 a period of mistaken and premature steps, though allegedly inevitable at the time.

At the end of that period, Soviet agrarian policy embarked on an entirely different course, known as the "New Economic Policy" (NEP). It was promulgated by Lenin at the Party Congress of March 1921 and found its first legislative enactment in the decree of 21 March 1921 on the agricultural tax (first to be paid in kind, later in cash) that replaced the food requisitioning of the preceding period. Under it, the peasants knew in advance how much they had to deliver to the state and what they could retain for their own consumption and for marketing to private buyers and traders. For the rest, they were free to till the land and keep livestock as they themselves saw fit. With a new currency of stable value introduced step by step from 1921 to 1924, they also were more able to buy needed commodities for their food sales. The NEP permitted private business in retail trade and small industry, but not in wholesale trade and large-scale industry, banking, and foreign trade, which the Bolsheviks considered the "commanding heights" of economic policy. For agriculture, NEP was gradually extended, allowing the employment of hired labor within certain limits and the leasing of land, especially in the land code of 1922 and in other legislation in 1925. Yet the state ownership of all land remained.

The economic success of NEP became obvious soon after its introduction. The sown area, yields, and livestock numbers recovered rather quickly, attaining and, in many indicators, surpassing the prewar level by 1928. Still, in the NEP period, the final goal of socialization was not forgotten by Lenin and his associates. However, under the conditions which forced NEP onto them, they had to behave with a certain circumspection. As Lenin put it in a speech on 17 October 1921:

> We thought that production and distribution would go on at communist bidding in a country with a declassed proletariat. We must change that now. . . . We tried to solve this problem straight out, by a frontal attack, as it were, but we suffered defeat. . . . Since the frontal attack failed, we shall make a flanking movement and also use the method of siege and undermining.

Lenin was not given time to see the attack, if only the indirect one, resumed. But the agrarian question—and NEP was in the main a retreat on this question—seems to have been much on his mind during his last years of life. One of his last articles, written on 4 and 6 January 1923, was "On Cooperation" (in agriculture). It was short but important. Here he was pointing to a new way to achieve collectivism in farming, namely by the way of cooperatives, which in the beginning were to be organised along the lines known also in Western peasant cooperation: consumers' cooperatives, credit, marketing and processing, land reclamation, farm machinery, livestock, seed and other cooperatives.[1] Such forms of cooperation, already well developed under the Tsarist regime, had been frowned upon and severely restricted in early Soviet years because they tended to strengthen the economic position of the individual peasant without making him collective-minded. Lenin now regarded such cooperation as a first step toward a comprehensive system of cooperatives so that, in terms of Marxist dialectics, by increasing quantity a qualitative

jump would be achieved. He envisaged, among other things, cooperation in all spheres of economic activity and by all people on the countryside, a markedly higher degree of general education, especially
among peasants, and some material surpluses as a security against
crises—all as preconditions for collectivization of agriculture and, im-
plicitly, of the economy at large. Lenin thought, and said in the article,
that this could be achieved in as short a "historical epoch" as one or two
decades, though he added cautiously, "at the best."

In its practical application Lenin's agrarian policy was far from following a straight course. Rather, it was one of very effective tactics for
seizing power, of wishful socio-economic thinking and of pragmatic
adaptation to economic necessities after the results of socialist zeal
proved disastrous in agriculture. Yet the theory remained consistent
throughout. Its guiding principles were: the "alliance" with the small
and middle peasants in a country where these formed the overwhelming
majority, until the Communists had seized power in the "socialist"
revolution; the persistent aiming at the socialization of agriculture
(together with the entire economy) as soon as such a revolution was
achieved; and at last, but not unequivocal in the earlier stages, the
recognition that the socialization of agriculture should not be imposed
by force upon a peasant majority, and that a certain level of economic,
cultural, and social development was required to make it viable.

Notes

1. The sequence is by the order of numbers and membership such cooperatives had
achieved in the Soviet Union by 1 October 1927. See Lazar Volin, *A Century of Russian
Agriculture* (Cambridge, Mass., 1970), p. 179.

2. Soviet Agriculture under Stalin: Collectivization and Its Results

THE LIBERAL Soviet policy toward peasant farming, disputed but once more confirmed by the 15th Party Conference in November 1926, gave way to a sterner attitude beginning in late 1927. Opposing trends of thought among Bolshevik leaders pertaining to the agrarian question became evident during these and the preceding years. The outcome of a protracted struggle within the leadership was Stalin's victory, first over the "Leftists" and then over the "Rightists." After that he took over the "Leftist" program in economic policy and proceeded to enforce the collectivization of agriculture.

The "Leftists" had stressed the socio-political dangers of letting the "class enemy," the "kulak," and his influence on the countryside grow stronger.[1] They wanted to squeeze as much as possible out of the peasantry, especially the "kulaks," without satisfying the demand of the village for nonagricultural goods. This meant unbalanced growth—unbalanced in relation to both agriculture and the consumer goods industry, in favor of heavy industry as the basis for economic development. Only after having accumulated capital by means of taxation and price policy, was the then strengthened industry to reestablish an equitable exchange between agriculture and the rest of the economy. This implied the use of force, if necessary, to quell resulting upheaval among the peasantry and also to prevent the peasants from reducing their production. Generally, it could be expected that the time needed to apply such a policy would not be unduly long, since Russia, as distinct from most developing countries of our time, had already passed the initial stages of industrialization before 1914 and possessed vast natural resources. The "only" condition was that the investment capital thus squeezed out was put to the most efficient use and soon rendered high returns—a condition granted, in Bolshevik eyes, by the very socio-political order and its planned economy. But this is not what actually happened.

Stalin's principal argument at the beginning of this policy was not collectivization as such, but rather the necessity to procure more grain, not only to feed the urban proletariat but also to export for the purchase of Western investment goods and technology necessary for rapid industrialization in Russia. Just as Lenin once upheld that the food crisis during the period of War Communism was mainly caused by "speculators," who therefore had to be fought mercilessly, Stalin and before him the "Leftists" held the "kulaks" responsible for withholding grain from the state and thus acting as enemies of socialism. Therefore the "kulaks" had to be the target of a fierce offensive, or as Stalin put it somewhat later, to be "liquidated as a class." Neither Lenin nor he ever admitted that the crises (especially that of declining grain exports in the late 1920s) were caused mainly by the regime's own policy, resulting in quite natural reactions to the price policy and to forced extraction of produce. It needed no "kulak" class consciousness to react as most (including middle and small) peasants did. The Marxist-Leninist (as it was soon called) bias toward class struggle and "building of socialism," combined with Stalin's assessment of the situation, contributed to make the expansion of the socialized sector of agriculture appear the adequate Bolshevik solution to what by then was called the "grain crisis." Under these circumstances, Stalin's policy of procuring grain by brutal force (as applied by himself in Western Siberia in January, 1928) became the focal point of Soviet agricultural policy in 1928/29 and soon grew over into a collectivization policy.

At the same time it was clear that the disappearance of the "kulaks" would eliminate exactly those farms which marketed an above average share of their output, be it through private trade or through state procurement channels. "Dekulakization" implied among other things, that the remaining small peasants would bring less to the market and consume more on their own. Also the extraction by force of sufficient quantities of grain and other produce from the twenty million small and middle peasants was impossible in the long run without ending in the same situation as in the early twenties, when the peasants either were ruined or reduced their production. Organizing large nonprivate agricultural enterprises seemed the most promising way out of the dilemma. In the first place, this would be in line with party doctrine; in the second, it would greatly reduce the number of units to be controlled directly by the state or the Party. It would also bring, it was hoped, the main advantages of large-scale production to agriculture: reduction of the labor force and increase of soil productivity by means of modern organization and methods.

The construction of more state farms might have solved the problem. However, they could be applied only in a small minority of cases, because turning all peasants into workers on state farms either would have cost the state too much in wages and capital inputs, or else would have made a mockery of what state farms were meant to be. Moreover, such "state socialism" was not in line with the ideology. Thus, collective farms were the obvious solution under the given goals and restraints. The XV Party Congress of December 1927 had already stipulated the "concentration and transformation of the small individual peasant farmers into large collectives" as the "main task of the Party on the countryside." Application of force was still not advocated, though the "decisive offensive against the kulaks," also proclaimed then, forebode that. It was the combination of this offensive and the drive for collectivization, with grain extraction as a background, that enabled Stalin to go on with collectivization at a speed that had been unthinkable a few years earlier, and also far surpassed the goals originally set in 1928/29.

It is not necessary to dwell on the "dekulakization" and mass collectivization that swept the country beginning in 1929 and ended in collectivization of almost two-thirds of all peasant households by 1932, and of the remainder, mostly in outlying, non-Slavic areas, during the following years. The process, with its ruthless force and cruelties applied against recalcitrant peasants, not to speak of those actively resisting, has been described in Western literature of that time as well as more recently, and in an indirect way has also been acknowledged in some Soviet writings of the Khrushchev period. Human suffering and political unrest among the peasantry during mass collectivization reached proportions that later made Stalin say, in a conversation with Churchill, that the strain on the Soviet system had been more terrible than that of the Second World War.[2] The great famine of 1932/33 was in part caused by drought, but it would not have been so catastrophic had it not been for the disruption of agriculture during the preceding years, especially the depletion of all of the peasants' reserves. Estimates of the death toll of "dekulakization," collectivization, and famine are of the order of ten million lives, not to speak of the suffering of other millions in prisons, labor camps, and through deportation.

Economically the losses were no less staggering. According to Soviet statistics, which tend to embellish the state of affairs, grain production stagnated: the three-year average of 1930–32 was 74 million tons, compared with 73 million in 1925–29. Yet state procurements of grain doubled and exports increased even more, leaving less to the peasants. There was less need for feed grain, it is true, because at the same time animal herds dwindled: by 1933, cattle numbers were slightly more than half of those in 1928, pig numbers were less than half, sheep and goats only one-third, and horse numbers were halved. Many peasants slaughtered their animals before being forced into the collective farm, and more animals were slaughtered or perished during the great famine. Other material losses were caused by arson, which desperate peasants used as a weapon against collective farms. To this must be added the socio-economic losses of skill and initiative, of personal interest in agricultural production, which was made still more severe by the decimation of leading cadres during the years of the Great Purge following collectivization.

It was not before the meteorologically excellent harvest year of 1937 that the diet of the Soviet citizen improved markedly, though the pre-Revolution output was barely reached (and not again reached from 1938 to 1940) for a greatly increased population, of which many more now lived in urban areas.

From the beginning of collectivization, the *artel'*, as distinct from both the commune (having wholly collectivized life, including consumption) and the T.O.Z. (*Tovarishchestvo po sovmestnoi Obrabotke Zemli*, i.e., Association for common Tilling of the Land, having a low degree of collectivization of production), was made the model organization of collective farming. Increasingly, the *artel'* was called *kolkhoz* (from *kollek-tivnoe khozyaistvo*, i.e., collective enterprise), which by now is the only name of the Soviet collective farm. Its structure was determined by a Charter. The Commissariat of Agriculture approved a first Model Charter on 5 May 1919; a new one was drafted in December 1929 and, after amendments on Stalin's instigation, was adopted on 5 January 1930. The latter was altered, though not essentially, in the Charter adopted on 17 February 1935 by the highest state and Party organs (with another amendment incorporated in 1938). For a whole generation of peasants, it became the "Basic Law" of collective farms. Each kolkhoz had to adopt its individual charter almost literally along the Model Charter's lines, with minor regional or local variants imposed—in practice—by the state and Party organs of the corresponding lower level. Although the Model Charter was changed in many ways by legislative and executive practice in the 1950s and 1960s, it remained formally valid up to 1969, when a new Model Charter was adopted. Some of its essential features deserve highlighting here (the numbering of the sections relates to the full text; see also Table 5.1.

II. The land is "governmental property of all the people," handed over to the kolkhoz free of charge for permanent utilization. Out of this land fund, the kolkhoz shall allot plots for private use to the member households.

III. Dwellings, personal livestock in restricted numbers, poultry, and other small animals, as well as stables for such livestock and minor implements for tilling the household plot are exempt from collectivization.

IV. The kolkhoz has to obey the production and delivery (procurement) plans imposed from above. Labor qualifications and the general cultural standard of the members are to be raised, and socio-cultural services are to be provided.

VI. From one-half to three-quarters of the property brought into the collective economy by a member is considered his share (without dividends being paid), to be returned in money when the member leaves. But land is to be returned only "outside the land enclosure belonging to the artel" (i. e., if there is free state land available). The collective crop and animal output in kind is to be disposed of in the following way (by order of priority): state delivery obligations and seed loans, payments for work done by the MTS (cf. below), other contract obligations, seed and fodder funds for the next year's production cycle and a reserve of these of 10–15 percent of the annual needs; funds in kind for social aid upon decision of the general meeting; over-plan sales to the state and on the free market; distribution of the remainder to members according to collective work done, measured by the number of labor day units earned. For cash income, the following order of assignment applies (according to the amendment decreed on December 4, 1938): payment of taxes; premia to the (state-owned and compulsory) agricultural insurance; repayment of loans; current production and management expenses (the latter to be not more than 2 percent of total cash income); assignments for socio-cultural purposes; replenishment of capital stock ("indivisible funds") and wages of hired construction workers (such wages are considered as part of capital investment), whereby the total replenishment shall amount to 12–15 or 15–20 (depending on regional characteristics) percent of total cash income; the remainder is to be distributed among members according to work done, measured in labor day units.

VII. All work is to be performed by the members; only specialists (and, in cases of exceptional need, other labor) may be hired from outside. Work is organized by brigades led by a brigade leader (*brigadir*); its remuneration is effected under a piecework system, whereby *trudodens* (labor day units) are allocated according to the fulfillment of work norms and the degree of skill required and physical difficulty of the work. Neglect of work or working animals, of implements or machinery, "shall be punished by the management in accordance with the rules of internal organization" (of the individual kolkhoz, i. e., not according to general labor or civil law); damage done to collective or state property is punishable under criminal law.

VIII. The general meeting of the members holds the highest authority; it elects the chairman and the members of the board of managers and of the auditing commission, and admits or expels members; it shall confirm the annual plans and rules for production, finance, building, and labor remuneration (including the labor norms), also the annual reports and current important reports of the management, the assignments to funds and the distribution of income, and the "rules of internal organization." Resolutions are passed by a majority of votes (if the quorum is fulfilled) through "show of hands" (i.e., no secret vote); "day-to-day direction of work" is executed by the chairman of the board (usually called kolkhoz chairman).

The central role in the affairs of a kolkhoz is that of the chairman. He must be elected by the general meeting of members. It may seem that this meeting thereby was assigned a decisive role and acted as a guarantee of democratic management in this allegedly voluntary cooperative. But the Soviet system as a whole, as it developed under Stalin and later, made any formally democratic institution a dummy, especially among the peasantry after the terror of collectivization. Moreover, the leading role of the Party organizations (article 126 of the Constitution of 1936) pervaded all spheres of Soviet, including rural, life. Where there were not enough members in a kolkhoz to form a Party cell, this role was assumed by the political organ of the MTS. Concerning production activities, the Model Charter itself provided for interference of state organs through the production and procurement plans and by means of prescriptions for the distribution of kolkhoz income. On the level below the chairman, the brigade or section leaders had command over the members, although they themselves depended wholly on the chairman. In most cases, there was a third and lowest level of command, that of the "team" (*zveno*, literally, "link") with its foreman (on this level, not infrequently a woman).

The regulations on private agricultural activities of members constituted an extensive and important part of the Charter. The plots were rather small (although in some regions exceeding 0.5 hectare), but as most of the grain needed was supplied by payments in kind for collective work, they could be devoted to what else was needed for human consumption, including potatoes. Even more important was the right to keep livestock in a number sufficient for the household's own consumption and even for some sales on the free market. The decisive supplement to such private animal husbandry was not only the payment in kind (grain also for feed, and hay, straw, etc.) but also the right to graze animals and to make some hay on public land. The plot alone would not have provided an adequate fodder basis.

The central characteristic of the kolkhoz charter was the underprivileged position of the kolkhoz member *(kolkhoznik).* Significantly, the Charter provided for expulsion of a kolkhoznik, but did not indicate the way in which he could leave the kolkhoz on his own will. In fact, he was not supposed to do so. What was not said in the Charter was that internal passports were issued to kolkhoz members only in exceptional cases. As this document was required for settling down in urban areas, most kolkhozniks could not legally do so. Similarly, a labor book was required for being employed outside the kolkhoz, and to kolkhozniks such books were not issued. To obtain an interal passport or a labor book, they needed special permit papers from the kolkhoz and village administration.

The Charter itself makes it clear that the kolkhoznik did not receive a fixed wage. Instead, he was credited with "labor day units," the value of which was determined as a residual at the end of the year by what was left of the kolkhoz income, divided by the total number of labor day units earned by all members. In view of the low producer prices paid by the state and all the other obligations that had priority, the remainder was very small indeed. In theory, this distribution was meant to divide the farm's annual profit among the "owners" (farm members) in accordance with work contributed, but in practice it was geared to making up incurred losses at the expense of labor remuneration. In this way, the individual kolkhoznik also paid for mistakes of the farm management and of official agricultural policy (plans and directives from above). As income taxes were paid by the kolkhoz as a whole prior to the distribution of labor incomes, the sum of the latter was reduced by the taxes paid. Thus, all members bore the burden of taxes equally, percentagewise, even those with the most miserable income. In contrast, workers of state enterprises had a tax-free minimum and a progressive tax rate for incomes above this level, though not beyond a certain percentage.

Soviet labor law applied and applies to workers and employees only; the labor relationships of kolkhozniks, including work norms, length of working day, and rules for juvenile labor are regulated by the charter and, mainly, the "rules of internal organization" of each kolkhoz. Most labor and other disputes, except for criminal and political offenses, may not be taken to court, but must be settled within the kolkhoz, where the position of the management is dominant. And although the kolkhoznik has the "right to work" guaranteed every Soviet citizen by the Constitution, the kolkhoz has no binding obligation to provide him with work during the slack season or where there is a labor surplus; but the kolkhoznik is obliged to work a certain minimum of days (fixed in later legislation, beginning in 1939), when required.

Concerning social aid and services, the kolkhoz was not under strict obligation to provide those rather modest ones indicated in the Model Charter. In a poor kolkhoz they were minimal indeed. Up to 1964, the state did not contribute to these funds.

As distinct from the rest of the population, the property rights of the kolkhoznik were vested with the household (in most cases identical with a family living together). The house and premises, animals, implements, and usage rights on the plot belonged to the household, not to individual persons. Of course, personal belongings such as clothes did not fall under this rule, but labor income in practice often did. This reflected old Russian and Tsarist peasant law, not the Soviet civil law.

So the kolkhoznik was a second class citizen in law and practice and this was emphasized daily in many ways. No wonder that he had to be prevented from leaving the kolkhoz, except when the state needed labor and therefore gave the necessary documents, mostly in an *orgnabor* (*organizovannyi nabor,* i.e., organized recruitment) action.

It is evident that Stalin's method of collectivizing agriculture contradicted the notion of a gradual approach without the use of force, as explicitly advanced by Lenin in his last years. In his article of January 1923, Lenin further mentioned two indispensable preconditions for collectivization: industrial progress (in general), and mechanization of agriculture, on which the superiority of large-scale over small-scale farming has to rest. Stalin reversed the sequence, making collectivization a means for attaining both, which, among other things meant regimentation and low prices (with resulting low wages) for the peasants. He put it this way in an article printed on 26 December 1929 ("Concerning Questions of Agrarian Policy in the U.S.S.R."): "The great importance of collective farms lies precisely in that they represent the principal base for the employment of machinery and tractors in agriculture, that they constitute the principal base for remoulding the peasant, for changing his mentality in the spirit of socialism." He quoted a similar dictum of Lenin in this connection, but hid the fact that for Lenin, machinery was the principal base for creating big collective farms and remolding the peasant. In place of the truly Marxist material basis of consciousness Stalin substituted a merely organizational one. Even if things went as hoped—which they did not—the peasants would have felt the advantages of modern large-scale production only at a later stage, not upon entering the kolkhozes. As things turned out, Soviet output of tractors and other agricultural machinery became sizable only after collectivization, and even then was inadequate for modernizing the farms, especially since most of animal horse power was lost during the process.

What was the motivation of the Stalinist approach? From the very beginning, the underlying ideology of the Bolshevik leaders prejudiced the discussion concerning collectivization and its eventual course because of the preference for social ownership and management of the means of production and also for methods of class struggle. No less important were the targets of practical politics: extraction of grain at low prices, de-kulakization, and organization of the remaining small peasants in large production units. As shown by the course of events, these practical targets were contrary to the wishes of the great majority of peasants. Therefore the application of force in the beginning, and the creation of a rural power apparatus in the longer run, became indispensable to pursue the targets energetically.

Stalin's fight for personal power may also have played a role, but seems to have been a side-issue. In any case, he started rapid and enforced mass collectivization only after he had defeated his rivals. Yet he seems to have set his mind rather early on carrying through industrialization as rapidly and as forcefully as possible, and this involved the question of capital accumulation. But was collectivization an adequate means to this end, and was it applicable without the proper industrial basis for modernized large-scale farming? The extraction of produce, especially grain, at low prices was a first and indirect answer of Stalin in the affirmative. It meant enforced capital transfer from agriculture to industry without regard for the capital needs of agriculture itself. Whether Stalin used and exaggerated the "grain crisis" as a pretext is a question of his tactics, not of his goals. At any rate he was not inclined to pay the peasants prices which would induce them voluntarily to produce and to market more grain. (In spite of a price increase in 1928/29, grain and other agricultural prices compared to industrial prices were still below the pre-Revolution parity, which itself had been considered unfavorable for agriculture.)

It has been debated in recent years whether a capital transfer took place at all. One may correctly make the point that the net transfer was actually much less than expected. On balance, it appears from the very low level of farm incomes during the thirties that as much as possible was transferred out of agriculture. However, there was also a considerable transfer in the opposite direction (e.g., industrial inputs, training of managers, high urban consumer food prices on the free, as distinct from the state-owned, markets, etc.). After all, state and collective farms required more capital than had the millions of small peasant farms formerly, although these small farms presumably would instead have consumed more. In sum, not too much could be transferred out of agriculture, because during the collectivization process this sector of the economy was ruined (above all, livestock herds were decimated), and the rural population was reduced to starvation.

It is hard to imagine that Stalin had wished for these negative effects from collectivization. However, the way in which then and later the campaigns for grain and other procurements were conducted under his leadership, and the way in which he slanted the prices further in favor of industry, drained labor from the villages and bound the remaining labor to the kolkhozes, hardly leave doubt about his general intentions. It is quite another question whether this scheme worked satisfactorily, whether industry had not in reality to accumulate most of its capital, mainly by means of paying low wages (but still not as low as kolkhoz remuneration) and by concentrating on producer rather than consumer goods.

What about the transfer of labor? Experience in any developing country with a rural labor surplus and a high birthrate (as was the case for Soviet Russia up to the Second World War) shows that the flow of labor out of agriculture is not dependent on the organization of agriculture, but on employment opportunities in the nonagricultural sector. Had it been otherwise in Russia, there would have been no need to make it so difficult for kolkhozniks to leave the kolkhozes of their own will. Kolkhoz organization, combined with administrative regulations (internal passport, labor book), was mainly a means to regulate, even to stop, labor outflow, if required regionally. For labor transfer as such, it was not necessary.

Thus, the motivation and effects of the Soviet collectivization are clear enough. On the whole, the goals were attained, though by far not fully and at extremely high social and material cost. The optimality of the operation is anything but evident. It was optimal only in establishing complete control over the peasant majority of the population. Thus, it became feasible to impose on this majority a long-term reduction of consumption, which otherwise would have been near to impossible to achieve. It enabled Soviet agriculture to put up with one otherwise fatal consequence of collectivization, namely the extreme fall of labor morale in collective production. The socialized large-scale farms had to feed the urban and industrial population at low prices, whereas feeding the peasantry was for the most part left to the private plots, where labor morale remained high.

A Soviet tractor industry was created under the first Five-Year Plan, 1929–32, and the importation of tractors was stopped in 1932, although the number of tractors available to agriculture far from satisfied the needs. Therefore, these machines had to be used to their fullest potential, and pooling was a logical means to achieve this. First pooling arrangements for agricultural machinery date back to the mid-1920s. After 1927 they spread on the basis of either state or cooperative property, most commonly under the name Machine and Tractor Station (MTS). During 1931 and 1932, when collectivization was on the whole completed, all these stations were transferred to the state and financed by it. State farms, being much larger on the average and state-controlled anyway, had their own machinery, but this was forbidden to kolkhozes, which had to be served by the MTS.

One MTS serviced a number of kolkhozes, depending on the latters' size. It is obvious that direction over all machinery gave the MTS an important economic lever over the production and decision-making process in kolkhozes. In addition, the MTS increasingly became the major instruments for channelling grain deliveries to the state, since their services had to be paid in kind, as a percentage of the harvest total of those fields, where MTS machinery had been used. Such payments amounted to more than half of all kolkhoz grain delivered to the state in

1937. No less important was the function of political control, as the Party had only few members in kolkhoz villages, and the local state officials, besides possibly being influenced by villagers' opinions, as a rule lacked the knowledge and ability to control the operations of large farms. The MTS had a special deputy director for political affairs on its staff, and toward the end of the collectivization period (1933–34) and again during the Second World War (November 1941–May 1943), a political section (*politotdel*) was added. The politotdel was to be the "Party eye and control" in agriculture, as it was said in a decree of 11 January 1933.

By 1940, the 7,069 MTS served more than 235,000 kolkhozes and had 435,000 tractors, 153,000 grain harvesters and 40,000 trucks. The permanent staff of MTS (537,000 employees in 1940) consisted mainly of agricultural and technical specialists and management personnel. The majority of MTS staff were seasonally employed so-called mechanizers who operated the machinery; they pushed the annual average number of MTS labor up to over 1.3 million in 1937. Most of these were kolkhoz members, and their remuneration was under the usual kolkhoz system of labor units. But during the time they worked for the MTS, they were provided a guaranteed minimum in grain and cash per labor day unit, and moreover were credited with more labor day units per day worked (rather, per norm fulfilled) than the average kolkhoznik. Thus, a new professional differentiation was introduced by the MTS, combined with social differentiation. Training of mechanizers was short and qualitatively inadequate in the earlier years, but improved later. Still, the level of technical skill and knowledge remained low by Western standards up to the 1960s.

A few agricultural operations became mechanized to a degree that was high for the standard of the time (especially in Eastern Europe). But this concerned only ploughing, grain seeding and harvesting, and growing of sugar beets. Other crops, such as potatoes, flax, and maize, were still largely tilled as well as harvested by hand in 1940, and almost all work in livestock farming remained manual.

Apart from the kolkhozes and MTS, the state farm (*sovkhoz*) constitutes the other organizational form of socialized farms.[3] This sector expanded during and after collectivization. Sovkhozes were mainly organized for grain farming in steppe regions, for technical crops such as sugar beet, for specialized animal breeding, and in later years for other branches of farming also. But on the whole their numerical importance remained small in comparison to the kolkhoz sector until the mid-1950s. They achieved greater significance in relation to state procurements, because they specialized on products for the state distribution and processing network, received more capital inputs, employed less labor per product unit, and therefore consumed less of their own food output. They were financed from the state budget, which received

ᵃᵈ to cover the losses. The farm workers and
eg ar wages and did not have to bear the risk of
ᵇ ᵌsidual wage system. By contrast to the
ᵥas owned by the "state of the people" and for this
' ᵌigher form of social property. The kolkhoz
tru but socialized only within the collective of its
property" and thus on a "lower" level of

ᵢs completed, Soviet agricultural policy turned
93ᵃ, these included the consolidation of the kolkhoz
systen., agricultural output, and diversification of production in accordance with industry's need for raw materials.

Agriculture urgently needed to recover—in terms of living standards
of its work force as well as of investment—after the turmoil and depriva-
tion of 1928–34. Net farm income per head, even in the record harvest
year of 1937, is estimated by Volin to have remained 10 percent below
that of 1928, and for kolkhozniks was even lower.[4] Kolkhoz labor morale
was very low, and there were not many managers and specialists on the
new farms who were competent agriculturalists and organizers. Dekulakization had wiped out most of such persons in rural areas, and
those brought in from urban areas, even if otherwise capable (which was
not always the case), often had insufficient training to operate large
agricultural production units. The subsequent years of the Great Purge
(1934–38), with their mass arrests and convictions and the resulting
frequent changes in administrative and management personnel and the
ensuing psychological pressure, worsened the situation. Such factors
were especially disturbing in view of the difficult natural conditions,
under which most of Soviet agriculture has to produce.

Among the first palliative measures to improve the state of the food
sector were the decrees of 1932 (6 and 20 May, and 22 August), which
reintroduced private marketing of agricultural output, though on a very
limited scale, by means of the so-called kolkhoz market. When the
compulsory deliveries to the state were met and seed and other reserves
were laid in, individual kolkhozniks as well as the kolkhozes themselves
now were allowed to sell on this free market what produce was left
beyond their own consumption requirements. However, they had to sell
their own produce only, trade by middlemen ("speculators") remained
strictly forbidden. Up to the 1950s, most cash income of kolkhozniks and
also of kolkhozes was derived in this way, and the supplies for the urban
population were improved, although at extremely high prices to the
consumer.

No less important in recovery was the introduction of a more logical
and viable system of state procurement quotas, initiated by a decree of
19 January 1933 and expanded during the following years. The pro-

curement quantities now had to be fixed in advance, so that kolkhozes could assess their possibilities and requirements. The irregular demands by state organs no longer were permitted although they remained current practice in many cases.

Slowly, agricultural production recovered, although livestock numbers remained below the 1928 level for more than twenty years. Total grain output, which had been between 75 and 80 million tons before collectivization, reached an average of 80 million tons again in the three-year period from 1938 to 1940, and less was now needed for feed in view of the smaller livestock herds. At the same time the share of government procurements (the various kinds of procurements, including payments to MTS and deliveries by state farms and private plot-holders) increased to 41.2 percent of total output of grain. The production of technical crops, which was favored by price policy and other measures and of which a much higher percentage had to be delivered, increased more. While retail prices of consumer goods rose sevenfold and the free kolkhoz market prices for foodstuffs rose by even more between 1928/29 and 1937, state procurement prices for grain climbed by only 5 to 28 percent (depending on the kind of grain).[5]

The slow development of animal farming demonstrated one of the essential shortcomings of socialized agriculture. It was possible, to a certain degree, to offset low labor morale and shrinking numbers of workers and draught animals in crop farming by the use of tractors and other machinery. But animal husbandry was not susceptible to such substitution, partly because at that time, mechanization of the stables (including the provision of appropriate buildings) had not yet reached the levels of crop farming, and partly because such machinery, where available, required more skill than was possessed by those using it, and more care than they were usually willing to devote. The kolkhoznik applied every effort to his privately owned livestock, but not so to public herds, and the official low prices for animal products made it impossible for kolkhozes to pay wages that might have changed this attitude.

Thus, Soviet Russia entered World War II (in 1941) with an agriculture still suffering from the results of collectivization and unable in peacetime to provide the population with more than a very modest diet. Military actions as well as the German occupation brought immense suffering together with economic losses to Soviet agriculture and its population, both directly through devastation and indirectly by depressing the yields of the remaining production potential. In 1945, the Soviet gross agricultural product had declined by 40 percent, compared to 1940.

The war also brought a further shift of people, resources, and production to the USSR's eastern regions. Agricultural labor potential

decreased drastically in numbers and deteriorated in the age-sex composition, as a consequence of not only the war and wartime industrialization, but also the failure of many demobilized men to return to
agriculture. Policy toward the private sector became more lenient, in
order to make use of its production potential in the war years of
emergency, to give rural people a chance for survival under near-famine
conditions, and possibly also in order not to raise anti-Communist
feelings. Yet the kolkhoz system as such, with the compulsory procurements, the directives from above, the labor enforcement, etc., was not
touched but rather strengthened by the war. Significantly, many of the
wartime regulations, especially those on labor organization and requirements as well as on urban kitchen gardening, were not repealed
after the war, or were only slightly changed. The kolkhoz system as a
means of extraction of produce proved to be well fitted for a war
economy.

The period 1945–52 was mainly one of restoring the prewar situation,
both in the organization of agricultural production and in its performance. As before the war, there were the socialized and the private sectors,
the latter comprising the remaining private peasants, the household
plots and livestock of kolkhoz members, and the growing numbers of
plots and animals held by non-kolkhoz workers and employees, urban
(in towns and on the outskirts of cities) as well as rural. An early
agricultural measure after the war was the decree of 19 September 1946
on "violations of the Kolkhoz Charter," aimed at restoring the kolkhozes'
use of land which had been misappropriated by non-kolkhoz organizations and—this was the politically relevant aim—by private persons. Even
with closer control, the private sector retained its importance to agriculture. According to Soviet data, the sector in 1953 still accounted for 45
percent of gross agricultural product (evaluated at 1958 prices), a
proportion not sizably less than in 1940 (percentages for the war and
immediate postwar years are not known, but probably were higher).[6] In
the socialized sector, state farms (including institutional farms) accounted for less than ten percent of the gross product in 1940 as well as
in 1950; kolkhozes accounted for the rest, which in 1940 and 1950 was
about the same volume as the contribution from the private sector.

In the newly annexed western territories (Baltic republics and the
western parts of Belorussia, Moldavia, and the Ukraine), collectivization
was not enforced immediately after the war, but a type of land reform
was carried out, fixing upper limits for the size of a private peasant
holding and expropriating the landlord estates. As to collective farms,
only a small number were organized there up to 1947. Then, in
1948/49, collectivization started in striking coincidence with the beginning of mass collectivization in Soviet-occupied Eastern Europe. By

1951, it was on the whole completed in the new Soviet territories, while in the rest of Eastern Europe it took more years and was interrupted in the mid-1950s (see chapters 3 and 5).

This last Soviet collectivization drive was soon superseded by a countrywide reorganization of kolkhozes, starting in 1950, entailing the amalgamation of several small kolkhozes into a larger unit. The merger campaign resulted in reducing the total number of kolkhozes from 254,000 to 97,000 during 1950–52 and in enlarging the average size of kolkhozes accordingly. The kolkhozes thereby ceased to be essentially one-village collectives. The process was accompanied by a further, indirect reduction of the private "mini-farms" of the kolkhoz members.

The slow output growth—in contrast to the organizational "successes"—during those years is not surprising in view of the bleak situation of Soviet agriculture after the war and of the small emphasis given to agricultural investment and incentives up to Stalin's death. For example, it took a long time to reerect the farm buildings and living houses in the war-stricken regions, because this had to be done mostly with the kolkhozes' and population's own scarce material and financial resources, and with little help from the state. Also, the production of fertilizer was only slowly stepped up. In addition, technical equipment, in part destroyed and in part worn out during the war, was not replaced at the necessary speed. It was not until 1950 that the machinery park markedly surpassed the level of 1940. This was clearly inadequate since labor, especially able-bodied male labor, was not as abundant as before the war and the number of horses did not regain the prewar level.

Perhaps the most important outcome of this development was that at a time when Western agriculture (primarily American, but during the 1950s, West European also) embarked on unprecedented mechanization, the Soviet counterpart remained far behind in this respect. Although machinery supplies to Soviet farms increased during the following decade, they did not catch up with what went on in the West. Before the war, the point could be made that large-scale socialized agriculture used more machinery per land unit than the peasant agriculture of many European countries, and thus seemed to fit into the picture of more rapid modernization, but this was no longer the case after the war. Now the "industrial revolution" in agriculture came into full swing on the medium and small peasant farms of "capitalism," not least by the introduction of labor-saving machinery in animal husbandry. The outflow of labor from agriculture, induced by rapid industrialization and temporarily enforced by the war, also took place in the Soviet Union but was not accompanied there by a corresponding increase of capital inputs. This time lag—a lag in comparison to the West as well as to the interior labor situation—makes itself felt up to the present.

The policy of industrial development at the expense of agriculture is most clearly mirrored in the prices, which are fixed by the Soviet government. Average agricultural prices actually paid by the state procurement agencies to kolkhozes in 1950, as compared to 1937, increased only slightly for grain and milk, even decreased for potatoes and meat, and increased from two- to four-fold for the main cash crops (sugar beet, cotton, flax fiber). These average prices were also depressed by the fact that the higher prices for above-plan deliveries to the state were abolished for grain, and increased delivery quotas for milk and meat prevented most kolkhozes from profiting from above-plan prices for these products. The free prices of the kolkhoz market more realistically reflected the demand and supply situation. They soared to nominally fantastic heights under the near-famine conditions during the war and again in 1946/47. Even after the currency reform in late 1947, with its sharp reduction of purchasing power, kolkhoz market prices still were four times as high as in 1937. Similarly, the index for all retail prices (including foodstuffs) in 1950 was nearly four times as high as in 1937. Among these, the consumer prices for basic food (grain and grain products) increased less, but did not reflect the low state procurement prices for grain because the state's profit margin on these remained great.

It is no wonder that agricultural incomes were much lower than wages in the rest of the economy. Including private subsidiary incomes and remuneration in kind, the kolkhozniks earned on average only one-third as much per annum as industrial workers; even in the exceptionally good harvest year of 1952, the figure was less than half. These are rough estimates, and they apply to kolkhozniks employed fulltime. For the many others the figures would have been even lower, though perhaps not as low as might be expected, because underemployment in the kolkhoz was in part compensated by work on the private plot and in animal farming.

The picture at the time of Stalin's death was of an agriculture that was just beginning to overcome the consequences of the war and of postwar policy. Yet it had to feed a population that had been increasing rapidly since 1949, the nonagricultural part increasing most and thus putting growing urban demands on agriculture.

Notes

1. A "kulak" by Bolshevik definition was a rich peasant or farmer, but in most cases the term referred to middle or thrifty small peasants by continental West European standards. The "middle" peasant, Soviet style, was considered a potential ally, and was a small peasant by Western standards. Later, during the collectivization, the word *kulak*, and also the term *podkulachnik*, that is, a kulak's henchman, was used to designate almost any peasant who tried to resist collectivization.

2. Winston S. Churchill, *The Second World War. Vol. IV: The hinge of fate* (Boston, 1950), p. 498.

3. The institutional farms will be disregarded here. These are farms of state-owned

nonagricultural enterprises and organizations, including the military, and research and educational institutions. As such they play a certain role regionally and at times or for training specialists and developing new varieties and breeds.

4. L. Volin, *A Century of Russian Agriculture* (Cambridge, Mass., 1970) pp. 257–58.

5. *Ibid.*, p. 251.

6. K.-E. Wädekin, *The Private Sector in Soviet Agriculture* (Berkeley/Los Angeles/London, 1973), p. 60; cf. V. A. Belyanov, *Lichnoe podsobnoe khozyaistvo pri sotsializme* (Moscow 1970), pp. 53–54, who, without indicating his price basis, gives 40 percent for 1953 and 43 percent for 1940.

3. The Early Postwar Period in Eastern Europe

AT THE END of World War II, the situation of agriculture in Soviet-dominated Eastern Europe was characterized by three preeminent factors:

1. With the exception of the Soviet occupation zone of Germany and the Bohemian and Moravian parts of Czechoslovakia, Eastern Europe was predominantly agricultural. Peasants constituted the majority of the population, and among them Communists made up a tiny minority, smaller still than among the population at large.
2. The disasters and destructions of war and German occupation had resulted in an acute food shortage, and the increase of food production to the prewar levels had to be a priority goal. Policies resulting in stagnation or even decrease of food production had to be avoided.
3. The hegemony of Moscow was a fact that only one country, Yugoslavia, succeeded in challenging at a later stage (and this was a special case, politically and strategically). So any policy, including agrarian, in these countries could deviate from Soviet orders and patterns only in matters of secondary importance.

Moscow—and politically speaking, Moscow was Stalin at the time—was determined not to give up what power and influence had been won during the war. To achieve this it had to prevent any resurgence of bourgeois political and social power in Eastern Europe, which by its very nature was likely sooner or later to become anti-Communist and West-oriented. Because the Communists were a minority in all the countries concerned, the political structure could for an initial period remain non-Communist, as long as it did not produce serious anti-Soviet actions. But the social and economic basis of such a structure had to be weakened, if not eliminated, if the war gains were to be consolidated. Afterwards it involved no great risk of disturbances if the social, economic and political life was gradually reshaped after a near-to-Soviet model which soon was to be called a "People's Democracy."

Because of the destruction of industry, the postwar economic life of Eastern Europe was based on agriculture even more than before. Any political power to be erected had to win over, or at least not antagonize, the peasants, who, with the exception of East Germany and Czechos-

31

lovakia, accounted for more than half the population and on whom the economy rested.

The social and economic structures in Eastern Europe had not been very stable formerly, again with the exception of those of East Germany and Czechoslovakia, and were much weakened during the war. The old upper classes could not count on strong support from the peasant masses, if these were not stirred by other measures. Denazification in Germany and punitive actions against wartime collaborators with German and Nazi authorities in other countries could be manipulated in a way that helped to destroy bourgeois and anti-Communist power and influence. Economic losses by such measures were not too great, because most losses had already occurred during the war.

For agriculture, this general situation dictated the Communist policy under the regimes in Eastern Europe, which were either newly installed or were being tolerated for a time by Moscow. Agrarian measures of a revolutionary character were feasible, and advisable from the Communist point of view, if they prepared the ground for a later Communist reshaping of agriculture and of the peasant class and at the same time seemingly corresponded to aspirations cherished by the peasants or a majority of them. Land reform was such a measure. It could be used to destroy the power of landlords and rich peasants and win sympathy from the majority consisting of small peasants. No pressure as yet was exerted towards the collectivization of peasant agriculture. To the contrary, any such intention was denied by the Communist leaders. In fact, one may assume that under the conditions of 1946–47, collectivization was not considered a goal of the immediate future, although it surely was already envisaged for the longer term, for socio-political reasons as well as for stabilizing Communist power once it was established.

A system of obligatory delivery of farm products was inherited in the previously German-occupied or -dominated territories. Although disliked by the peasants, its necessity could hardly be disputed under the impact of latent postwar famine. Slightly changed, it could be used as a means to oppress the remaining wealthier peasants by disproportionately greater delivery quotas without unduly arousing the small peasants, whose quotas were made less onerous.

Peasants' cooperatives of the traditional type and of various forms were favored, and although some new forms were introduced, this was nothing essentially new to the East European peasant. As a rule, the pre-1945 cooperatives continued to exist or reemerged. However, they were gradually changed in character, becoming unified on a nationwide scale, being given the monopoly for their respective activities, and becoming subjected to the general economic plans. A temporary sideissue were peasants' associations for mutual help, the formation of which was inspired from above. Before and during collectivization, the various cooperatives were also used to prepare the ground with those peasants who were not yet collectivized. They continued to exist, of course, where the collectivization process was reversed later on, as in Poland and Yugoslavia.

Political organizations of peasants were also tolerated or even reanimated in the first postwar years, although sometimes in other forms than previously. In most countries (Poland, Hungary, Bulgaria, Yugoslavia, and Rumania) peasant parties were even represented in parliament and government. But care was taken that these would not gain decisive influence on important issues of general policy.

Almost a generation earlier, in the 1920s, a wave of land reform had swept Eastern Europe. The October Revolution had demonstrated that in agrarian countries there was a revolutionary potential where small or landless peasants existed on the one hand, and large estates on the other. In some parts of the region (mainly Bulgaria and the Serbian and mountainous parts of Yugoslavia) land distribution was no issue, as there existed practically no latifundia. In other areas, not much was achieved in the way of a land reform, especially not in Hungary (where Bela Kun's Communist revolution of 1919 had been quelled by the conservative powers), and in Poland and Albania. Yet in the Croatian and Slovenian parts of Yugoslavia, in Czechoslovakia, and in Rumania, millions of acres were redistributed, although sizeable estates and domains remained. In part, the peasant demand for land could be satisfied by expropriating landowners of other nationalities, often of the nobility, thus turning the social issue into a national one. This applied first of all to Czechoslovakia, Rumania, Croatia, and Poland.

The new peasant holdings originating from these reforms of the 1920s were not really viable, and the rural population continued to increase, causing further fragmentation of holdings. In addition, the world economic crisis, which depressed agriculture most of all, made life for East European peasants very difficult after 1929. Underemployment was widespread among them, the agricultural labor surplus in 1937 being estimated at 20–50 percent in most of these countries (again with the exception of East Germany and most of Czechoslovakia).

Thus, another land reform became the first and most spectacular measure of the Soviet-inspired governments. Its implementation was similar to that carried out in Soviet-annexed territories in 1939/40 and resumed in 1945. Not only estates, but also peasant holdings exceeding a medium size of between 20 and 50 hectares in most countries, were affected (in Hungary, the upper limit was 115 hectares, and in Poland in certain cases, up to 100 hectares). As distinct from Russia in 1917, this was only rarely brought about as a spontaneous movement of the peasants, except where German landowners and peasants were expelled; it was a measure ordered from above. In an extremely short time a total of 20 million hectares (50 million acres) of land were redistributed to 65 million (including dependents) small peasants and landless agricultural workers. Only rarely did this result in economically viable holdings, and in the light of later events one may confidently say that this was intentional. If anything, the speed in itself precluded the formation of rational production units and the orderly registration of new ownership rights. No compensation was granted to the former owners.

Not all land thus expropriated was given over to peasants or workers. Some was kept to organize state farms on the basis of former estates. This is precisely what had been intended in Russia twenty-seven to twenty-nine years earlier and had been largely thwarted there in the beginning by spontaneous peasant partitioning. As in the early Soviet period, state farms were to serve several purposes: to act as model and research farms for new production methods; as seed-growing and livestock-breeding enterprises; and as nuclei for state (which later became Communist) influence on the countryside. Their share in total agricultural land remained small in the initial period. Only in Poland and Czechoslovakia, with their large stretches of expropriated and deserted land on former German-settled territory, did it amount to as much as 11 percent by 1948 and 1949, while in Rumania it was 5 percent, and in the other countries less than that.

Such land reform tended to increase the social tension between small (including the new) peasants, and big peasants, who had remained and were deprived of only part of their land. As in early Bolshevik policy, the wager was on the economically weak and on leftist or radical peasant organizations, where such already existed, and the regimes aimed at winning these over politically. Often, too, owners of medium-sized holdings, who were hostile to the regime, were the target of "class warfare" incited by the regime.

With all that, it must not be forgotten that governments in Eastern Europe, once they had become fully Communist controlled and once they had decided to pursue collectivization, had less fear of active peasant resistance than had the Soviet government and Bolshevik Party twenty years earlier. The Soviet troops, garrisoned over most of the region, were potential and reliable organs of suppression if the regimes were in danger. On the other hand, the regimes in most countries were not entirely Communist during the first postwar years and therefore were not wholly reliable from a Communist and Soviet point of view.

The political turning point, as far as agriculture is concerned, but also in other respects, came in 1948. By then the Communists were firmly in power in all East European countries of the Soviet sphere of influence. In June of that year, the Soviet representative, A. Zhdanov, pointed out in a speech at the Cominform Bureau that the Soviet road to socialism was to be regarded as the model of development for all. This, of course, also implied collectivization of peasant agriculture, though it was not publicly stated at the time. The simultaneous rift between Moscow and Belgrade apparently speeded up matters on this front.

In the early postwar years, Yugoslavia and Bulgaria were quickest to undertake early steps toward collectivization. Bulgaria had a few dozen peasant production cooperatives even under the old regime in 1944 and before; by autumn 1944 their number had increased to 579. In Yugoslavia, most of the collective farms of 1945 were so-called working cooperatives of war veterans and were mainly based on expropriated land. In late 1946 Yugoslavia already had 454 peasants' working cooperatives of various types. In 1946, the Yugoslav government decided to favor only those that were similar to the Soviet kolkhoz, and the 1947–51 economic five-year plan provided that at its end, half of all arable land should be in collective farms. Albania, at that time under strong Yugoslav influence, began to create collective farms by 1946. In other countries, comparable early attempts were either wholly absent or concerned only a few cases.

As can be seen, there was no reason to reproach Tito's Yugoslavia for not wanting to collectivize the nation's agriculture. And in fact, the published documents of the famous Soviet-Yugoslav controversy of 1948 contain no such assertion; rather, they contend that the Yugoslavs precipitated collectivization without adequate social and economic preparation. Nor does the published anti-Yugoslav Cominform resolution of 28 June 1948 contain a decision to proceed with collectivization in the rest of Eastern Europe. But it remains a fact that immediately afterwards, all the East European governments took official steps in that direction, and this was preceded by a Soviet collectivization campaign in the newly annexed Western territories of that country, starting in late 1947.

By far not everything has been published or otherwise clarified on these events, but it is obvious that beginning in the second half of 1948 what almost amounted to a race among the East European states ensued for the prize of first achieving the socialization of agriculture. Yugoslavia seemed to be in the leading group, with Bulgaria and Czechoslovakia. East Germany, where in autumn 1949 the Soviet Military Administration was replaced by a German Communist government (which in practice was very much under Soviet orders), was an exception. Obviously with a view to Soviet foreign policy goals, collectivization there was officially started only after Stalin's last effort (in his letter of March, 1952) had failed to initiate a development toward a neutralized German state comprising the whole nation. The other governments and Communist Parties pressed on with collectivization, and especially in 1949, the voluntary nature, adhered to in theory, was obviously infringed upon in practice rather often. This seems to have been less so during 1950–52, but considerable indirect pressure continued to be applied at any rate.

It is impossible to draw a clear line between application of force and indirect pressure, if such pressure is exerted by the state and goes beyond what in the West are considered legally and morally acceptable means. This is especially so where para-governmental organizations (like the Communist Party) hold great power and are more or less openly supported in their actions by the state apparatus; under Communist regimes, they in practice often have even greater power than the public administration. Without trying to draw a distinction between the application of force and that of indirect pressure, I shall very briefly outline the measures most commonly used in Eastern Europe to make hesitant or resisting peasants join the collective farms.

Most important was the general system of delivery quotas for agricultural products. As the prices for these were set by the state, and in an inflationary environment did not cover real production cost, nor equal a free market price, compulsory deliveries were a burden on the peasant. Their quotas, usually per hectare, were set disproportionately higher not only for the larger farms, but also for individual peasants in general as compared to collective farms. Moreover, the public authorities had possibilities to manipulate their assessment.

Peasant incomes could be improved by sales at much higher prices on the residual free market, which could be done legally after delivery quotas were met. The peasant had a vital interest in such sales, even before or without fulfillment of his delivery obligations, and therefore had a strong incentive to evade the regulations imposed upon him. To some degree almost every peasant did so and thereby became liable to fines, imprisonment, or expropriation. It was, of course, up to the authorities to decide how thoroughly they wanted to investigate individual actions, and how they would evaluate them.

Taxes, too, were disproportionately higher for well-to-do peasants and were lowest for collective farms. As they were mainly based on total income (including non-monetary income) and as on a farm such income has largely to be estimated, the authorities were able to discriminate heavily against a peasant, if they wanted to.

Inputs—machinery, fertilizer, quality seed, etc., and the corresponding services rendered by supply co-operatives and machinery stations—were in great shortage, and their supply was almost totally a state monopoly. State and collective farms were given first priority, and only what remained was distributed to individual peasants. Because of the shortages and the fixed prices this was more a matter of distribution practice than of the financial capacity of a given peasant. Again, discrimination was easy to apply when this was to serve a purpose approved by the authorities.

"Class warfare" against "kulaks" (rich peasants) was a declared policy, as it had been during Soviet collectivization. As then, the definition was very loosely applied. Once a peasant was declared a kulak (and, by definition, an enemy of the regime and of progress), almost any measure could be taken against him, up to deportation for some minor offence. Even a small peasant could be considered one who was on the side of the kulaks, especially if he spoke out against collectivization or against the regime in general.

This set of restrictions, measures, economic and administrative levers, and last but not least, the whole political atmosphere of Stalin's time, offered many ways to make life difficult for peasants who actively or passively resisted collectivization. The preceding enumeration is not complete, but it comprises the most important aspects. Under these circumstances, what is surprising is not the progress of the collectivization as such but rather its slowness. Quite possibly, the regimes did not want to apply their powers fully at once, as this might have meant a disruption of food production beyond what was tolerable even in the somewhat improved economic situation of four to five years after the end of the war.

By 1952 or 1953, the share of total agricultural land in farms of the kolkhoz type was minimal in East Germany, ranged between 5 and 10 percent in Poland, Rumania, and Albania, and amounted to roughly one-fifth in Hungary (for a very short interlude in Hungary, in summer 1953, they occupied three-quarters of all arable land). Kolkhoz-type cooperatives accounted for more than one-quarter of agricultural land in Czechoslovakia and for more than one-half in Bulgaria. In Yugoslavia, where in 1951 every sixth peasant household had been in a collective farm, the process suffered a reverse, and by 1954 only an insignificant number of such farms remained.

Several factors combined to effect this change in Yugoslavia. Being under heavy outside pressure from the Soviet bloc, and having to rely heavily on the loyalty of the nation, Tito could not afford to ignore peasant resentment against collective farms. Two years of severe drought (1950 and 1952) had made the country's food situation precarious, and any measures that acted as disincentives on the peasants' willingness to produce and to go to market were bound to render it worse. Moreover, the laws of 1949 had assured the Yugoslav peasants that they would be allowed to leave the collective farms three years after entering, if they so wished. Since many had joined in 1948 and 1949, this trial period elapsed by 1951/52. Under the politically more liberal atmosphere of that time, it was simply not feasible to abandon that commitment. At last, in a decree of 30 March 1953, the legal land ownership rules were redefined, the failure of the previous collectivization policy was admitted, and leaving the collectives was permitted even if the three years had not yet elapsed. The great majority of Yugoslav collective farmers took advantage of this possibility to quit the collectives.

Beginning in 1949, a process of revision of earlier Model Charters started in Eastern Europe and the policies emerging during the last year of Stalin's life showed a number of common characteristics. For example, the words *collectivization* and *collective farm* (= kolkhoz) were avoided everywhere, to be replaced by words such as *reorganization, cooperation,* etc. For the collective farms themselves, various terms existed, conveying the meaning of a production cooperative based on common agricultural work and ownership, except for the remaining subsidiary plot and animal farming of member households. Apparently, it was feared that an open reference to the Stalinist model of the 1930s would have a deterring effect. Nevertheless, without being labelled as such, this model in fact applied. Yugoslavia, Hungary, Poland and (since 1958) Bulgaria, provided in the Model Charters for the possibility of peasants leaving the collective farm of their own will, and this departed from the usual pattern in an important formal aspect at an early stage. The Model Charters of the other countries kept silent on this question, as did the Soviet kolkhoz charter up to 1969.

Except for Bulgaria, where since 1950 only one form of collective farm (similar to the Soviet kolkhoz) was provided for, and for Rumania, where there were two forms up to the mid-1950s, three or four types of collective farms existed in all countries. But in their essential contents, these were actually two types, with the addition of one or two intermediate types, which were considered as transitory forms. This pattern emerged more clearly when the collectivization drive accelerated in 1949 and 1950, and in all countries new obligatory Model Charters were officially promulgated, replacing earlier, less uniform charters.

The Model Charters for Type III (or III and IV) farms almost wholly copied the Soviet kolkhoz charter in content, if not in form. (For a description of the internal organization on the basis of the Model Charters, see chapter 5). These Type III and IV farms represented what was called the "higher type" of collective farming. At the other end of the scale, Type I farms (the "lower type" of collective farming) were production cooperatives where only the fields were cultivated in common and the results distributed among the members according to their labor and their land contribution. In most such cooperatives, the distribution was effected after deducting the collective's obligation towards the state and other agencies and for investment, but according to some charters, the gross result, that is, before deductions, was distributed. This resembled the Soviet precollectivization T.O.Z. (see p. 17) and other forms that had existed in the Soviet Union up to 1929. There were even looser forms of peasants' associations, especially in Rumania, Yugoslavia, and, at a later stage, in Hungary.

In the intermediate Type II farms, working animals, implements, and machinery were also collectivized, but most of the productive livestock still remained the private property of the members. In the early years, this was called the "Bulgarian" type, although it soon disappeared there.[1] It was closer to Type I farms than to Type III, and was considered to be a "lower" type. Both lower types were devised explicitly to make it easier psychologically for individual peasants to change over to a collective form of production. Accordingly, the number of such collectives was great in the initial stages of collectivization. Later on, they either were tranformed into, or joined with (they usually were of smaller size) a farm of the "higher" type.

It is hard to assess the impact of the early collectivization years (1949–52) on agricultural output in quantitative terms. Figures for Hungary indicate that grain output stagnated, in spite of expanding areas, and that potato output went down, as compared with 1948 results. Part of this effect was due to the diversion of valuable arable land to cash crops needed for industry and demanded by the top planners. In Rumania grain output had not regained the prewar levels by 1947 and decreased sharply during the following years, while potato output increased threefold after 1949 and more than doubled the prewar level. Comprehensive data for the years 1945–48 are rare and not reliable. On the whole, it may be said that animal production was most affected by collectivization, but less so in the incipient stage, when part of the collectives were of the "lower" types, where the productive animals remained in private ownership. The indices of gross agricultural production (in constant prices), as reported by the United Nations' Economic Commission for Europe, on a multiannual average rose by roughly 3 percent per year in East Germany (where there was no collectivization as yet), Rumania, and Bulgaria; they remained more or less stagnant in Poland and Yugoslavia and went down by one percent per year in Czechoslovakia during 1951–54.

This was a slow overall development in any case, especially in view of the preceding wartime setbacks (which had not yet been made up by 1948) and the faster recovery in Western Europe. It is also a fact that with the relaxation and reverse or slowdown of the collectivization drive in Poland, Hungary, and Yugoslavia, the output growth there became faster again after 1955. But this does not prove much, as it may also have been due to the increased investment growth rates. Moreover, Bulgaria, where the collectivization went on during 1955–57 and was completed by 1958, had a comparable output growth. (For more on the development of output during collectivization, see chapter 8.)

It has sometimes been asserted that collectivization in Eastern Europe was carried out in accordance with national conditions rather than after the earlier Soviet model, especially since less brutal force and haste were applied, and the peasants were allowed to join various "lower" forms of collectives, while only the "higher" equalled the Soviet kolkhoz. To the degree that this is true, it is highly probable that it was Stalin and his advisers who wanted it this way, apparently because they had learned some lessons from the years 1929–34. If the more cautious, yet still forceful, initial drive for collectivization in Eastern Europe had not been devised in Moscow, it would be hard to explain the almost uniform approach by means of essentially two types of collective farms through-out all of Eastern Europe (except for Bulgaria since 1950) and even in China (at that time) and North Korea. Not only did the "lower" forms have precursors in pre-1929 Soviet Russia, but there was also the pro-vision in all model charters (except for some Type IV farms) to distribute part of collective income according to property brought in, as had been practiced, though on a limited scale, during the early years of Soviet collectivization.

If one takes a closer look at the speed of collectivization, one finds that it was not as dissimilar from the Soviet case as might appear at first. One has to bear in mind that the standstill or, in some countries, partial reverse, of collectivization, during 1953–56, and its long-term delay in Poland and Yugoslavia, was not part of a planned policy, but is attributa-ble to external events: the Cominform-Yugoslav rift and Stalin's death with consequent unrest in both Eastern Europe and the Soviet Union. Abstracting from these events, it may be said that there was a short period (early 1946 to mid-1948) comparable to the New Economic Policy (1921–27) in the Soviet Union. Looking at the intensity of the collectivi-zation drive after 1948 and the successes achieved in Bulgaria, Czechos-lovakia and Hungary by the time of Stalin's death, one may conjecture that its completion was envisaged for the mid-1950s, that is, within seven to nine years, compared with the four to seven years it took in the various parts of the Soviet Union after 1928. The above-mentioned external events then retarded it by another five to six years.

A distinct similarity to the Soviet model was the creation of machinery stations (state-owned pools), fulfilling the same functions as their Soviet counterparts, the MTS (see p. 23 above), and often given the same or a very similar name. Those agricultural machines confiscated from former estates and the bigger peasant farms, and those newly produced, were in their hands, if not in possession of the state farms, which had priority in supply. In total, there was not much machinery available, partly because there had not been much before the war, and partly because some of it had been destroyed or worn out during the war. Out of current production, nothing was sold to individual peasants. In the beginning, with few collective farms existing, the MTS served mainly the peasants, in the first place the small peasants at preferential rates. This differed from Soviet procedure only insofar as machinery (and other industrially produced inputs, such as fertilizer) were available to a relatively greater extent in postwar Eastern Europe than before and during collectivization in Soviet Russia. Thus, a material basis, although a small one, was available for large-scale farming already in the initial stage of East European collectivization.

As in the Soviet Union, the overall economic policy was to give industry, especially heavy industry, as much as possible of available capital resources. Although such a policy was not taken to the same extreme as during Soviet collectivization, demand for consumer goods and for capital inputs for agriculture was consciously neglected. However, for Czechoslovakia and East Germany, and to a lesser degree Poland, capital accumulation at the expense of agriculture was of secondary importance. There, agriculture did not play such a dominant role in the composition of the gross national product, and accumulation by industry itself was more feasible. Nevertheless, it remains true that a certain amount of capital was deliberately transferred to industry by means of the fixed prices, not to speak of labor. Where labor was already short, large-scale, modernized farming was a way of releasing it for industry, as seen mainly in Czechoslovakia (and in East Germany after 1952, when collectivization started there). In these respects, the situation differed from that of Russia during collectivization because a higher degree of industrialization had already been attained. Yet such national variants did not affect the policy of capital transfer as such.

A major deviation from the Soviet model was the fact that land was not nationalized (formally, socialized), as had been done immediately after the October Revolution in Russia. Nor was it juridically made collective property upon the creation of a collective farm (except for the "higher" type in Albania). Although in the early postwar years, the policy, first enunciated in the Bulgarian Model Charter,[2] was not without ambiguity (in Rumania it was not made clear until as late as 1956), it applied equally to all East European countries: formally, the peasant remained an individual owner of the land he had brought into the collective farm. In actual practice, this title to the land became almost irrelevant as time went by, because there were so many restrictions on selling or inheriting land holdings that the right of ownership lost its economic importance. The collective farmer could sell, bequeath, or donate land only to the state or to the collective, or to a comember of the same collective, and in this way, the public sector acquired much of the land. Yet the fact remains that juridically, land ownership differed from the Soviet pattern.

Several reasons can be discerned why peasant land, including that received during the land reforms, was not formally socialized or nationalized at once. First, there was the century-old striving of the peasants, among them the small peasants on whom the regimes tried to rely, for land ownership. Second, the precarious position of the tiny Communist minority in the villages in spite of their political domination from their urban strongholds had to be considered. Also, the aim was not to let the population and the Western former allies realize too early that the installation of purely Communist power and the collectivization of agriculture were the ultimate goals from 1945 on. In later years, such reasons lost their importance, but by then the progress of collectivization and the practical restrictions on the exertion of land property rights had brought about the socialization of land without a formal legislative act.

Looking at agrarian policies in postwar Eastern Europe up to the time of Stalin's death, it must be kept in mind that prices and costs of production were secondary questions. Prices were fixed for compulsory deliveries, the rest of output being left to the peasant to consume or to sell on the free market as "surpluses." An increase in the volume of production (of crop production in the first place) in order to satisfy the food requirements of the urban population, and also of the Soviet armed forces in the respective countries, was a primary goal. With necessarily few technical inputs in an overwhelmingly small-peasant agriculture, labor was the main input. It was available in most cases, the question being only how to organize it, or rather not to disorganize it by disproportionately low returns and/or by precipitated political measures. Peasants' real incomes were sufficient under the circumstances, when enough food for the family out of its own production was an important advantage of agricultural compared to other labor, and when the free market for "surpluses" provided at least some cash, not to speak of possible barter operations.

With collectivization beginning, other elements influenced such a policy. A greater share of the output of the new large-scale producers was being delivered to the state, and the marketing channels for it could be more easily controlled. For the collectivized peasant it became more difficult to withhold produce for his own consumption or for free marketing, and thereby he lost part of the incentive to produce in spite of low official prices. Early attempts to close the legal free markets (the counterparts to the Soviet kolkhoz markets) were soon given up; in Bulgaria, for instance, the Party Congress of July 1949 found it necessary to officially readmit such markets. It was inevitable to pay higher producers' prices to the collective farms than to the individual peasants, not only because they needed more capital inputs than small farms but also because they had to make up by labor remuneration at least part of the incentives which were avilable to the individual peasants. Although in practice such remuneration proved far from sufficient to achieve good labor morale, the fact of higher prices being fixed for collective farms was a general feature.

From the Communist point of view, collectivization as a socio-political action was justified in itself. The only restriction, apart from considerations of not destabilizing Communist power, was that a transition period of disorganization, with a corresponding decrease of the volume of agricultural output, had to be avoided. Economic efficiency of the new large collective farms could be neglected for some time, if there was reasonable hope that total gross output would soon increase fast enough. Given these goals and conditions, the first stage of collectivization in Eastern Europe may be called a success. Things began to look different when the regimes became politically unstable in the years after Stalin's death, when during the subsequent restabilization period (after 1957) economic efficiency and, in some countries, scarcity of agricultural labor (relatively, in relation to capital available for substitution) became major problems, and when, finally, industrialization and urbanization put increasing demands on the quality of food and on the availability of agricultural raw materials for industry.

Stalin's death, in early March 1953 was followed by an almost immediate change in Soviet agricultural policy. After a few months of infighting, the change found its reflection in published decrees. N. S. Khrushchev became the policymaker for agriculture until his demotion in October 1964, though he was not undisputed at the beginning and again toward the end of this period.

Khrushchev's impact not only on Soviet but also on East European agricultural policies can hardly be overestimated. The eleven years of his reign in the Soviet Union and of his dominance (though not a dictatorship like that of Stalin) over the other bloc countries ushered in a new period for the whole region. It was characterized by Khrushchevian reforms of the Stalinist system and at the same time by the completion of collectivization outside the USSR, with the exception of Poland and, of course, Yugoslavia. The shape, which the agrarian system took in the bloc countries was influenced by these reforms already while collectivization was implemented. Most of the changes were more or less strict emulations of Moscow's policies, and only a few minor differences emerged in the East European countries. Therefore to understand the bloc-wide process, it helps first to show the major developments within the Soviet Union up to the early 1960s, and on this background to outline the collectivization and the simultaneous reform process in Eastern Europe.

Notes

1. Jerzy Tepicht, *Marxisme et agriculture: Le paysan polonais* (Paris 1973), p. 55.
2. *Ibid.*, pp. 51, 54.

4. A New Deal for Soviet Agriculture: The Khrushchev Era

General Features

THE FIRST MEASURES of Soviet agrarian policy after Stalin's death were aimed mainly at eliminating extreme or economically absurd severities of the preceding period. Only then did a new line evolve, whose underlying ideas were:

1. Increase of output, to be achieved
   a) by expansion of the agricultural area and
   b) by more capital inputs
2. Modernization of the production process not only by increased capital inputs but also by means of more efficient organization, improved agricultural techniques, and better spatial allocation
3. Increase of incomes in agriculture as an incentive (instead of the former coercion) for better performance by managers and workers
4. Increase of the share of the socialized sector and the meeting in full of the country's food demand by this sector so as to make the private sector's contribution dispensible

The relevant measures were meant in the first place to increase the volume of output, combined, where possible, with better quality, and, in the second place, to reduce production costs. As time went on, the second aim, that of economic efficiency, gained in importance. But the increase of labor cost far outran any reduction of other costs, so that prices went on increasing, and the state had to renounce much of its profit margin on food and eventually even subsidize the procurement system in order to keep consumer prices stable. Such a policy was directed at socialized farming, while the private sector was favored only at the beginning and was soon restricted.

With all that, three facts must be kept in mind. First, in spite of setbacks, the output growth was respectable. Calculated in Soviet gross output figures (at 1965 prices, as published in 1971), it increased from a 1950–52 three-year average of 36.6 billion rubles by 56 percent up to 1956–58, then slowed down but still achieved a total increase of 78.5 percent up to the 1962–64 three-year average. Second, if such growth turned out to be unsatisfactory, the causes must be sought in a wider context. The pent-up demand for food was great, industrial expansion and population growth made the purchasing power of the nonagricultural population increase at a faster rate than food production, and Stalin's agrarian policy had economic and psychological effects beyond its actual application. Third, Khrushchev was under heavy time pressure not only because of the above-mentioned factors, but also because his policy change against some internal resistance compelled him to demonstrate quick successes. It is not by accident that the word "campaigning" *(kampaneishchina)* is not infrequently found in retrospective Soviet and Western writings on this period. Haste and exaggeration, which may well have been also an outflow of his personality, made many of his measures have harmful effects, although they aimed in the right direction. His main mistake seems to have been that he overestimated the chances for quick success, and too early thought that the fast production growth of 1953–58 would continue. In this belief, he reverted to ideologically determined targets as early as 1957, probably because he himself was not ready to renounce them.

For an overview of agricultural policies under Khrushchev and their logic and shortcomings, it is useful to examine them in two ways:

1. Those features receiving varying but recurrent emphasis throughout the period (mechanization, labor and investment, cropping patterns, expansion of the livestock sector, prices, farm sizes, and rural resettlement).

2. Those which can be more meaningfully ranged chronologically, and which yield a picture of the shifts of policy mainly in organizational terms (procurement system, planning, remuneration and mobilization of labor, restriction of the private and expansion of the state farm sector, liquidation of the MTS, start of the fertilizer and irrigation programs, and shortlived administrative reforms).

Paramount for an understanding is the fact that Soviet agriculture was still not highly mechanized by the standards of the time, in spite of what the propaganda contended. If by nothing else, this is evidenced by the great share of the country's labor force that was (and is) employed in agriculture. According to the results of the January 1959 census (the first after the war), it amounted to 38.8 percent, among whom 24.3 were classified as workers "without special skills"; including the private sector the total was about 44 percent.

Efforts were made to increase the share of trained cadres (technicians and managers) in the agricultural labor force either by transferring available staff from urban and administrative posts or by training new personnel. Some success in this respect was achieved, but not to the desired degree. Thus, the number of machinery operators ("mechanizers") increased from 1.7 million in May 1953 to 2.95 million in April 1964.

By 1964, mechanization encompassed not only ploughing, grain seeding, and harvesting, but also the cultivation of maize, potatoes (except, of course, on private plots, which still comprised more than half of the potato area), sugar beet, and cotton as well as harvesting of sunflower and silage crops and mowing of hay. However, most other activities were mechanized only in part, especially in animal production. Because of the "progressive" label attached to mechanization, the policy was in some cases pursued in spite of the fact that vast underemployed labor resources were available locally, for instance in the western Ukraine or in the Asian cotton-growing areas. Another problem was that the standardized very big machines could not always be used efficiently on the relatively small fields in the western and northwestern parts of the Soviet Union.

Annual state investment in agriculture, mainly in sovkhozes, increased six fold from 1953 (985 million rubles) to 1964 (5,786 million). The most conspicuous increase, almost a doubling, occurred in 1954, and was largely due to investment in the newly ploughed Virgin Lands and the state farms there. From 1957 to 1959, annual state investment decreased somewhat, but then went up again, in part because of the mass conversions of kolkhozes into sovkhozes in 1959–60. The quantity of machines supplied to agriculture fluctuated in a similar way. Not all of the capital inputs were put to good use, partly because of inefficient handling on farms and partly because of erratic changes in the centrally directed production policy. (On kolkhoz investment, see p. 57).

All this adds up to the contradictory picture of an agriculture where mechanization is greatly emphasized for economic as well as sociopolitical reasons, but is not really achieved. The problem was aggravated by Khrushchev's later political course—in itself fundamentally sound—of intensifying agricultural land use, because this increased the labor requirements while the available work force decreased.

Maize is a quite labor-intensive crop, and its cultivation was greatly advocated. On Khrushchev's instigation the Central Committee adopted a resolution on 31 January 1955 for expanding the area sown to maize from 2.5 million hectares in 1954 to 28 million by 1960. Although Khrushchev's recommendations for planting maize already transgressed reasonable limits, they were followed later on and were surpassed indiscriminately by lower echelons. The total maize area reached 36.5 million hectares in 1962, accomplished partly by a severe reduction (by almost two-thirds) in the area devoted to oats. Maize was sown in climatically unsuitable northern and eastern regions, allegedly as a profitable silage crop. However, apart from bringing low and sometimes no yields there, it also lacked digestible protein, especially where the silage contained few or no cobs of milk-wax ripeness. So it represented an unbalanced animal feed ration in most cases, and depressed animal productivity in spite of large amounts being fed. Moreover, specialized machinery for sowing, weed control, and harvesting of maize was totally lacking in the beginning. Its production, though increasing, rose far more slowly than the rate of expansion of the maize area. Thus, the maize program contributed to the worsening of the labor situation.

Khrushchev was undoubtedly right, when, beginning in 1954, he opposed the expansion of Lysenko's grass-rotation system with its large tracts sown to perennial grasses in dry farming regions unsuitable for such production. But in later years the campaign was directed indiscriminately against all grasses, including clover, which is a valuable crop in the more humid, western parts of the Soviet Union.

Another of Khrushchev's campaigns was against summer fallow. To do away with what practically was a traditional three-field rotation in the non–black earth region was basically sound, although it contributed to exacerbating the already strained labor shortage there and deprived the population of some possibilities for grazing privately owned livestock. But the eradication of summer fallow also in the dry-farming regions of the south and east was against agronomic reason. It contributed greatly to the declining yields there, and to dangerous wind erosion on vast former steppe, now grain, areas.

Toward the end of his rule Khrushchev's emphasis on maize lessened somewhat, and this resulted in a decrease of the maize area by roughly one tenth. Instead he emphasized fodder legumes, especially peas, as a crop that would supply the protein lacking in maize silage. At that time he also advocated the growing of sugar beet for feed and had the area for this crop expanded. This, however, was a short interlude brought to an end by his removal from power. Similarly, once he departed, the area sown to maize declined drastically to 23.4 million hectares in 1965.

One of the lasting changes in the overall cropping pattern was the expanded area of cotton on irrigated land. Stalin's venture of growing cotton on unirrigated land in the south of the European part of the country was dropped. Cotton output increased even more quickly than the cotton area because, with increasing amounts of fertilizer applied to cotton, and with improved cultivation and varieties, yields increased considerably. But there was also the danger of cotton monoculture, with consequent plant disease, instead of a rotation system with lucerne as a valuable alternative.

The changes in the overall cropping pattern were intended mainly to expand human consumption of crop and also livestock products, the latter through enlarging the feed base for more animal production. The task of providing bread grain was solved by the expansion of agriculture into the dry steppe regions. Except for catastrophic years, e.g., 1963 and 1975, bread grain supply was no longer a problem. Difficulties in providing foodstuffs such as vegetables and fruit were, and are, more a problem of transport, storage, and distribution than of output volume. Things look different, though, when it comes to animal production. Most of Khrushchev's programs, indirectly even the Virgin Lands campaign, were aimed at increasing the feed supply for the animal industry. They were paralleled by efforts to increase the livestock herds. Numerically, this was achieved until 1963, but then the drought caused extensive emergency slaughtering and thus a reduction in herds, above all in pig herds. But the achievement was less when looked at in terms of actual meat, milk, and wool output. The increased herds were not always of good breed. Moreover, the fodder basis grew less than expected and less than the animal numbers required, and the nutritive value of much of the feed was unbalanced. Therefore, annual milk yields per cow remained low, though they increased somewhat, and pork production per sow was very inadequate.

Agricultural producer prices were raised throughout the Khrushchev period, and later, and the great disparity towards industrial prices was lifted. As early as the decree on agriculture of 7 September 1953 considerable price increases were provided for basic procurements as well as for over-plan sales to the state. During subsequent years further increases were applied and were extended to other products (see Table 4.1).

These enormous price increases must be seen against the abysmally low level of the preceding time. Still they were substantial enough to result in producers' prices above world market prices (at the official exchange rate) and near those of small peasant agriculture of Western Europe by 1962. Prices for grains and potatoes, and also sunflower seed rose more than average for crops, while those for cotton and sugar beet increased less because they had already been privileged in Stalin's time. Similarly, eggs and wool remained below the average increase for animal products. On the whole, it may be said that at the end of the Khrushchev era, meat was overvalued relative to grain. For example, the ratio of the procurement price for grain to that for live pigs more than doubled, from 1:7 to roughly 1:15, during 1952–62.

It must be borne in mind that these are average prices of all kinds of procurements, including basic procurements, over-plan sales, and the so-called contract sales of cash crops. However, they do not include prices paid to state farms, which as a rule were lower than those for kolkhozes, but which they began to approach after 1958.

The price differentiation by regions became pronounced with the 1958 reform of the procurement system. On the other hand, prices of industrial goods for agriculture, such as machinery, trucks, lubricants, etc., which now the kolkhozes (and no longer the MTS) had to buy, did not remain stable but increased markedly during 1958–61.

In spite of the great procurement price rises, the increasing produc-

Table 4.1   Indices of Average Soviet Procurement Prices Actually Paid to Collective Farms and Individuals (1952 = 100)

Product category	1953	1956	1958	1960	1962	1964
Total of agricultural products	154	251	296	299	332	354
All crops	132	207	203	202	219	...
Grains	236	634	695	717	840	840
Potatoes	316	814	789	885	1043	1500
Cash crops (unginned cotton, flax, hemp, sugar beet, tobacco)	115	147	143	139	143	158
All livestock and livestock products	214	371	546	562	639	...
Meat animals	385	665	1175	1243	1523	1609
Milk	202	334	404	405	434	455

Source: S.G. Stolyarov, *O tsenakh i tsenoobrazovanii v SSSR*, 3rd edition (Moscow, 1969), p. 121; all crops and all kinds of livestock and livestock products up to 1962, according to A.N. Malafeyev, *Istoria tsenoobrazovania v SSSR* (Moscow, 1964), pp. 412-13.

tion costs of kolkhozes made themselves felt. This applied predominantly to animal production, where, owing to a low degree of mechanization, labor inputs increased with the expanding livestock herds, and wages also increased. In most cases, kolkhozes produced the required animal products at a loss. Much against their will, the leaders, especially Khrushchev, felt compelled to raise the procurement prices for animal products by 35 percent for meat and by 10 percent for butter in June 1962. This time retail prices, too, were raised, on the average by 30 percent for meat and 20 percent for butter. All the same, livestock production has remained an unprofitable branch of kolkhoz agriculture.

Concerning the organizational development of agriculture there were two outstanding features of the Khrushchev era: the liquidation of the MTS and the expansion of the state farm sector. In addition, the process of increasing farm sizes continued, as illustrated by the figures in Table 4.2 (one kolkhoz household was equivalent, on the average, to roughly 1.2–1.4 fully employed workers, this ratio moving to the lower margin during the 1960s).

The increase in sizes was caused by mergers of enterprises. For example, kolkhozes would be amalgamated (and sometimes then be converted into a new sovkhoz), and often one or more kolkhozes would be absorbed by an existing sovkhoz. Also, the expansion of agriculture into dry-farming areas was accomplished through the creation of new farms (as a rule, sovkhozes) far larger than average. Although it was at times

Table 4.2   Increase in Size of Average Soviet Farm

Year	Average size per kolkhoz		Average size per sovkhoz	
	Hectares of agricultural land	Number of households	Hectares of agricultural land	Number of workers (annual average)
1937	1534	76	12200	285
1953	4211	220	13100	352
1965	6100	420	24600	663

realized that some farms were too large to be manageable, instances of their division were by far outnumbered by the mergers. Only at the end of Khrushchev's reign did the increase of average farm sizes come to a temporary halt.

With the extremely large farms the task of subdividing them by sections became more acute. Only a few farms in the Southern steppes did not comprise more than one village, while most kolkhozes comprised three to seven villages, and sovkhozes even more. The subunits were brigades for kolkhozes, and sections (*otdelenie*) for sovkhozes. Often some of these were specialized for a certain branch of production, such as cattle, pig-fattening, wine growing, etc. (in sovkhozes more so than in kolkhozes). When the MTS were incorporated in the kolkhozes, a widespread form of subunit was the combination of a tractor brigade of the former MTS with a kolkhoz brigade of manual workers. The degree of autonomy of such subunits varied according to circumstances. Although there was much discussion on this question, the policy never was to grant them full independence, which would have been equivalent to splitting up the oversized farms.

Below this level, the question of a smaller unit, the so-called link (*zveno*) came to the fore again after 1962. It had been an issue in the late 1940s for grain growing, but at that time, the idea of giving such small groups a degree of their own responsibility was rejected. For manual work, especially with labor-intensive crops, the *zveno* has always existed, but was wholly subordinate to directives from the brigade leader. The new *zveno* controversy concerned links which had machinery at their disposal and would be granted a considerable degree of economic independence. Some such links emerged, tentatively, in 1962 and later, but the issue remained hotly disputed, and no decision was made under Khrushchev.

Connected with mergers and enlargement of farms was another scheme, that of the *agrogorod* ("agrotown"), which had always been a pet idea of Khrushchev. He had suffered a rebuke under Stalin when he advanced it in 1950 and 1951, but took it up again in 1957, when he was the master of Soviet policies. The essence of the idea was to concentrate the scattered villages into settlements of an urban type. This correponds to the persistent Marxist goal of adapting agricultural production to industrial methods, and of changing the rural life accordingly. At times, Khrushchev advocated such settlements to comprise only a few huge houses for several thousand inhabitants. Of course, the agrotown also implied greater concentration of agricultural production and partial or total liquidation of private agricultural activities of the workforce. All economic and cultural facilities were to be provided in these houses or adjoining to them, and buses or trucks would bring the workers to the more distant fields and livestock premises during the day. But usually an "urban-type settlement" on the countryside was not envisaged in so extreme a form. Although Khrushchev had to scale down these plans toward the end of his rule, he never abandoned them entirely.

Major Measures in Chronological Order

The first measures of the new policy after Stalin, apart from reducing the agricultural tax in August 1953, were contained in the decree of 7 September 1953, "On measures for the further development of agriculture" and those of 26 and 29 September, on animal farming and on the increase of potato and vegetable output and procurements, respectively. The latter two granted substantial increases of procurement prices. Contract sales, which were already in general use with cash crops, were enlarged in order to encourage other increased over-plan sales to the state. In April 1954 a fixed proportion of such sales was set relative to procurements.

A new feature of the September 1953 decrees was the introduction of "permanent hectare norms" of procurements for whole districts (raiony). They were devised, on the one hand, to exclude annual fluctuations of per hectare procurement norms, which in practice was a logical precondition of contract (i. e., advance) sales. On the other hand, they were to exclude unequal procurement burdens for the individual kolkhozes. In years of good or record harvest, the over-plan sales accounted for a large share of the total and thus raised the average price level (e. g., in 1956 and 1958). By reducing basic procurement obligations (percentagewise, if not in absolute volume) in favor of over-plan sales, the government increased the average price actually paid, apart from increases of the basic procurement prices. This is what in fact was effected by the various measures of 1953 and 1954.

The amount of grain going to the MTS for their machinery work was indirectly somewhat reduced by the introduction in 1953 of fixed rates of payment, in place of a percentage of the yield of the respective fields. Even so, such payment in kind still accounted for the bulk of kolkhoz grain coming into the state granaries: in 1956, the figure was 47.2 percent, against 17.2 percent from basic procurements, 23.9 percent from above-plan and contract sales, and 11.7 percent from other means of procurement.

With the liquidation of the MTS in 1958, a fundamental overhaul of the procurement and price system became necessary because the single most important source of grain procurements now had been cut off. The new systems were enacted, on Khrushchev's instigation, by the decrees of 18 and 30 June 1958. The different kinds of procurements were now unified, and were all called state purchases. Purchase quotas were obligatory on the farms, as the procurements had been, so their essence did not change. However, above-plan and contract sales were abolished, and the new unified prices were close to the old above-plan prices.

With the reform of the procurement system, the quotas were, on average, fixed somewhat lower than before for grain, and higher for animal products, potatoes, and vegetables (among others). The procurement (now purchase) quotas were in principle set for a longer period of time, and so were the prices, but they were subject to annual reviews in case of bountiful harvest. What actually happened was that quotas increased and prices fell in the record harvest year of 1958, without a revision of quotas and prices during the following, less bountiful years.

One main defect remained: procurement quotas were still too high for many unprosperous kolkhozes, which therefore had to cut down on seed and other reserves (buying seed later from the state) and on payment in kind for kolkhozniks. The essentially sound venture of reducing or abolishing grain procurement quotas for regions where grain production was disadvantageous for climatic reasons was dropped three years later, in 1961, under the pressure of generally disappointing harvests and the state's urgent need for grain.

Another one of the early and most conspicuous post-Stalin measures was the expansion of crop farming into the dry steppe regions of Southeast Russia, Western Siberia, and, above all, Kazakhstan, initiated in spring, 1954.[1] Roughly 40 million hectares were involved, though not all of this area could be kept under cultivation. Almost all of these new lands were sown to spring wheat. As this was an extensive, highly mechanized method of farming, with little accompanying animal production, not much of the grain output was needed on the farms for feed and for the workers' consumption. The share of the new lands in government grain procurement therefore was great. It amounted to almost half of total procurements, as distinct from total output, during 1954–62.

The new lands, or Virgin Lands, program alleviated the delivery strain on the traditional grain producing areas and put them in a position to expand fodder and row crop together with feed grain production, and thereby to develop livestock farming. But it also meant that for a while, the bulk of capital investment was redirected to the new lands, and less was available for the traditional areas and for the animal sector.

In the short run, the campaign was successful, favored by two meteorologically propitious years, 1954 and 1956. But the unmitigated continuation of spring wheat monoculture during the following years, almost without summer fallow, rapidly depleted the accumulated fertility and moisture reserves of the virgin soil. Beginning with the drought year of 1957, this resulted in several bad or disastrous harvests, heavy wind erosion, and widespread weed infestation.

By the end of 1954, the new course of Soviet economic policy, with its accent on consumers' demand, was brought to a halt, though not fully reversed. Production of producer goods again became the undisputed favorite of planners. For agriculture, this implied that sooner or later a brake had to be put on the increase of the (still miserable, though increasingly improved) remuneration of labor, if it was not greatly surpassed by productivity gains. Otherwise inflation or rationing of food was inevitable, not only because agricultural earners would have more money to buy industrial consumer goods, but also because the increasing numbers of workers in the expanding heavy industries would have more money to demand more and high quality foodstuffs.

In the first two post-Stalin years, private agricultural production far outpaced the planned increase, especially in livestock production, and thus greatly helped to balance the growing demand. The socialized animal sector did not quite attain the set targets. Although this was compensated by the latter sector's expanding crop production, the shift of weight seems to have caused concern to Khrushchev. Yet before resorting to an increasingly restrictive policy toward the private sector (cf. below), he tried to increase socialized, that is, kolkhoz and sovkhoz, production.

An important and promising move in this direction was the reform of the planning procedure for agriculture. A decree was issued on 9 March 1955 (published on 24 March), which was to give farm managers more freedom in planning and some influence in the fixing of their procurement obligations. The procurement plans still were set by the government and its agencies, but the kolkhozes and sovkhozes were to decide themselves in which way, by which sowing pattern and kind of livestock farming, these could be met best. Moreover, they were to have some say in the initial stage of the formulation of procurement plans. But the decree remained a dead letter, like its timid predecessor of 28 December 1939. "Administrative planning" in practice continued to be followed for the next nine years, as was officially acknowledged by another decree, of

20 March 1964, which reemphasized the 1955 decree and threatened punishment to all state officials infringing on the still restricted planning autonomy of farms.

The year 1956 brought two important events: two decrees on kolkhozes, both published on 10 March, and a record harvest. The two decrees reflected the necessity to revise the 1935 Model Charter and, anticipating a kolkhoz congress and a new charter, gave the kolkhozes some leeway to adapt their charters to present needs. They dealt with three main issues, the first of which was higher and more regular labor remuneration. Increased cash wages, to be paid in part before the end of the year, were recommended to kolkhozes, and for this purpose, the advance payments on procurements and contract sales were increased. Similar recommendations had been part of the decree of 7 September 1953, but without great avail then. Now they were reemphasized and expanded, but again, this did not essentially change the situation. The labor day unit *(trudoden')* remained the predominant form of accounting and remunerating kolkhoz labor, although it had gained more real value during 1953–55 and again in 1956 as a consequence of that year's bountiful harvest.

The second issue of the decrees was to make more labor available for kolkhoz production. Kolkhozes were empowered to fix minimum labor day requirements per kolkhoznik and the daily work norms in accordance with needs. Such requirements were fixed generally by legislation in 1939, were stepped up during World War II, and had been revised upwards again in a central decree of 19 April 1948. The new right for kolkhozes to fix labor day quotas autonomously was devised as a right to increase both the minimum requirements and the norms, and that is what actually happened wherever this right was exerted. Thirdly, it was recommended to kolkhozes that they make the sizes of individual private plots dependent on fulfillment of the labor requirements: for non-fulfillment, part of the plots was supposed to be withdrawn.

There was also the decree of 27 August 1956 on restrictions of livestock holdings of the urban population and the prohibition on buying grain and grain products from retail stores for feeding to privately owned livestock. In addition, there was the announcement, by decree of 4 July 1957, that beginning in 1958, no more state deliveries would be exacted from private producers. This was interpreted as a measure of liberalization toward the private sector, which in fact it was not. On the contrary, it meant that the state, through increasing its own output, hoped to be freed from the necessity of receiving foodstuffs from the private sector, and thus also be freed from the necessity to pay attention to this sector's needs.

Spring 1957 saw the beginnings of the transfer of the MTS into kolkhoz property. At the same time, the mass conversion of kolkhozes into sovkhozes began. Stalin in his last years (according to Khruchchev much later, in his speech of 28 February 1964), had even considered the liquidation of the state farm sector. However, things had developed in the opposite direction. Beginning in 1954, the Virgin Lands campaign led to the organization of hundreds of gigantic sovkhozes there, and especially after 1957, most kolkhozes on those lands were converted into sovkhozes or annexed to the new sovkhozes. A related case is that of sovkhozes on newly irrigated land in Central Asia. Another motive for conversions was to organize large enterprises of truck and dairy farming near to big cities or other industrial agglomerations. But the conversions were not restricted to such cases.

Above all, the underlying idea was that sovkhozes present a "higher" type of socialist farming. Yet to become effective as a stimulus for conversions, such a theory needs practical impulses at a given time. It is true, the sovkhoz, being state owned, is more manageable in the framework of a centrally planned economy, but on the other hand, it has the disadvantage of often being a financial burden to the state. Under certain circumstances, in a more primitive stage of economic development, for instance, central planners may feel that their purposes are best served by an organizational form such as the kolkhoz, which enables them to get the maximum amount of produce from agriculture for nominal prices, leaving it to the agricultural population and labor force to make their living as best they can. This applies even more where the work force is abundant and can afford individual labor inputs, apart from collective work, on private plots. Such was the case up to the early or middle 1950s. But the outlook must change, when and where labor is no longer abundant, necessitating technically skilled labor for the more modern methods of agricultural production, and when the planners think they can afford to pay regular (though still not high) wages for such labor. Their intention thus became one of integrating agriculture into the overall economy and getting more out of it (with greater productivity and less workers to be fed on the farms). When capital inputs for modernizing agricultural production in general, and for opening up new farm land in particular increased, control over such inputs, which is more easily enforced in state-owned enterprises under direct state direction, became more important. Thus, both ideology and expediency played a role.

The available evidence suggests that various motives of both the central leadership and local authorities interacted to promote conversions of economically strong, as well as of weak, kolkhozes; and that Khrushchev at times tried to prevent economically unfounded conversions, but on the whole took a benevolent attitude towards them. This does not mean that he wanted to dispense with kolkhozes in the foreseeable future. Rather, he wanted to strengthen both sectors economically while increasing the numerical weight of the state sector. Thus, the expansion of that sector did not contradict the "reform" (actually liquidation) of the state-owned MTS, begun in 1957 and officially inaugurated in May 1958, which strengthened the position of the kolkhozes.

The MTS held most of the capital stock other than livestock and buildings of kolkhoz agriculture. Their role had been emphasized by one of the first post-Stalin agrarian measures, when (as part of the decree of 7 September 1953) the machinery operators who were kolkhoz members and only seasonally attached to the MTS were made permanent MTS personnel and thereby state employees. This brought the total work force in MTS to three million in 1954. By the same decree, an important part of Party control over kolkhoz agriculture was vested with the MTS by assigning to them instructors under a *raikom* (district Party committee) secretary specifically responsible for agriculture. This organizational setup was revoked by a Central Committee decree of 19 September 1957.

The MTS "reform" of 1958 was a sharp reversal of the preceding policy toward the MTS and meant that they were incorporated into those kolkhozes which before they had serviced, in fact supervised and directed. Some nuclei remained under the name of Repair and Technical Stations (RTS), but the number of these diminished, and in 1961 the rest were taken over by the newly organized *"Sel'khoztekhnika"* organization. The former brigades of the MTS became tractor brigades of the kolkhozes, with the MTS personnel becoming kolkhoz members (as many had been before 1953).

The essence of the liquidation of the MTS consisted of letting the kolkhozes as producers dispose of the technical means on their own and thus enabling them to use these means in the way they deemed most rational. By contrast, the MTS had to serve several kolkhozes each, and moreover to act in the state's interest, which often did not coincide with the economic interests of the individual holkhozes. A side effect of the reform was to free the state budget from financing the MTS, which were deficit enterprises in most cases. The kolkhozes had to pay for the machinery they took over, and these purchases, originally said not to be enforced on the kolkhozes, but then in fact "voluntarily" speeded up, put a heavy burden on kolkhoz finances.

The MTS reform indirectly provided an additional impulse for converting economically weak kolkhozes into sovkhozes. As the latter were financed from the state budget, conversion freed a kolkhoz of its financial burden, among other things of that imposed by the MTS reform. In short, the motives for converting kolkhozes into sovkhozes accumulated by the mid-1950s and became interrelated with the liquidation of the MTS.

But another side of the coin was the increasing burden on the state budget because of the increased numbers of sovkhozes. This sector on the whole had always been subsidized. The incorporation into the sovkhoz system of many undercapitalized former kolkhozes brought with it the transfer from the method of extremely low labor-unit remuneration to one of state wages (not high, but still much higher than former remuneration in such kolkhozes). Consequently the deficit of the sovkhoz sector grew to enormous totals, from less than 100 million rubles in 1952 to 1.1 billion in 1965.[2] All told, it was not surprising that the wave of mass conversions ebbed by 1961.

In the good harvest year of 1958, cash incomes of kolkhozes continued to increase, and procurement plans were in general fulfilled, so the remuneration of kolkhoz labor continued to rise as well. However, in 1958 and 1959, Khrushchev and his associates warned over and over again that not too much was to be distributed as individual remuneration. Instead, they stated, the money should be used for "social consumption," that is, for fringe benefits and transfer payments (schools, kindergartens, canteens) as well as for village reconstruction (agrotowns, among other things). As to the method of individual remuneration, it was advocated that payments in kind be reduced and cash wages increased, these to be paid at intervals before the end of the year and combined with a guaranteed minimum wage of 70–80 percent of the expected final remuneration. There was also a strong recommendation to make labor remuneration more dependent on actual productive performance rather than on work norms fulfilled. The norms themselves were to become "progressive" and "technically well grounded," which in most cases was tantamount to higher norms.

The framework for such changes had been established by the decrees of 1953 and 1956, and after the production successes of 1956 and 1958, it apparently was thought that the time had now come for their large-scale implementation. The reform of the procurement system in 1958, with its higher average prices to producers, was to help in this.

But the production results of agriculture were disappointing again during 1959–61. Grain output remained below the 1958 record, and total gross output, according to Soviet statistics, increased only slowly. This depressed kolkhoz incomes, while the payments for MTS machinery and the growing inputs of labor and capital, as required to comply with Khrushchev's demands for a change in the cropping patterns and for an increase of livestock numbers, implied assignments greater than ever before for the kolkhoz "indivisible funds," or investment funds.

Kolkhoz investment, which had amounted to 2.1–2.2 billion rubles in 1955 through 1957, increased to 2.8 in 1958 and 3.5 billion by 1959, in spite of the decreasing number of kolkhozes (due to conversions). The proportion of kolkhoz gross income assigned for investment increased from 17–18 to 23–24 percent of the total. As a consequence, the rise in labor remuneration came to a stop, sizable advance payments in cash did not become widespread, and in fact were discontinued in many kolkhozes. The reform of kolkhoz labor remuneration was not actually put into effect.

Had the intentions been realized, kolkhoz wages would have become similar, in quantity and in system, to sovkhoz wages. On the other hand, the trend for wages to become more dependent on production results also made itself felt in the sovkhoz sector through a tendency to deprive sovkhoz wages of some of their former stability. This was not only a question of gearing incentives more toward productivity goals. It was also a practical necessity in view of the greatly increased number of sovkhozes and the resulting burden on the state budget, out of which sovkhoz wages had to be financed in the numerous cases of deficit sovkhozes. A reform of the wage system in the state agricultural sector was enacted in autumn 1961 and was rapidly implemented during the following twelve months. The new system was called *akkordno-premial'-naya sistema* ("group piece-work and bonus system"). Basic wages now were guaranteed at the level of the former wages, and a bonus wage was paid in correspondence with production results. At the same time, the daily work norms, too, were revised, mostly upwards.

The new sovkhoz wage system put into reality what had been envisaged, but only in part attained, in the kolkhoz sector since 1956 and 1958. Beginning in 1962, kolkhozes were recommended to introduce the sovkhoz wage system either fully or in the form slightly adapted to their special needs and financial possibilities. With the sovkhozes as with the kolkhozes, the higher work norms and the disappointing production results of 1959–61 prevented an appreciable increase of wages actually paid under the new system. It was only in the better harvest year of 1962 that labor remuneration began to increase again in kolkhozes, in part owing to their improved terms of trades (produce and input prices).

There was yet another side to the wage reforms. The insistence on increased cash wages and reduced wages in kind was a blow to the private production of the agricultural population. This affected not only kolkhoz members, because a certain amount of payments in kind, mostly in the form of bonuses, had been extended to sovkhoz workers as well. Reduction of payments in kind impaired private consumption of kolkhozniks and sovkhoz workers unless it was compensated for by sales of produce to them, which was done only to a limited extent. Most importantly, private livestock production was adversely affected, since private owners of livestock have to rely heavily on feed grain and other fodder, as well as on grazing and haying rights, provided by the public farms or organizations.

The generally favorable attitude toward the private sector (except for urban livestock owners) was somewhat but not decisively tempered down by the measures of 1956–57. By 1959, the then greatly intensified campaign against the private sector, combined with a disappointing harvest, began to show results.

Action against the private sector was taken in different ways. One was to reduce the plot sizes on various administrative grounds: nonfulfillment of the minimum of workdays, village reconstruction, conversion of kolkhozes into sovkhozes—the effects naturally depended on how the cuts were implemented in practice. A second way was through the principle that the general meeting of members could decide to have the plot sizes reduced generally in a given kolkhoz, which was very much a matter of manipulating the assembly. The same applied to statutory upper limits for livestock holdings, but with these, fodder was more important. The management could greatly influence private animal farming by granting or withholding feed—as payment in kind, and/or as grazing and haying rights. Also, the agrotowns implied by their very nature that household plots and livestock holdings had to be reduced in favor of a more compact outlay of the settlements. The plots would then either totally disappear or be located outside the settlement, which made access more time consuming. Last, restrictions, as opposed to a liberal handling of the rules, on the free market (the kolkhoz market) could greatly influence the turnover and profits on these markets, for example, by fixing or not fixing upper price levels, or by making access to marketing (including transport) facilities easy or difficult. These are effective incentives or disincentives for the sale of private output beyond the needs of own consumption of the producers. Some or all such measures were applied increasingly in most of the country in the Khrushchev period after 1957, especially in 1959–60, with the result that private agricultural production stagnated. The fact that it did not contract sizably up to 1962 bears eloquent witness to the tenacity with which the population clung to this resource of food.

The restrictive policy toward the private sector had negative results for the socialized sector as well. It not only reduced the incentives for work (payments in kind) and the food supplies for the urban population through the kolkhoz market, but also affected socialized animal production. A delicate division of labor had long established itself between the two sectors: private owners received or bought young animals, first of all piglets, from their public farm and in turn sold fattened or half-fattened pigs and calves to these or directly to the state and consumer cooperatives' agencies. There is good reason to assume that part of the milk, meat, and wool which was accounted for as kolkhoz and sovkhoz production in fact originated from private producers and was purchased from these by public farms, thereby enabling the latter to fulfill their procurement quotas. Moreover, the private sector utilized the labor of housewives, old people, youngsters, and invalids, who were not (or only to a minor degree) obliged to work for the kolkhoz or sovkhoz. Such labor in most cases did not become available to the public farms when its application in the private sector was prevented.

By 1960, a feeling of latent crisis must have been prevalent among those concerned with agriculture, from lowest to highest levels. The negative consequences of the various policy measures, together with the still precarious achievements of Soviet agriculture and the less favorable weather after 1958, all combined to curb output growth. In 1961/62, the mass conversion of kolkhozes, the campaign against the private sector, the propaganda for agrotowns, and the call for kolkhozes to invest more, were all toned down, though not completely dropped. The reform of the sovkhoz wage system was enacted and its application in kolkhozes recommended, as outlined above. Higher advance payments on procurements and more bank loans were granted to kolkhozes. Prices for agricultural machinery were lowered, and a program for supplying more electrical energy to kolkhozes from the state grid was announced.

By that time, Khrushchev appears to have become defensive concerning his agrarian policies. He delivered more speeches than ever on agriculture and took new measures for stepping up its production and making it more efficient. In the few years before his ouster in 1964, Khrushchev gradually came to the conclusion that the best way to increase output was no longer to expand further the tilled and sown area, but to raise yields through more intensive farming methods (e.g., fertilizing, irrigating, tilling with more care, etc.). At the same time, he seemed increasingly to be again leaning towards a more consumer-oriented economic policy.

The most important indication of the new approach was the great production increase for fertilizer envisaged in his Seven-Year-Plan (for 1959–65). The increase between 1957 and 1960 had not been spectacular and amounted to somewhat less than 500,000 tons of effective nutrient. But then production went up in bounds, to 6 million tons in 1964, almost double the production of 1960. Moreover, a program for yet more accelerated increases in fertilizer, pesticide and herbicide production was announced in December 1963.

Up to that time fertilizer application was almost wholly restricted to cash crops like cotton and sugar beet. Nor was manure applied to a great degree in public farms, because much of the livestock was held privately. Even the manure derived from the small (in proportion to sown areas) public herds was not always put to use, because labor and transport for bringing it to the immense and distant fields was often lacking. Small grains and row and most other crops were badly deficient in nutrients, which is one of the main explanations for their low yields. But even with rapidly increasing fertilizer output, Soviet fertilizer and manure application remained far from the level prevailing in Western agriculture. In any case, a start for improvement was made during the early 1960s and the need for more fertilizing was clearly realized.

A related issue is that of intensified land use by means of expanding the irrigated area in the southern parts of the country. Stalin's earlier plans in this respect had been postponed, presumably because the Virgin Lands campaign promised quicker results with less capital inputs. The change toward intensification now found its expression in the resumption of great irrigation projects and was heralded by Khrushchev in his speech at the January 1961 plenary session of the Party's Central Committee.

On the organizational plane, the remaining Repair and Technical Stations (successors of the MTS) were abolished as such and incorporated into an all-Union agency for supply and repair of agricultural machinery, spare parts, etc., called *Soyuzsel'khoztekhnika* (with agencies and stations *sel'khoztekhnika* on regional and local levels) by decree of 10 February 1961. For a number of years this organization did not live up to expectations. The procurement system was again changed. Kolkhozes as well as sovkhozes obtained the right to negotiate procurement quotas and to sell their produce on the basis of contracts so concluded. But this remained a formality: the state agencies had their purchase plans to fulfill and were able to set amounts and delivery conditions, especially so after their position was strengthened by a reorganization enacted on 26 February 1961.

The procurement system was again reshaped only one year later in connection with a major restructuring of rural administration. This too was shortlived and was suspended soon after Khrushchev's removal two years later. The underlying idea of the 1962 reform was to unify the administration of both the kolkhoz and the sovkhoz sector by means of the so-called kolkhoz-sovkhoz production administrations, which were to direct all agricultural activities, including procurements. By the mid-1950s the Ministery of Agriculture with its hierarchy down to the district administrations was responsible for practically all fields relating to agriculture. Before that, at different times there had been various other central agencies, such as for sovkhozes, for special agricultural branches, for procurements, etc. Beside the Ministry, as beside any other state organization, was the Communist Party, which exercised indirect influence. At least formally the Party's influence was indirect, through the Party members in the state administration (and, up to 1958, in the MTS), but in fact it was rather direct and very powerful. Now, by a decree of 21 February 1961, the Ministry of Agriculture lost most of its functions, retaining only research, advice, and information activities for agriculture. *Soyuzsel'khoztekhnika*, as well as the new state committee for procurements became independent central agencies. By 1962, the new local production administrations were entities entirely separate from other administrative organs not related to agriculture. The same structure applied to the higher, provincial (*oblast'*) level, but not on the Republican and all-Union level. The Party organization was divided along analogous lines.

All these measures were of little avail, although the better weather in 1962 resulted in an improved though not spectacular harvest. Then followed the catastrophic drought of 1963. The Soviet Union was compelled to buy 13.7 million metric tons of grain from the West during 1963 and 1964. One of the reasons the drought had such adverse consequences was the heavy reliance, for grain procurements, on the recently ploughed dry-farming regions, which suffered most.

By the time the 1963 crop failure became evident, Khrushchev had but one more year to govern. He stepped up the fertilizer and irrigation programs, but did not inaugurate fundamentally new measures in agriculture, except for one: the introduction of old-age pensions for kolkhozniks by a law of 15 July 1964. Although these pensions were far lower than those introduced for state employees and workers long before, they heralded significant progress in a previously disregarded field.

Khrushchev's sanguine ways of reforming the system of socialized agriculture came to an end with his removal in October 1964 and gave way to a more patient and considerate approach under his successor L. I. Brezhnev. But the general direction of agrarian policy, as it had developed during the last years of Khrushchev's rule, was not basically changed, with two notable exceptions. First, the administrative reorganizations of 1962/63 were repealed. The restrictions on the private sector, already attenuated since 1962, gave way to a policy that not only reinstated this sector in the role it had played up to 1958 but also declared its contribution to the nation's food balance indispensible for a certain period to come. This did not mean legitimizing it for the longer term, though. It was not until toward the end of the 1960s that, with the development of "agro-industrial integration" (see chapter 12), a basically new move was inaugurated by Khrushchev's successors, although even this had been previously outlined in theory in the Party Program of 1961.

Notes

1. An interesting analysis of how Khrushchev pushed through this bold venture against resistance within the leadership was given by Richard M. Mills, "The Formation of the Virgin Lands Policy," *Slavic Review* 29:1 (March 1970), pp. 58–69.

2. K.-E. Wädekin, *Die sowjetischen Staatsgüter* (Wiesbaden, 1969), pp. 58ff.

5. Collectivization in Eastern Europe: The Common Pattern and the Deviations

General Outlines of Agrarian Policies

IN THE SOVIET UNION, the years immediately after Stalin's death, 1953–55, were those of the New Course (political relaxation) and of a political power struggle, the outcome of which had not yet been definitely decided. For Eastern Europe, the period brought political relaxation, uncertainty, and in some countries popular unrest, especially after a serious uprising in the GDR in June 1953. Collectivization made slow progress in the GDR, Poland, and Rumania, suffered stagnation, although much had already been achieved in Bulgaria, and setbacks in Hungary (where it was resumed with some vigor in 1955 for an interlude until mid–1956) and in Czechoslovakia. In Albania, it was fully resumed already in 1954 and for the most important agricultural parts of that country completed in 1956; only the mountainous areas of Albania remained predominantly individual peasant and tribal for another ten years.

In 1955–56, efforts at carrying through collectivization were again made in Bulgaria and Rumania, while the standstill in the GDR continued. The events of autumn 1956 in Poland, where an open uprising was just prevented by an important change of government, and in Hungary, where such an uprising could only be quelled by Soviet military action, led to renouncing enforced collectivization in Poland and to its severe setback in Hungary. These events apparently influenced government attitudes not only in the GDR but also in Rumania, where the December 1955 decision to speed up collectivization was applied hesitantly, and where a bad 1956 harvest may also have had some delaying impact. Only Bulgaria did not seem to care about the disturbances in other countries and completed collectivization during 1956–58. In Czechoslovakia, collectivization was continued with renewed emphasis beginning in 1957.

The final collectivization drives began in 1959 (in Hungary, in the second half of 1958) and resulted in full collectivization in Czechoslovakia and the GDR by 1960, in Hungary by 1961, and in Rumania by 1962 (four years earlier than Bucharest had anticipated in June 1961).

63

Thus, the process in the Soviet bloc restarted and reached its completion in the late 1950s and early 1960s. Of the nonbloc countries, Albania took the final step only in 1966, when in her mountain areas, too, almost all peasants and herdsmen were made collective farmers. In Yugoslavia, collectivization after the Soviet model was no longer an issue after summer 1953, and only a few collectives remained by 1954.

Although the process was not wholly uniform throughout Eastern Europe, a common chronological pattern, with deviations due to political events outside the agrarian sector, emerged in those countries which remained within the Soviet-dominated bloc: hesitation or setbacks during 1953–56, and resumption of the collectivization campaign after 1957, that is, after the restabilization of the political regimes. Of the bloc countries, only Poland remained one of predominant private peasant agriculture, and Bulgaria (except for the stagnation period of 1952–56) and Albania were hardly affected by the post-Stalin political instability.

The model charters of the collective farms in the various countries and their revisions during the years 1953–60 demonstrate an organizational convergence up to the crisis of the mid-1950s and some differentiation afterwards. The application of the Soviet model becomes more apparent when one compares those charters not only with the kolkhoz Model Charter of 1935, but also with its later revisions (especially as contained in the two decrees of 6 March 1956) and with the other Soviet policies during that decade. Through these alterations, the actual organization and processes in the Soviet kolkhozes were modified without changing the Model Charter as such until its thorough revision in 1969. In other aspects of agrarian policy, too, the Soviet model was applied, although with more cases of either deviating from or anticipating Soviet measures. The greatest number of differences of course are found in Poland and Yugoslavia, since the main part of their policies had to be geared to a small-peasant private agriculture, and Yugoslavia moreover was not a bloc country.

Roles were transposed in the field of social insurance for collective farm members, where the Soviet Union followed the lead of several East European countries with some delay during the 1960s. In the GDR, collective farmers were incorporated into the comprehensive social security system, long established in the rest of the economy, from the beginning of collectivization. It is not quite clear, though, in which forms and to what degree this was done on the individual farms that were affected. When the social insurance system for East German collective farmers was unified in 1959, its main feature was that the premia were paid only by the farmers themselves, with no equal employer's share being contributed by the collective farm since juridically, it was not an employer. Correspondingly, most of the benefits, too, were only half of those enjoyed by employees in other parts of the East German economy.

Czechoslovakia's Model Charter of 1953 (Type III, which soon became the only one) provided for health insurance, with the premia to be paid by the collective farms. Bulgaria introduced old-age pensions for collective farmers in 1957. This move preceded full collectivization and the subsequent buying or otherwise acquiring of the members' land titles by Bulgarian collective farms. Significantly, the argument was advanced by collective farmers at the time that the "rent" payments they received for their land contribution constituted a kind of old-age insurance. The logic of the collective farm acquiring the land title and creating an old-age insurance at almost the same time is convincing. (For the further development of social security in Eastern Europe, see pp. 178–79.)

A common feature, probably brought about as much by general economic necessity as by the first post-Stalin measures in the USSR, was the raising of producer prices for agriculture (for the USSR, see Table 4.1). It continued, at various intervals, in all collectivized countries throughout the 1950s, less so in Poland and Yugoslavia. As pointed out earlier, the creation of large socialized farms has its own logic in favor of price increases, as these farms no longer belonged to a class considered alien or even hostile to the Communist regimes. In addition, another kind of logic became effective when the MTS were abolished and collective farms had to buy out of their income the machinery and more industrially produced inputs in general. This also contributed to raising the prices.

Price increases were not necessarily effected only by raising the officially fixed prices. They could also be brought about, in terms of average prices actually paid, if the shares of compulsory deliveries, purchases, and contract procurements were changed. This was so because different prices operated for each of these marketing prices. A shift toward more contract procurements, together with the raising of a number of producer prices for vegetables (including potatoes) and animal products had been decided in the Soviet Union in September 1953. The East European countries followed the Soviet example, some simultaneously and some after a few years. Interestingly, East European countries also set the example for the Soviet Union. Even earlier than in the Soviet Union, compulsory deliveries were wholly abolished in Hungary and Rumania in 1956 (effective for 1957), whereas Moscow took this final step in 1958, and the other countries then followed. The idea was to do away with the system of multiple prices and to give the farms enough incentive to increase their output and marketings (procurements) on their own initiative. It did not work out in practice, as farm costs (for labor and capital) also increased at a fast rate. The planned purchases became compulsory, much as the former deliveries had been, and the bonus prices for above-plan procurements recreated a price differential which did not fundamentally differ from the earlier one, although on a higher level. On the whole, the rise in producer prices for collective farms was greatest during 1953–57, and then levelled off for a number of years.

State farms for a time continued to be paid special prices below the level for collective farms, but Hungary was the first to equalize their prices to those paid to collective farms and individual farmers, beginning in 1958. The Soviet Union and other countries gradually raised state farm prices in order to bring them closer to the collective farm level, but this move took a long time, into the 1960s and even 1970s, to be more or less completed. Poland, mainly in her restructuring of the whole system during 1957, did not deviate from this overall pattern.

The share of agriculture in overall gross investment, financed in part by collective farm incomes and in part directly by the state (mainly on state farms), showed a similar common trend, but deviated from the Soviet case insofar as in the beginning, most of the additional incomes were spent on improving the miserable labor remuneration. During 1949–55, agriculture generally received less than 10 percent of gross investment, while in the late 1950s and early 1960s this share was on the whole well above 10 percent (in Bulgaria even 19, although in the GDR only 9 percent).[1]

The Soviet example of abolishing the MTS in 1958 and of transferring their machines to the kolkhozes was followed by the GDR, Czechoslovakia and Bulgaria. This happened during 1959–61 (with some planned delay in the execution of the measure), although a number of Czechoslovak and Bulgarian MTS were retained in outlying and agriculturally poorly endowed regions. Albania stuck to her MTS. Rumania kept her MTS but changed their organizational position so that within larger collective farms the tractor brigades, although still being sections of the MTS, became firmly installed on the farms themselves. Quite apart from political considerations, it was probably useful to retain the MTS as a means of concentrating scarce available agricultural machinery in countries like Albania and Rumania with their low overall level of economic development.

In Hungary and Poland the events of autumn 1956 led to the disappearance of the MTS even earlier than in the Soviet Union. The remaining Hungarian collective farms were permitted to own machinery, and when collectivization was resumed there in 1958, no new MTS were formed. In Poland, where only a tiny collective farm sector remained after 1596, and did not again expand for a decade, it was equally natural that the state-owned MTS was turned into servicing and repair stations for individual peasants as well as for collective farms. Most of their machines were sold to the latter and to small cooperative machinery pools of the peasants, the Agricultural Circles, in which old Polish traditions of peasant cooperation were continued and reanimated, beginning in 1957.

Yugoslavia preceded the Soviet move to dissolve the MTS by more than seven years, in 1951–52, that is, when collectivization came to a halt there. One wonders whether the Venzher/Sanina proposal for abolishing the Soviet MTS, energetically rejected by Stalin in 1952, had not been influenced by the Yugoslav precedent.[2] The functions of the Yugoslav MTS were in the main taken over by the General Agricultural Cooperatives in their relationship with the individual peasants.

Thus, it seems that the lack of a uniform pattern with reference to the abolition of MTS was brought about by various circumstances in the other countries. In Yugoslavia, Poland, and Hungary, these were a result of internal political events, whereas in Rumania and Albania they may have been a function of their economic stage of development.

The movement toward bigger and bigger farm sizes, as demonstrated for the Soviet Union, on principle applied to the socialized sectors of agriculture in the East European countries as well. Yet statistically it did not evidence itself as clearly there as long as collectivization was not completed. During that process many collective and also a number of state farms were being organized. Their size not only depended on the leadership's intentions but also on local circumstances and on whether "lower" or "higher" type collective farms came into being. The Bulgarian case is telling. In August 1958, when collectivization was on the whole achieved there, the leaders started a "campaign for the more rapid economic development of the economy of our country." It paralleled Khrushchev's simultaneous ambitious Seven-Year-Plan for the Soviet economy, but also seems to have been influenced by the Chinese "Great Leap Forward" of the same time. Economically it ended in failure, like the Soviet and the Chinese ventures, but in the process many Bulgarian collective farms were amalgamated into huge farms of, on the average, four times the previous sizes. With 4,000 hectares of agricultural land and 1,300 households, their sizes even exceeded those of kolkhozes in comparable Soviet regions, like the Moldavian or Georgian Union Republics.

In all Communist countries, state farms are meant to be large or gigantic agricultural production enterprises, exclusively or predominantly concentrated on only a few kinds of crop or animal production. From state farms proper, one has to distinguish other state-owned farms, often called institutional farms in Western literature. These are experimental, training, research, and model farms, or else farms attached to industrial firms or military units for supplying their canteens with food. Institutional farms may also be large and specialized, but often are of medium size. In state and institutional farms all the means of production are owned not by the respective production collective but by the state as the representative of society as a whole, as understood in a Communist system. While collective farms have a work force composed of farm members (who do not have the status of "workers"), the state and institutional farms have wage-earning blue and white-collar employees (workers) with the state as the employer. The directors and managerial personal are appointed by the state organs to which these farms are subordinated. They act within the parameters prescribed by their affiliation and are under the direct orders of the state authorities—and in practice, indirectly, also of the Party organs.

In these respects state farms were fundamentally different from the collective farms, although the real differences were less than the formal ones. More important were the economic consequences of the fact that state farms were, in a way, sections of one all-state agricultural enterprise, their profits going to the state and their losses being covered out of the state budget, so that labor and capital costs were in the end borne by the state. Mainly for this reason they could exist with the low prices the state paid them. This kind of financing also implied that their success indicators differed somewhat from those of the collective farms and that regular wages, as in nonagricultural state enterprises, could be paid. (All this underwent some, though not a total, change in the next decade.)

The Yugoslav state farms acquired a certain economic independence after the general introduction of the system of self-administration and in this respect became similar to the General Agricultural Cooperatives. Beginning in 1950, this type of farm was increasingly called an agricultural estate *(poljoprivredno dobro)*, with no reference to the state, or else was called a social economy, or a combine. Most of them were located in the more fertile parts of the country, especially in Croatia and the Vojvodina. Their endowment with capital inputs was far above average, as were their per hectare yields.

In the Soviet case, many state farms proper (i.e., excluding institutional farms) were not in fact so much larger on average than those in other countries. To a great extent, the larger state farms were located in areas with very extensive land use (dry farming, pasturing), while those with patterns of intensive land use did not fundamentally differ in size from their comparable counterparts in Eastern Europe, especially toward the early 1960s. The GDR, Poland, and Yugoslavia were still exceptions to this general picture at that time. More special to the Soviet Union was the fact that during the 1950s the state sector expanded to a degree that made its share in agricultural land almost equal to that of the kolkhozes (because of the Virgin Lands campaign and for other reasons). In no other Communist country did this come about. Expansion as such of the state farm sector was not peculiar to the Soviet Union, though. It happened, to varying degrees, in all East European countries during the decade of collectivization and more or less parallelled it, most markedly in Poland and Czechoslovakia. By 1956, the Polish state farm sector reached a peak of expansion, occupying one-sixth of the total agricultural area, but shrank again during the second half of the decade.

In most countries, state farms proper occupied by far most of the state-owned land. However, in Rumania, institutional farms accounted for the greater part, and in Czechoslovakia and Bulgaria, as well as in Hungary up to the early 1960s, their share was quite sizable. The moderate expansion of the state farm sector was in part a substitute for collectivization. In the GDR, Bulgaria, and Yugoslavia this sector remained smallest, and was in most cases composed of experimental, training, research, and model farms, or very specialized farms, as indeed had been Lenin's notion. In the other countries, the expansion led to a situation where many state farms did not correspond to this notion. They became simply large-scale enterprises, differing from collective farms mainly by the organizational setup and subordination, and by greater capital intensity of production.

The Statutory Setup of Collective Farms

By the middle and toward the end of the 1950s, the Model Charters for collective farms were revised, in most countries more than once. During the collectivization process, the "lower" type collectives numerically retained their importance and even increased their share for a time because of the formation of new ones. Yet once collectivization was more or less completed, most of them were converted into or amalgamated with the "higher" (kolkhoz) type collective farms sooner or later. They gradually either wholly disappeared or accounted for only an insignificant part of the collective farm sector ten years later. For this reason, only the "higher" type collective farms are dealt with in some detail below.

In some countries, a special legal setup is used to put the collective farm sector into a more comprehensive framework than a model charter alone could provide. The latter concerns only the internal farm organization and some obligations towards the state, whereas laws also regulate the relationship to other organizations and enterprises. Such a general law has been under discussion in the Soviet Union for a number of years but has not been issued up to the present. By contrast, Yugoslavia adopted her Fundamental Law on general agricultural, not only production, cooperatives as early as 6 June 1949 and later, under changed circumstances, complemented it by the decree of 26 January 1954 on the General Agricultural Cooperatives. Czechoslovakia issued her first specific law on production cooperatives (i. e., collective farms) in 1949 (with a second version in 1959, and a third in 1975); the GDR followed in 1959, Poland in 1961 and Hungary in 1967 with an amended version in 1977.

Table 5.1 contains a comparative selection of the more significant social and juridical features of the "higher" type collective farms in the Soviet Union and Eastern Europe, including the rules of membership, land ownership, private plot and animal holdings, distribution of collective income (for labor and for land contribution), and working requirements. Consequently these aspects are dealt with here only briefly, along with other aspects where great uniformity is shown. The table does not include Yugoslavia, since the situation there was basically different after 1953, with Soviet type collective farms almost wholly disappearing. Nor is the Polish new model charter of 1957 (later incorporated into the public law of 17 February 1961 on collective farms) taken into account. That charter cannot be compared with those of the other countries because it did not contain detailed and obligatory rules, but rather gave a framework of fundamental principles for organizing collective farms. It provided for only two kinds of collective farms, basically corresponding to the "higher" and the "lower" types. Of the earlier Polish model charter one feature deserves special mention as it had no parallel in the other bloc countries: the possibility of dissolving a collective farm by the members' will was implied by indicating the formal procedure demanded in such an event.

Most of the many revisions of the Model Charters during the 1950s are accounted for in our comparative table. Generally, the rules applying at the time of Stalin's death or shortly after are specified first, and relevant later changes are added with an indication of the year of their introduction (by decree or by revising the charter).

Permitted private plot sizes for collective farmers everywhere roughly equalled those applying in the Soviet Union, while the scope left for private animal farming remained somewhat broader in the GDR, Albania and, most of all, in Hungary and Poland. During 1954–56 a relaxation on private animal farming took place in Hungary, Rumania, and Bulgaria, in contrast to a certain tightening, beginning in 1956, in the Soviet Union, Albania and, by 1959, in the GDR. With full collectivization being attained or closely approached, the limits on private plot activities were generally again set narrower, or handled more strictly in practice. The exception was Hungary, where a more flexible limitation was spelled out in the 1959 version of the Model Charter (art. III/4): Produce from the private plot and exceeding the family's own consumption needs must be put at the collective farm's disposal—to be paid for, as it was understood.

Apart from the features outlined in the comparative table, the following major characteristics of the collective farms were more or less uniform throughout the Soviet bloc:

1. All Model Charters mentioned nonagricultural, complementing farm activities, especially in the area of processing and building. The 1935 Soviet Charter referred only to rural crafts and subsidiary home activities, but not to subsidiary enterprises of a more or less industrial type. These were explicitly rejected at the 1952 Soviet Party Congress, and it was only after 1960 that they were favored, while the more traditional rural crafts were repressed under Khrushchev. The somewhat greater emphasis given to these activities in the Model Charters of the East European countries anticipated the later change in the Soviet attitude.
2. Production buildings and agricultural implements of peasants, except those small ones which were needed for private plot and animal farming, were uniformly socialized in the "higher" type collective farms.
3. The general assembly of members was the highest organ of collective farms everywhere. It was not concerned with current affairs, but met at more or less infrequent intervals for deciding on (often merely rubber-stamping) long-term questions. Generally, voting at these assemblies was by lifting hands, not by secret ballot. For some fundamental decisions, a majority of two-thirds was required. Following the Soviet example of 1956 (the March decrees), Bulgaria and Albania provided for assemblies of elected farm section representatives in the big farms, where it was physically difficult to bring all members together. This practice became general in subsequent years of increasing farm sizes, without its being incorporated in the charters.
4. The management of current farm affairs, where it was not left to the chairman, was taken care of by the elected board. Usually the latter consisted of five to nine persons, or of three to five in smaller collective farms (specially mentioned in the East German, Hungarian and Polish charters). As distinct from the Soviet charter, the possibility to depose the chairman by vote was explicitly mentioned in the East German, Polish, Czechoslovak (but no longer after 1957), Bulgarian (since 1955), and Hungarian (since 1959) charters.
5. An auditing commission was provided for in all charters, in Bulgaria under the name of a "control council." Its main duty was to control the finances and the adherence to economic regulations by the management. Protecting the members' rights as one of its duties was mentioned only in Poland and the GDR, whereas the Hungarian charter of 1955 (article 12e) stated that members may appeal to the executive committee of the district administration if a complaint can

Table 5.1 Comparative Major Characteristics of Soviet and East European Collective Farm Charters (kolkhoz and "higher type" of the 1950s)

	USSR	GDR	Czechoslovakia	Poland (until 1956)
Membership:	"Voluntary" admission at end of sixteenth year of age; provisions exist for exclusion, but not for voluntary withdrawal.	As in Soviet Charter of 1935, but voluntary withdrawal also mentioned (only after harvest and upon written application).	As in Soviet Charter of 1935; since 1957, admission at end of fifteenth year of age.	"Voluntary" admission at end of eighteenth year of age (for 14-18 year olds, working but not voting rights). Apart from exclusion, regulations for leaving at will, upon application one year in advance and after discussion in members' assembly.
Land ownership (cf. distribution of income):	Nationalized (i.e., state-owned); land handed over to kolkhoz for use "forever," without rent; not to be alienated. If a member leaves, his land physically remains with the kolkhoz; in compensation, he may receive another piece of land, if available, from the state land fund.	"Remains in the ownership of the peasants"; alienation only to the collective farm or to other of its members or (Law on Agricultural Production Cooperatives of 1959) to the state. Upon leaving, another piece "of equal size, and considering the quality, at the border of the cooperative's land" to be returned to the leaving member.	Remains in the ownership of the member. Upon leaving, same provisions as in GDR. Otherwise the collective economy would suffer.	Remains in the ownership of individual member, not physically, but as an ideal share, recalculated and registered on the basis of a uniform land quality.

Private plot and its size per household:	Indirect right of use (given by the kolkhoz out of its land fund); may be reduced (since decree of 6 March 1956) if work minimum not fulfilled. Maximum 0.25-0.5 hectares, regionally up to 1.0 (in practice, much less on irrigated land).	"In personal ownership for own use." No more than 0.5 hectare.	In personal ownership for own use.	If a member leaves, he receives land of equal value on the borders of the collective farm as a compensation. The plots adjoining the farmsteads are not part of the collective land fund; the collective farm gives plot land to those who have none. Plot sizes range from 0.3-1.0 hectare.
			0.5 hectare; up to 1.0 (of which, arable up to 0.5) in pasture regions; no more than 0.1 admitted for "special crops" (vegetables, wine, etc.).	
Private ownership of livestock (maximum):	1 cow, 2 calves; 1 sow (2 in some cases) with young; 10 sheep and/or goats; 20 beehives; no maximum for poultry and rabbits. Regional variations: 8-10 cows with calves; 2-4 sows with young; 100-150 sheep; 10 horses; 5-6 camels; 2 mules; 2 donkeys. (Beginning officially in 1956, and in practice earlier, these regional maximums were greatly reduced by decrees and "recommendations").	2 cows with calves; 2 sows with young; no limits for sheep, goats, poultry, rabbits, and other small animals; 10 beehives; 1 horse with 1-2 colts, or 1 ox. *Charter of 1959:* sheep, up to 5 with equal number of young (up to 11 months of age) no limit; ownership of a horse, colt, or ox no longer permitted.	1 cow, or up to 3 goats; 1-2 pigs for fattening; 5 sheep; 10 beehives; small animals and poultry (not explicitly stated) numbers are unlimited. *1957:* no maximum for sheep, goats, and other small animals, but said that size of private economy must not jeopardize the collective economy (as in Soviet decree of 6 March 1956).	2 cows with young; no limitation for all other kinds of animals.

Table 5.1 Continued

	USSR	GDR	Czechoslovakia	Poland
Distrubution of collective income or labor remuneration:	Remuneration according to labor day units earned and in total sum as a residual of kolkhoz income, paid at end of year.	Labor day units as in Soviet Charter of 1935, but 10 percent increase or reduction not mentioned. Generally possible to reduce remuneration or payment for land brought in if work minimum not fulfilled. Principle of group or of individual performance, with additional remuneration possible if plan overfulfilled.	Labor day units as in Soviet Charter of 1935.	Labor day units as in Soviet Charter of 1935.
	The number of labor day units credited to the individual brigade as a whole may be increased or reduced by 10 percent according to productive results.		Up to 50 percent of over-plan production may be paid, in cash or in kind.	The number of labor day units credited to the brigade may be increased according to production results; moreover additional remuneration for better and/or faster work; yet total of such increase must not exceed 10 percent. Of collective farm income available for distribution, 20-25 percent (1953: 5-25) to be distributed for land shares brought in, up to 15 percent for implements and livestock brought in. No land rent in Rolnicy Zespol Spoldzielczy (which corresponds to type IV).
	No payment for land shares brought into the kolkhoz.	Rental payments for land shares brought in, maximally and in total up to 20 percent of produce or cash available for distribution.	Rental payments for land shares brought in, maximally and in total up to 15 percent of the sums available for distribution; no rental payments in type IV collective farm. Charter of 1957: rental payments "may" be made, without further specification.	
Advance payments toward end-of-year calculation of remuneration:	Possible, to the amount of 50 percent of expected kolkhoz cash receipts and to 10-15 percent of grain in kind at time of threshing.	Advance payments in cash and in kind possible (no further specification).	50 percent of expected remuneration in kind, although in exceptionsl cases, and with consent of the district administration, up to 70 percent.	Up to 60 percent of the remuneration fund available at a given time.

	Decree of 6 March 1956: in total, no less than 25 percent of current cash receipts, and no less than 50 percent of advance cash receipts of the kolkhoz from delivery and procurement agencies.	1959: Advance payments for credited labor day units at up to 70 percent of their planned value.	(1959: **"may"** be up to 80 percent). Advance payments in kind also possible: as long as delivery and/or procurement obligations are not fulfilled, only up to 15 percent of the grain already delivered, and up to 10 percent in the case of potatoes.	
Work requirements:	All able-bodied members obliged to work on kolkhoz; since World War II: also juveniles in twelfth to sixteenth year of age. Nonmembers to be employed only in exceptional cases. Minimum of labor day units per year introduced in 1939, later put higher. After 1956 to be fixed by each kolkhoz individually. Work norms (up to 1948 and since 1956) to be set by each kolkhoz individually.	As in Soviet Charter of 1935; minimum of 150 labor day units per year; after 1959: to be fixed by each coll. farm. "Announcement" of 2 February 1955: if minimum unfulfilled without good reason, the person's share in his brigade's premia has to be cancelled, and general reduction of remuneration and land rental payments also possible. Work norms to be set along the guidelines issued by the Ministry of Agriculture and Forestry (these guidelines no longer mentioned in 1959 Charter).	As in Soviet Charter of 1935. Minimum of 150 labor day units per year. Work norms to be set by each collective farm individually.	As in Soviet Charter of 1935; minimum requirement 100 "days per year" (if not fulfilled, land rental payments are not made). Work norms are to be fixed by the collective farm (after 1953, on the basis of unified model norms), and are valid after approval by the Ministry of Agriculture.

Table 5.1 Continued

	Hungary	Rumania	Bulgaria	Albania
Membership:	"Voluntary"; admission at end of sixteenth year of age. Since 1959: at sixteenth to eighteenth year, if parents or legal guardian agree. Apart from excluding, also regulations for leaving on own will, but no earlier than 3 years after joining, and after members' assembly's decision on application.	As in Soviet Charter of 1935; since 1956: admission at end of eighteenth year of age. "Leaving" the collective farm mentioned in another connection (cf. land ownership).	"Voluntary"; admission at end of eighteenth (since 1953/54: sixteenth) year of age. No regulations for leaving at will, but entrance fee to be returned "in case of leaving or exclusion." 1958: Application for leaving mentioned as being subject to voting in members' assembly.	As in Soviet Charter of 1935.
Land ownership (cf. distribution of income):	Fields are brought in for "common utilization" (wording of 1955) and their owners registered in cadaster; the regulations for leaving, land rent, and private plots also imply actual individual ownership. Upon leaving, land to be returned out of the unconsolidated fields of the collective farm or from the state land fund.	1953: All land in "dispositional" ownership (patrimoniul) of collective farm; 1956: "constituting private property . . . brought into the cooperative for common utilization"; may be inherited. As in Hungary about leaving, but in 1956, the state land fund no longer mentioned.	Land remains in ownership of member, may be alienated, but only within collective farm. 1958: Not physical piece of land, but ideal share in the land of the collective farm remains in individual ownership. Upon leaving, or if non-members of the collective farm inherit, land is returned out of unconsolidated collective farm land.	Land becomes "social property" of the collective farm; other provisions as in Soviet Charter of 1935. Upon leaving, land may be assigned on the borders of collective farm's land in exceptional cases.

Private plot and its size per household:	The plots adjoining the farmsteads are not brought into the collective farm (if received from the farm, they are returned upon leaving). 1959 also: free disposition (to sell, exchange, donate, etc.). Plot ranges: 1951: 0.29 hectare (maximum); 1955: 0.29-0.58. 1959: if vineyards or orchards, only 0.14-0.22; and within the upper limits the local conditions, family size, and contribution to collective work have to be taken into account.	Concerning plots adjoining the farmstead, as in Soviet Charter of 1935, but 1956: "remain in personal ownership." Plot range: 0.2-0.3 hectare.	As in Soviet Charter of 1935, including the 1956 amendments. Plot range: 0.2-0.5 hectare. Since 1958 (as in Soviet amendment of 1956), it may be reduced, but not to less than 0.1-0.2 hectare, if labor minimum is not fulfilled.	As in Soviet Charter of 1935, including the 1956 amendments. Plot range: 0.05-0.3 hectare, depending on contribution to collective work (1954-56: Up to 0.3 hectare).
Private ownership of livestock (maximum):	1 cow with 1 (1955: 1-2) young; 1 (1955: 1-2) sows with piglets; 1-2 (1955: 3-4) pigs for fattening; 5 sheep or goats; no upper limits for poultry, rabbits, beehives. Beginning in 1957, the restrictions on private livestock holdings other than cattle were almost completely lifted in actual practice.	1 cow with 2 calves of up to two years; 1 sow with piglets; 3 pigs for fattening; 10 sheep and/or goats; 20 beehives; no limits on poultry and rabbits. Regionally, up to: 2 cows with young (1956: 2 cows and 4 young); 1 sow with piglets; 2 pigs (1956: 3) for fattening; 20 sheep and/or goats; 20 beehives; no limits on poultry and rabbits.	1 cow with calves, or 1-2 Saane goats with young; 1-2 pigs, of which 1 sow with young (since 1953/54: up to 2 sows with young); 3-5 sheep with young (since 1953/54: 5-10 sheep and/or goats with young in mountain areas); 10 beehives (1958: unlimited number); 1 donkey (1958: or mule); no upper limits on poultry and rabbits.	Since 6 December 1956, up to: 1-2 cows and 1-2 calves; 15-20 sheep and/or goats; 1-2 sows; 10-25 beehives; 1 horse or donkey or mule; no upper limits on poultry and rabbits. Earlier (1954-56) norms varying by regions and according to family size: 1-4 cows and 1-4 calves; 1-3 sows; 15-40 sheep and goats; 15-30 beehives; 1 horse; 1 mule or donkey; no upper limits on poultry and rabbits.

Table 5.1 Continued

	Hungary	Rumania	Bulgaria	Albania
Distribution of collective income or labor remuneration:	Labor day units as in Soviet Charter of 1935, but no mention of increase or reduction of number of units credited. Premia for over-fulfilling plan. 10-20 percent, in cash or in kind, of total collective farm income to be distributed as rent in proportion to the fields brought in (so defined since 28 July 1955). 1959: Up to 25 percent of income available for distribution (similar to provision in 1951); but in 1959: to be paid in cash, also to land owners who are not members of the collective farm.	Labor day units as in Soviet Charter of 1935. Increase of number of units credited by 20 percent if plan fulfilled, with reduction by 10 percent if not (no longer mentioned in 1956). Piecework rates. 1956: If plan fulfilled or less labor spent, additional remuneration according to criteria recommended by the Ministry of Agriculture. Rental payments for land must not exceed 25 percent of total collective farm income available (in kind and in cash, to be accounted separately).	Labor day units as in Soviet Charter of 1935. No mention of increase or reduction of number of units credited; but additional payment according to fulfillment of plan, as to be fixed by the collective farm as a percentage increase. Work and remuneration for piecework by large groups, small groups or individuals. Of all collective farm income available for distribution in cash and kind, up to 30 (since 1953: 25) percent to be distributed as rental payments for land shares brought into the collective farm, to the amount of 1-5 labor day units per 0.1 hectare (since 1953: no more than 4 labor day units). Beginning in 1958, the rental payments were to be fixed by the charter and the members' assembly of each individual collective farm.	Labor day units as in Soviet Charter of 1935 (including its 1956 amendments). Increase or reduction of units credited according to degree of fulfillment of plan, with no percentages indicated. No rental payments for land shares brought into the collective farm.

Advance payments toward end-of-year calculation of remuneration:	1951: it may be decided to make advance payments; but 1955: Members "may demand advance payments" of up to 50-60 percent of planned remuneration; in kind advance of grain to be paid immediately after the harvest, cash depending on the actual receipts of the collective farm (but for both, no absolute figures or percentages indicated).	As in Soviet Charter of 1935, including its amendments of 1956, but percentages deviating: up to 50 percent of the expected remuneration in cash; up to 15 percent of the grain already delivered, and up to 40 percent (1956: up to 50) after all obligations and the collective farm's own needs are met.	As in Soviet Charter of 1935, the percentages being 50 percent of the planned cash remuneration in monthly advance payments (the same as in 1950), and up to 10-15 percent of the delivered quantity of grain, which must not amount to more than 60 (since 1955: 80) percent of all grain that remains after all obligations and own needs of the collective farm are met. 1958: Permitted to form a special fund of the collective farm for paying a guaranteed minimum remuneration; sums of advance payments out of this fund equal those stipulated in the 1956 Soviet amendments.	As in Soviet Charter of 1935 (since end of 1956, including the March 1956 Soviet amendments). But in kind payments of grain calculated as 10-15 percent not of the harvested quantity but of that already delivered or procured.
Work requirements:	As in Soviet Charter of 1935. Minimum requirement of 80 labor day units in 1951, and 120 in 1955. Since 1959, to be fixed by collective farm itself, but not below 120. Work norms to be set by each collective farm individually, but on the basis of nationwide guidelines (these guidelines no longer mentioned in 1959).	As in Soviet Charter of 1935. Minimum requirement mentioned (but by whom fixed?); in 1956, a figure of 80-120 labor day units was indicated. Land share rental payment cancelled if no family member fulfills work minimum. Work norms recommended by the Ministry of Agriculture; after fixing by collective farm, to be submitted for approval.	As in Soviet Charter of 1935. The changed text as of 11 March 1955 provided that the minimum work requirement be fixed by each collective farm individually (no figure indicated); if the minimum is not fulfilled, the advance payments and—where such is given—the land rent payment is to be halved, and the premia for overfulfilling the plan is to be cancelled.	As in Soviet Charter of 1935, but from the very beginning, including a minimum requirement for work to be fixed by each collective farm individually.

not be settled by the general assembly of the farm. This was an exception to the theory and practice of the Soviet Union and the other countries. The Hungarian charter of 1959 dropped this stipulation and instead provided for an arbitration commission of the collective farm.

6. The internal farm organization was uniformly characterized by sections, mostly called brigades, and subsections under various names and corresponding to the Soviet *zveno*. The latter were not mentioned in all charters (and not in the Soviet charter either, although in the 1956 decrees) but they seem to have existed everywhere, at least in the more labor-intensive production. The brigade leader was appointed by the collective farm board, and in the GDR this appointment had to be approved by the general assembly.

As can be seen by comparing the Model Charters, the Soviet system of labor remuneration (labor day unit) as a residual claimant on collective farm income was gradually changed into a more stable kind of remuneration. This was effected by introducing advance payments during the year and fixing these as a percentage either of current and/or expected collective farm income, or of the planned or expected final value of the labor unit in a given year. In principle, this had been provided for from the beginning (in the Soviet Charter of 1935 and its 1938 amendment), but the actual impact had been negligible up to the mid-1950s, except for grain given in kind. Only in the GDR was a guaranteed minimum (7 marks per labor day unit) paid at an early stage (1952). By 1959, the remuneration in German collective farms was more than that, and collective farms which then paid less than 3,120 marks per full-time worker and year (i. e., roughly 10 marks per day), received public credits to make this new minimum remuneration possible.

From 1955 on, more emphasis was put on advance payments. By the end of the decade these played a considerable role and, as time went by, practically turned into minimum wages. In the actual implementation the Soviet Union was not the leader in this movement; Hungary and Bulgaria, apart from the GDR, were. On the whole, premia and other additional remuneration were rarely dealt with in detail in the charters. Their system largely depended on orders and "recommendations" issued by the state and Party authorities. Therefore the Model Charters present only a rudimentary picture of the ongoing changes in this sphere.

It was similar for work norms. The Soviet Charter of 1935 did not mention centrally prescribed norms. However, in 1948 such were introduced in the USSR, together with uniform evaluation scales for the various kinds of work (the 1956 decrees again left their fixing to the individual kolkhozes). The Soviet changes were reflected in the relevant provisions of the model charters of other countries, depending on the time of their publication. At a later stage, the trend toward centrally regulated and unified work norms again became prevalent.

With regard to the remuneration of labor and other distribution of collective farm income among members, two things must be pointed out. First, since the remuneration fund under the labor day system was a residual claimant on total collective farm income, increases of the numbers of labor day units credited to individuals or work groups as premia or bonuses changed the distribution in their favor within the collective farm, but did not affect the total sum to be distributed. Thus, other farm members received correspondingly less. Conversely, a reduction in the number or value of labor day units credited to groups or individuals raised the average value on the farm. Second, the differentiating effect of the rental payments for each member household's share of land brought into the collective farm (such payments were nonexistent in Soviet kolkhozes) was less than has often been considered in the relevant Western literature. The land reforms had levelled out many of the former differences in the sizes of holdings. Moreover, the state or the collective farm often attributed deserted or otherwise free land to collective farm members who formerly owned little or no land, designating it as their land contribution (not property) brought into the collective farm. In effect, the land shares as credited to members, and correspondingly accounted toward the "land rent" distribution, did not differ greatly in size, and tended to level out. These payments, where made, in actual practice constituted a kind of basic remuneration for all, including the nonworking (disabled) farm members, and did not differ very much among the households.

The overall picture emerging from the comparative analysis of the collective farm model charters is one of predominant uniformity based on the Soviet model at least up to two or three years after Stalin's death. Some differentiation, which set in later, does not essentially change this picture. However, it seems that uniformity then was effected more by mutual consultation and coordination, although still much under Soviet guidance. In a few aspects, the changes originated from, or at least were anticipated by, individual smaller bloc partners. This perhaps also applies to the enumeration of rights—not only the duties—of the collective farmers, as they were given in some of the East European model charters but absent from the Soviet charter until its revision in 1969.

The time for developing further the collective farm system, and with it the state farm system, had come by the early 1960s. At that time, agriculture was restructured in the whole of Eastern Europe, fundamentally along Soviet lines, except in Poland and Yugoslavia. The ensuing organizational and administrative changes were started in Khrushchev's time, but are more a feature of the following period of reforms, which for most of Eastern Europe may be termed the post-collectivization period.

Notes

1. *Ekonomika stran sotsializma, 1964 g.* (Moscow, 1965), pp. 60, 270–71.
2. J. Stalin, *Economic Problems of Socialism in the U.S.S.R.*, English ed. (Moscow, 1952), pp. 97–101.

6. The Development of the Sectors of Farming

The Collective and State Farm Sectors

APART FROM SOME fundamental issues, the present chapter is mainly concerned with land ownership and usage and with the production by socio-economic sectors. Most of the characteristics of the two main categories of socialized farms were given in the preceding chapter; their further organizational development will be dealt with in Chapters 11 and 12 on the reforms and agro-industrial integration.

The distinction established in Communist countries of three sectors of land ownership—state, collective, and private—is meaningful in terms of actual usage rights, but less so for juridical ownership. Especially with regard to the private sector, formal property rights lost most of their importance during collectivization and the following period. The individual property title of members to their former land, as a share in the collective farm's land, has been made irrelevant in practice, though it is still upheld formally in Poland, Czechoslovakia, Hungary, and the GDR. The East German reorganizations of the 1970s have made it impossible to attribute even a theoretical land share to individual member households, and in Hungary, most of the shares have been bought up by the collective farms since 1968, or otherwise passed into their possession. In these two countries, however, part of the members' income is still being paid as a sort of "rent" for the land once brought in, which is no longer done in Czechoslovakia. In Bulgaria (Party Program of 1972) and Rumania (Constitution of 1965), collective farm land—not the household plots—was officially declared collective property, while in the USSR, private land ownership ceased juridically to exist as early as 1918. In Albania, too, the land is nowadays considered state property. Where the individual title to land and other shareholding property still exists, a member receives payment for his share upon leaving the collective farm, but only in Hungary is it not merely nominal or zero in amount.

In state farms, naturally all means of production belong to the state. By contrast, in the collective farms of countries in the CMEA (Council of Mutual Economic Assistance), and also in Albania, the means of production (other than land) formally are the common property of the farm members. Of course, this does not apply to state-owned resources, among them some land from the state land fund, which are given to the collectives for utilization, nor to the livestock and small implements in private ownership of farm member households and in the few remaining "lower type" cooperatives. Yet the actual disposition of the common shareholding property is very much dependent on state orders and recommendations. Moreover, dissolving and returning to private ownership a collective farm is practically impossible in the existing economic and political conditions of the Soviet Union and Eastern Europe. Even in Poland, where such a possibility is explicitly provided for, the state and Party organs have many ways to prevent it.

Collective decisions on how and what to produce are also limited. This is so not only because of the general rules of law and the specific rules of the collective farm statutes, but also because procurement and, in part, even production plans are imposed by the state, which also prescribes prices. Equally important is the fact that the collective farm managers, although elected by the members, in reality are more or less strongly "recommended" by the superordinate state and Party organs, thereby turning the "election" into a subsequent formal approval in most cases. In their decisions, the managers have to comply much more with the wishes of these organs than with those of the farm members. In practice, the decision-making processes are for the most part directed, although in a less obvious way, by the actions of the Communist Party and other "social" organizations within the farm or village. As far as there remains some room for informal influences and relationships within the farm and between farm and public organs, it is hardly discernible, although its existence should not be wholly denied. In sum, the distinction between state and collective farms with regard to the ownership of land, fixed assets, and turnover capital is not of great importance for the state's factual disposing over such property. Even a Soviet author, more than a decade ago, considered it a mere formality for the socialist farms of his country.[1]

Over time, one discerns a pronounced change in the composition and also in the farm sizes of the socialized sector (see Tables 6.1 and 6.2). In Bulgaria and the Soviet Union, the state sector expanded mainly at the expense of the collective farm sector, while in Poland, both sectors expanded by taking up former private peasant land. In Czechoslovakia and the GDR, it was predominantly the collective farms which took over private peasant and plot land during the 1960s as well as the 1970s. Socialized farming increased its predominance in animal production, too, as is demonstrated by the shrinking shares of privately owned cows and pigs in all the countries (Table 6.1).

Generally speaking, collective farms initially were run on a primitive level of technology and organization, in spite of their large size, even in the Soviet Union. Each of them produced almost every kind of agricultural output of the given climate, without much regard to local comparative advantages or disadvantages. As time went by, this situation improved, especially in connection with the economic reforms of the 1960s and later. The level of mechanization rose and a certain specialization took place, although in both regards, collective farms continued to lag behind state farms.

With increasing supply and application of machinery, and correspondingly declining numbers of workers, the subunits of both collective

Table 6.1 Percentage Share of the Main Farm Categories in the Total Agricultural Area and of the Private Sector in Cow and Pig Holdings, 1960, 1970, and 1978

	Collective farms of all types (excluding private plots)	State-owned agriculture		Private sector (individual peasants' plots, other private users)[a]		
		State farms proper	Institutional farms	% of agr. land	% of cows	% of pigs
USSR						
1960	56.4	37.4	4.8	1.4	47.3	26.2
1970	37.5	56.4	4.6	1.5	39.0	24.5
1978	31.9	65.0	1.4	1.4	31.0	20.1
				(1.7?)[b]		
Albania						
1960[d]	69.5	9.0		21.5	75.7	28.9
					(1966)	(1966)
1971	85.2[c]	14.1		. . .	53.4	13.0
1971[d]	78.7	15.8	1.9	3.7	53.4	13.0
1973[d]	78.7	. . .	. . .	. . .	. . .	. . .
Yugoslavia						
1960	5.7[e]	4.8	0.6	88.9	96.3	89.9
1970	5.0[e]	10.6	1.8	82.6	96.1	85.7
1977	0.9[e]	17.6		81.5	97.2	83.2
Rumania						
1960	50.2	11.8	17.5	20.4	87.9	71.5
1970	54.1	14.0	16.1	15.8	53.6	36.5
1978	54.3	13.6	16.4	15.7	46.7	27.2
Bulgaria						
1960	79.9	6.6	4.3	9.1	45.7	35.8
1970	68.0	15.6	5.7	10.7	34.4	27.4
1978		90.3		9.7	31.0	25.7
Poland						
1960	1.1	11.2	0.4	87.3	93.3	92.7
1970	1.2	14.0	0.4	84.4	89.9	90.6
1978	2.8	17.1	1.1[f]	79.0	87.1	74.9

Table 6.1 Continued

Hungary						
1960	48.6	12.1	7.2	32.1	67.3	67.7
1970	67.6	12.8	2.5	17.0	45.9	62.0
1978	70.3	12.6	2.6	14.5	32.7	52.3
Czechoslovakia						
1960	62.1	15.5	4.9	17.6	33.0	28.5
1970	55.7	20.2	9.2	14.9	18.7	22.2
1978	62.1	20.0	10.6	7.4	5.6	7.8
GDR						
1960	72.8	6.2	1.9	19.1	54.8	46.4
1970	78.2	7.0	1.1	13.7	20.9	21.1
1978	81.3	8.6g	0.8	9.3	2.3	9.5

a Derived as a residual.

b The striking expansion of the residual area to 1.7% is due to the land of non-private interfarm agricultural enterprises and organizations.

c Including plots of collective farmers.

d Harvested area only.

e Including General Agricultural Cooperatives; for 1977, excluding those cooperatives which were incorporated into other "organizations of associated labor" (i.e., state and comparable farms).

f Presumably, including the State Land Fund.

g The share of state farms in a stricter sense of the work (VEG = Volkseigene Guter) was only 1.9 percent in 1976, and again 5.9 in 1978 *(Statistisches Jahrbuch der DDR, 1978,* p. 157).

Sources: For Albania: *Vjetari Statistikor i RPSh, 1971-1972* (Tirana, 1973), pp. 74, 84-85, 106-107; *30 Vjet Shqiperi Socialiste* (Tirana, 1974), pp. 104-105; for Yugoslavia: *Jugoslavija 1945-1964* (Belgrade, 1965), pp. 105, 116; *Statisticki bilten,* no. 1065, p. 17; *Statisticki godisnjak SFRJ, 1978,* pp. 224, 237; for all other countries: *CMEA statistical annual, 1979,* pp. 228-229, 257, 261 (for livestock in the GDR, in combination with *Statistisches Jahrbuch der DDR, 1978,* pp. 179-180).

and state farms became smaller in manpower and larger in area or livestock per worker. During the 1970s, brigades or sections of farms have more and more been organized and managed according to the production or branch principle (i.e., by specialization of output or work process instead of by location). Crop and animal production often became completely separated, within the farm or between farms, most conspicuously so in the GDR, where by 1977, practically no mixed farms existed any longer.

In Yugoslavia, no fundamental distinction has been made since the early 1950s between state and collective (group) ownership, although for a different reason. The state farms, as well as the cooperatives and all other socialized enterprises, became considered socialist in the

Table 6.2   Average Sizes (hectares of agricultural land) of State Farms Proper (excluding institutional farms) and of Collective Farms (excluding private plots)

	Year	State farm	Collective farm
USSR	1960	26,104[b]	6,463
	1970	20,529[b]	6,094
	1978	17,548[c]	6,599
Moldavian Republic	1960	2,033 (1,295[d])	4,286 (2,995[d])
	1976	.. (1,440[d])	.. (3,020[d])
Estonian Republic	1960	3,596	1,855
	1976	4,542	4,333
Georgian Republic	1959	1,626	1,055
	1969	3,581	1,432
Albania[a]	1960	2,060	229
	1970	2,105	736
	1973	...	1,091
Yugoslavia	1960	1,320	179[e]
	1970	4,911	561[e]
	1977	1,032[f]	174[e, f]
Rumania	1960	3,071	1,495
	1970	2,881	1,728
	1978	5,149	1,660
Bulgaria	1960	5,627	4,865
	1970	6,000	5,496
	1978	ca. 29,000[f, g]	
Poland	1960	398	117
	1970	509	214
	1978	1,894	258
Hungary	1960	2,598	734
	1970	4,906	1,658
	1978	6,394	3,097
Czechoslovakia	1960	3,104	421
	1970	4,265	631
	1978	7,663	2,426
GDR	1960	592[h]	238
	1970	867[h]	490
	1975	1,024[h]	781
	1978	760[i]	4,820[i]

[a] Tilled land only.

[b] Because of very extensive land use in mountainous and steppe regions, the average sown area per state farm was "only" about 9,000 hectares in 1960, 6,100 in 1970, and 5,600 in 1977.

[c] Beginning in 1973, the feedlot farms of the Ministry of Agriculture of the USSR were statistically included among the state farms proper, which was one of the reasons why the average size, in terms of land, declined.

[d] Sown area only.

Table 6.2 Continued

e Including the General Agricultural Cooperatives.

f The 1977 sizes are not comparable to the earlier ones, as categorization and counting of
  the socialist farms were changed.

g Beginning in 1973, Bulgarian statistics no longer gave the land used by state and collective
  farms separately, but only that of the so-called agro-industrial complexes which comprise
  both sectors and cover 80-90 per cent of the agricultural area.

h Including experimental farms of universities.

i For state farms only the VEGs, for collective farms only those specializing in crop
  production; those specializing in animal production had less than 50 hectares of land on
  average.

Source: For Soviet Union and Union Republics (Estonia, Moldavia, Georgia): *Narodnoe
khozyaistvo SSSR za 60 let* (Moscow, 1977), pp. 355-356, 371; *Narodnoe khozyaistvo
Estonskoi SSR v 1976 g.* (Tallin, 1977), pp. 120, 124; *Narodnoe khozyaistvo Moldavskoi
SSR v. 1960 g.* (Kishinev, 1961), pp. 80, 100-103; *Sovietskaya Gruzia za 40 let* (Tbilisi,
1961), pp. 71, 81; *50 let Sovetskoy Gruzii* (Tbilisi, 1971), pp. 132, 142-143; for Albania:
Vjetari Statistikor i RPSh, 1971-1972, pp. 74-75, 81; *30 Vjet Shqiperi Socialiste*, pp. 104-
105; for Yugoslavia: *Statisticki godisnajak, SFR, 1978*, pp. 495-496; *Statisticki bilten*, no.
720, p. 16, no. 816, pp. 11, 15, and no. 881, p. 9; for all other countries, *CMEA statistical
annual, 1979*, pp. 224-225, 229-230, 278 (for the GDR in 1978: *Statistisches Jahrbuch der
DDR, 1978*, pp. 156-157).

---

framework of Yugoslav "self-managed socialism", and therefore as being
the property of their members and workers and/or of a communal or
regional socialist organization. The farm organization in the socialist
sector of Yugoslav agriculture is differentiated not so much along formal
categories, but according to output mix, land relief, dispersed or compact location of fields, weight of nonagricultural farm activities, extent
and kind of cooperation with private peasants, etc.

The main subdivision of a Yugoslav socialist farm is the working unit,
.which may have varying degrees of self-administration. In very large
farms, these units are in most cases formed according to the production
principle. In other farms and in those whose territory is dispersed over a
wide area, organization by location prevails (the territorial principle).
Sometimes special units are formed for cooperation with private peasants. Depending on farm size and dominant production, sections on an
intermediate level may also exist, and these have a greater degree of
self-administration than the units below them.

In many aspects of their actual economic activity and organization, the
Yugoslav General Agricultural Cooperatives differed from the collective
farms of the Soviet or CMEA type, the last of which in Yugoslavia
seem to have disappeared in the late 1960s. Farm production on its own
has increasingly made up the smaller part of the economic activities of
General Agricultural Cooperatives. By contract and other cooperation
with members and also non-members, they now engage not only in
agricultural production but in related servicing and also in processing
the produce. It would be more fitting to characterize them as multipur-

pose cooperative entrepreneurs. It is similar for other socialist farms. On the whole, this sector derives almost half of its income from nonagricultural activities and devotes more than one-third of its assets to them. (For the sake of simplification though, the General Agricultural Cooperatives continue to be treated as representing collective farming in some statistics of the present book.) In the practice of Polish collective farming, one finds similar activities, although to a lesser extent, but the theoretical reasoning on ownership relations there corresponds to that of the other CMEA countries. At any rate, the majority of Polish collective farm members of today did not own land formerly, and most of the land farmed collectively was given by the state.

After a prolonged period of stagnation, collective farm hectarage in Poland expanded again during the 1970s. Collective as well as state farms received land through the State Land Fund. Such transfer was favored by the law of 1968, amended in 1973 and again in 1979, on rents to be paid to retiring peasants who hand their land over to the Fund or directly to a socialist farm. The activity of the State Land Fund reflects the proclaimed (or not proclaimed) official policy and its changes. During 1970, the Fund received 111,200 hectares of agricultural land from private peasants (of which 63,100 were in return for an old-age pension); by 1975, this figure tripled (358,900 hectares, of which 314,600 were for a pension). The figures declined again in 1976 and 1977. The land sold by the Fund to private peasants dwindled to 16,700 hectares during 1975, but by 1977, reached 77,700 hectares and slightly surpassed even the 1965 level. And while up to 1976, collective and state farms received 88 percent of land handed over by the State Land Fund, the three sectors' shares in 1977 were almost equal. Even so, peasant farming on balance lost 281,200 hectares during 1977.[2]

The socialist sector in Poland in 1974 owned more agricultural land (25.1 percent of the total) than it utilized (19.0 percent), whereas it was vice versa for the private sector (74.9, as against 80.2 percent). The discrepancy was even greater in 1965 (no such data are available for earlier and later years).[3] Some of the land which the social sector, mainly the State Land Fund, cannot utilize at a given time is leased to private peasants.

Under the conditions of predominant individual peasant farming in Poland, there have also been agricultural cooperatives of a more traditional kind, based in part on private and in part on collective property. The Polish state assigned them a considerable role in sponsoring collective economic activities of peasants and in the transactions between these and the state. In 1973, about 70 percent of the public sector's purchases from peasants, and roughly three-quarters of the supply of investment and other production goods for peasants, were channeled through agricultural trade cooperatives. In the other CMEA countries, consumer cooperatives account for part of the purchases (in addition to the state procurement agencies), and for most of the rural retail trade, but not for the supply of production means, except for sales of livestock feed and of some small implements to private plot holders.

The division into three categories of land ownership often, but not always, coincides with differences of farm size and organization (private peasant farms and plots are discussed later). During the 1960s, it was the prevalent opinion of Communist leaders and economists in Eastern Europe and the Soviet Union that the "optimal" size for modern large-scale farming is 3,000 to 5,000 hectares of arable land per farm (implying even larger farms in terms of total agricultural land). Much of the increase in average collective farm sizes in the GDR up to 1975 resulted from incorporating the usually smaller "lower" type farms into those of the "higher" type. Similarly, in Hungary and Czechoslovakia, the aim was to eliminate the smaller ones. Soviet, East German (for crop production), and Hungarian collective farms have attained the "optimal" average size, while the Bulgarian agro-industrial complexes far surpass it (see Table 6.2). Through horizontal cooperation, production units larger than the nominal sizes of the components have also been achieved in the Czechoslovak collective farm sector. In many cases, such cooperation or integration, which has been going on in most countries, sooner or later makes the individual collective or state farm lose its economic and juridical identity, thereby forming agricultural enterprises of a new order of magnitude.

The state farms in most countries had already by the mid-1970s attained the above-mentioned size, except in Poland and East Germany. State farms used to play the leading part in the integration process, including the formation of agro-industrial complexes. Thereby, the state sector also surpassed the formal average farm sizes. In the Soviet Union, state farms by 1964 had in a number of cases exceeded what at the time was considered an optimal size, and a significant reduction was then discernible. The all-Union averages hide divergent regional sizes, though. Under comparable natural conditions, Soviet farm sizes and their development have been similar to those in Eastern Europe (see the data for Moldavia, Estonia, and Georgia in Table 6.2). Recent agro-industrial integration, where it came about in the USSR, again resulted in gigantic production units.

Average sizes of socialist farms in Yugoslavia, including the General Agricultural Cooperatives, did not increase after 1970. Yet the data do not disclose the changes that actually occurred in the most recent years because a number of former farm sections now are considered farms on their own. It remains an open question, though, whether they really have economic independence from those other farms which act as guiding enterprises for them.

In Albania, a strong campaign to increase farm sizes set in after collectivization was completed. A new kind of "higher type" of collective farm was introduced on 15 July 1971. These farms in some respects resembled state farms. Their machinery continued to be owned by the state's MTS, but one MTS was affiliated to each collective farm of the new variety, and rendered technical service exclusively to that particular farm. Long-term investment was financed by the state, and such assets remained state property within the collective farm. By 1976, 23 percent of Albania's arable land was in such farms. The new "higher type" collective farms of 1971 had more than double the tilled area (41 such farms, averaging 2,700 hectares of tilled land by the beginning of 1975) than the still existing old "higher" type collective farms (428, averaging 1,200 hectares of land at the same time.).[4]

In Yugoslavia, and to some increasing degree also in Poland recently, data on land area owned and farmed by the collective farms did not show their true size as production units. They often farmed more land in loose or in close cooperation with individual peasants, and this sometimes came near to a practical, though not formal, incorporation of such peasant farms.

The Polish and Yugoslav average sizes of collective farms were largely determined by the local conditions of their formation on a voluntary basis. Therefore, they do not tell what sizes the agrarian policy-makers would have preferred. It is different for the state farms, where the Yugoslav average of 1970 corresponded to that of other Balkan countries, and the Polish average almost to that of the GDR at the time. Since then, Poland also made efforts to increase the size of state farms. But the area of Polish state farms was often unconsolidated, and if for no other reason, their management was not so strongly centralized as in other countries. In Albania, the tilled area is located mainly in the coastal plains and the lower hilly ranges adjoining it, and that is where one finds most of the state farms. The average size of these farms was about the same as in Hungary or Czechoslovakia and may be assumed to have increased since 1970, the last year for which data are available, when it stood at 2,105 hectares.

A few words are necessary on organizations and enterprises servicing agriculture. Only in Poland and Yugoslavia have they existed, apart from the cooperatives characterized above, as cooperatives or joint ventures of the peasants themselves. Yet they, too, mostly were organized on state initiative and made to fit into regulations deemed appropriate for an otherwise socialist economy and society. Important in Poland were the Agricultural Circles, which were complemented by the state-owned Machinery Centers. The latter took over the repair and maintenance of agricultural machines where such exceeded the capacities of the Circles or of associations of them. Apart from technical services, the Agricultural Circles sometimes also took crop farming upon themselves, especially on former peasant land of the State Land Fund.

In the other countries, organizations or enterprises with comparable functions were either state-owned or were joint ventures of several collective farms. After the abolition of the MTS, the Soviet Union, GDR, Czechoslovakia, and Hungary sold their machinery park to the collective farms. The state provided stations for major repairs, which also acted as agents for selling machinery, spare parts, fertilizer, and similar inputs to the farms. Beginning in 1970, the transformed Rumanian MTS combined such functions with those of the MTS under the name of "Stations for the Mechanization of Agriculture," and the system expanded somewhat. The GDR combined both sorts of functions in its County Enterprises for Agricultural Technology, as did Czechoslovakia in her remaining MTS, renamed Construction and Tractors Stations. Yet in both countries, most of the machines are in farm ownership. The giant Bulgarian agro-industrial complexes as a rule each formed an Agro-Chemical Center and a machinery repair enterprise of their own.

A new and special kind of servicing enterprise was the Agronomic Center or Agro-Chemical Center, which came into being in the GDR in the late 1960s, and in Czechoslovakia in 1971. They soon began to play a role also in various regions of the USSR, although initially they lacked a unified form. In Hungary, it was envisaged in 1976 to create such centers, but except for a few units and a national Center of Agro-Chemistry and Plant Production, not much seems to have come out of it so far.[5]

Usually the agro-chemical stations or centers were state, or joint collective and state farm, enterprises. Their main functions were to help and advise with the application of fertilizer and plant protection chemicals, but they also carried out mechanized field work and transport for the collective farms. In the Moldavian Republic of the USSR, kolkhozes in 1973 began to form inter-farm Associations for Mechanization and Electrification, to which they handed over their machinery, repair shops, gasoline stations, etc. These stations, or technical centers of these associations, remind one of the former MTS, although they are bigger and not state owned. A Soviet decree, published on 30 August 1979, inaugurated the formation of an all-Union self-accounting association for agricultural chemical services, "Soyuzsel'khozkhimia," with regional associations and stations for servicing the farms. Associations for soil improvement were another new venture in the GDR, while in the Soviet Union such works, when undertaken on a great scale, were carried out by a specialized state organization. Such new organizations and enterprises have come into existence under various names and forms and are part of the so-called industrialization and agro-industrialization of agriculture, which became the new policy formulation by 1970 (to be dealth with in chapter 12).

It is accepted official doctrine in Eastern Europe and the Soviet Union that at some future time, collective property (i.e., that of the peasants who once formed a given collective farm, and of their descendants and of later admitted members) will merge with state property into a new kind of property of all the people of a country, with the characteristics of the state farms predominating. The process is expected to "require a prolonged period and undoubtedly will last beyond the current and, for a number of [CMEA] countries, also the subsequent decade," that is, beyond 1990.[6]

Opinions differ as to how long the collective farm sector will continue to exist. The Rumanian Party Program of 1974 indicated a period of twenty to twenty-five years during which the sector will have to be strengthened economically, which implies a continued existence at least up to the year 2000. The XIth Hungarian Party Congress (1975) also emphasized that for the immediate period, both collective and state property will have to be strengthened and developed. Earlier Hungarian legislation, and a number of subsequent acts, referred to various kinds of cooperatives, among which collective farms are the most important and numerous. The East German Party Program of May 1976 emphasized the strengthening of the collective farms. However, subsequent measures, including the proclamation in 1977 of Model Charters for the new specialized collective farms of crop and of animal production, showed that the type of farm favored for the immediate future has much more in common with state farms than with the previous collective farms.

Soviet policy under Brezhnev in principle seemed to adhere to a course of continuing the existence of the collective farms, but in practice did not prevent the further expansion of the state farm sector at the expense of the kolkhozes (see Table 6.1). The new Soviet Constitution of 1977 explicitly stated that collective property will "come closer" to state property. Most of all, it was the Bulgarian Communist Party, especially at its March/April 1976 Congress, which emphasized the merging of state and cooperative property into a unified property of all the people, starting during the 1976–80 quinquennium. In fact, Bulgarian statistics since 1972 have no longer differentiated between collective and state farms. Although similar views have been expressed in all CMEA countries, it was the Bulgarians who first formulated them as official doctrine to be applied in present-day practice.

Generally, however, the warning is often given that hasty decisions in this matter would do harm. Two Soviet authors stated, with a view to all CMEA countries, that "urging on and accelerating the liquidation of these cooperatives would essentially mean to make the cooperative sector run the farms 'at the expense of society', which would put an extreme burden of additional expenses on the state, without providing new sources for covering such expenditures."[7] In other words, guaranteeing incomes and investment capacities equal to those of the state farms would cost more than the public budget can afford. The special, and still disadvantaged, position of the collective farms in most cases remains necessary for some time to come in order to prevent a still greater cost increase of food production.

The Private Sector

In the countries with collectivized agriculture, the private sector consists of plot mini-farms and animal holdings, and usually of a few remaining small and dwarf peasant holdings. Exceptions are the commercial gardening cooperatives and the remnants of church land property in the GDR, the specialized production cooperatives in Hungary, and a few other remaining "lower type" agricultural cooperatives in the CMEA countries. Quantitatively, they have become almost negligible and usually are medium-size rather than large-size producers.

The land used for plot farming may be in formal private ownership (of remaining individual peasants and, depending on the country and the local situation, plots of employees and collective farmers); in state ownership (mostly employees' plots, but in the Soviet Union, also collective farmers' plots since all land there is nationalized); or in collective ownership (left for individual usage to farm members, but often also to employees residing in collective farm villages). Apart from these differences of formal ownership, the more essential criterion is the private usage right, which is founded on a legal claim and can be withdrawn or expire only under fixed legal conditions. If so, it is possible that the individual or family household can exert this right on a defined basis, although under certain restrictions, for maximizing private income. The essence of this state of affairs is not changed by the fact that in Communist countries, the notion of "private" ownership and production is usually said not to apply, and that "personal" is the official term. Even where the land is owned by the state or a collective, the other means of agricultural production (livestock, small premises, and working implements) of the private sector as a rule are private property not only in actual practice but also juridically.

The extremely small proportion of land in private use in the Soviet Union would amount to more than 3 percent if instead of total agricultural land, only sown area and perennial crops were taken into account, and if some additional land under vegetable and fruit gardens, which only formally is registered as communal or belonging to nonagricultural enterprises, were added. In Albania, grazing rights are an important basis of the private sector, although no relevant data are available (but cf. the share of livestock holdings in Table 6.1). Comparing the data of 1970 with those of 1960, one must not forget that in the earlier year, collectivization in Albania, Rumania, and Hungary, and to some degree even in the GDR and Czechoslovakia, had not yet been completed.

Most collective farms are very large units, but their production, especially animal production, depends on a more or less important economic symbiosis with their members' plot farms. The latter deliver some of their produce to the collective farms, or through them to the state, and in exchange receive not only cash payment, but also young animals, most of all piglets, feed, etc. This is strikingly illustrated by the fact that during 1976–78, the private sector of the six CMEA countries with collectivized agriculture (USSR, Rumania, Bulgaria, Hungary, Czechoslovakia, and the GDR) held only 7.6 percent of all breeding sows, but produced 32.2 percent of all pork.[8] Moreover, the household plots form the basis of existence for some labor, which otherwise could hardly be available for collective production at peak times.

It is not always clear whether the collective farms (often illegally) register as their own the produce bought from their members and delivered to the state or kept for restocking their own animal herds, thereby inflating their productive performance. It seems to happen sometimes, and therefore, apart from other reasons of possible under-reporting, data on the private sector's share in total agricultural production must be looked at with reservation. For Hungarian collective farms, the practice is neither denied nor forbidden, and this obviously is why data on pork production there are no longer given separately for the farm and its members.

As far as the individual peasant is concerned, his means of production, including most of his land, are private property anyway. But he may in one way or another cooperate with the public sector and use that sector's means (machinery, fertilizer, production services) in return for payment, entering into a share-cropping arrangement or delivery contract, etc. This public input in the private sector is most widespread in Poland and Yugoslavia, and is considered there an element of a gradual socialist transformation of agriculture which these governments hope to achieve in the long term.

A recent development in both Yugoslavia and Poland is the cooperation, on their own initiative, of small numbers of peasants for specific production purposes, not necessarily committing the whole farm. Earlier efforts in Yugoslavia to win peasants over to closer and permanent cooperation with the social sector had very limited success, and more recently "basic organizations of cooperators" and "working organizations of cooperators" were advocated. Their juridical basis is the Federal Constitution of 1974 and, more specifically, the "Law on the Associated Labor" of 11 November 1976. Beyond this general framework, no strict rules were set, and a variety of such organizations developed, in animal more than in crop production. It does not seem that in the near future this will become a general and economically viable form of peasant cooperation. A comparable cooperation arrangement in Poland, given legal status and privilege in late 1972, may comprise a small group of farms, but no less than three. Usually, these cooperating peasants have delivery or other contracts with public processing or trading enterprises. The important point in the Yugoslav as well as in the Polish ventures is that the participating farms retain their separate juridical identity, and that the state exerts only indirect and limited guidance. Therefore, they are part of the private sector, and must not be classified among the collective farms, although some Polish land statistics do so.

In five of the countries with collectivized agriculture (USSR, GDR, Bulgaria, Hungary, and Albania), the main part of private land use and animal production is that of the collective farm members on their plots, and in Czechoslovakia and Rumania, it forms at least an important part. State farm workers in the GDR are not supposed to have private plots. However, the new Model Charters for collective farms in 1977 permitted the increasing numbers of these farms' wage-earners (workers), in addition to member labor, to run plots. While the number of collective farmers is decreasing, the number of such farm employees' plots may increase. As far as state farm workers in the GDR are concerned, their number, including those of other agricultural organizations of the state, amounted to some 150,000 in 1978, roughly one-fifth of the total annual average workforce in agriculture.

Private plot farming serves several functions, for the collective farmers themselves as well as for the economy at large. It complements the income in kind and in cash to or above the subsistence level where such income from collective work is inadequate. It utilizes labor reserves which in the large-size socialist farm cannot, or can only in part, be put to productive use, so that otherwise there would be underemployment for most of the year. It contributes to the supply of urban areas with food, and in the countryside relieves the state and cooperative retail network from the supply of the population with food, as this network in many rural places exists in only a rudimentary form. Moreover, it contributes to the productive performance of the collective farms through the symbiosis outlined above. As time went by, these functions were no longer needed as urgently as before. Nevertheless, they could not yet be dispensed with, as has been admitted in many official statements of more recent years, in contrast to some statements and actions during Khrushchev's time. In most CMEA countries, a certain benevolence towards private subsidiary farming became official policy during the early 1970s and was re-emphasized by the end of the decade.

Land utilization (predominantly for vegetable, including potato, and fruit growing) and animal farming in the private sector are very intensive. This is why in spite of the small share in the agricultural area, this sector in the second half of the 1960s still produced roughly 30 percent of the total gross output of agriculture in the Soviet Union, Hungary, and Rumania, more than 25 percent in Bulgaria, and about 28 in Czechoslovakia (where there still were a number of individual peasants on hilly and otherwise marginal land at that time). In Polish collective farms of that time, 46 percent of the income of collective farmers originated from their private plots and livestock, and 21 percent of state farm workers' incomes.[9] Since then, the private sector's relative share has gradually declined, except for Hungary, but not always the absolute volume of its output. The share in Bulgaria was 20.9 percent in 1970, and probably amounts to slightly less than one-fifth by now. In Hungary, the private sector contributed 32.2 percent in 1974, and 31.15 percent in 1978, while in Czechoslovakia, it was a mere 13.5 percent by 1977. For the GDR, it may be assumed to have amounted to still less, not much more than 10 percent.[10] A Soviet author indicated 33.7 percent in 1960, and 25.3 percent in 1975 for the USSR,[11] but 1975 was a catastrophic drought year which hit the private sector less than the socialist sector's cropping results. A Hungarian estimate for the year 1974 put the shares at 21 percent for the Soviet Union (of which 46.3 percent was from households other than those of kolkhoz members), 16 percent for Bulgaria, 15 for Czechoslovakia, and 31 for Rumania.[12] Judging by the recent numerical development of the shares in land and animals (see Table 6.1), it may have further declined during 1978 in Rumania (where it seems roughly equal to the Hungarian percentage), Czechoslovakia, and the GDR. Already in 1975, 37 percent of Czechoslovak collective farms no longer had any private plots,[13] and a similar development is likely for the GDR (for private plot incomes, see pp. 181–82).

In Albania, the upper limits for private plots and animal holdings were reduced by half after completion of collectivization (1967), and the workers of state farms were called upon to renounce their plots. The Albanian Model Charter of 1971 for the new "higher type" collective farms permitted plots of only 0.03 hectare (one-thirteenth of an acre).[14] However, it seems that the rules were not always adhered to and that actual developments, as well as policies, were full of contradictions. In 1975, complaints were published that many plot-holders exceeded the limits for privately-owned livestock, and that they used machinery of the socialist sector and even employed hired labor.[15] It was also stated, in a leading Albanian newspaper early in 1976, that the time for the total abolition of private land use was approaching.[16]

The share of the private sector in total livestock herds gives an indication of its weight, at least in the animal part of agricultural production. It also shows different intensities of private plot farming by countries, and the importance of pig versus cattle farming according to the economic conditions as well as to the national traditions. Thus, the private share in pig herds was rather small in Albania (it was much greater for goats and sheep), whereas in East Central Europe, most of all in Hungary, it exceeded that in cows. Because of the character of small peasant farming, with its greater labor intensity, the peasants in Poland and Yugoslavia also had a greater share in livestock holdings than in land.

In Poland and Yugoslavia, the share of the private sector is by far dominant, but is gradually shrinking. By 1977, the private peasants still produced 77.9 percent of overall ("global") output of Polish agriculture, and in Yugoslavia they contributed 77.0 percent to the national income generated in agriculture.[17]

The peasant farms in Yugoslavia in 1969 (census results) numbered 2.6 million, averaging 3.9 hectares (including nonagricultural land) per farm. Since 1960, their number had slightly increased, because those with less than one hectare grew in number. The latter had a total of two million family members permanently employed outside agriculture. During the same nine years, the number of Yugoslav peasant farms with more than three hectares declined. From the Yugoslav literature (some of it as yet unpublished) it emerges that the total number of individual peasant farms increased since 1969 and that the average size continued to diminish and the annual outflow of labor from private agriculture rose to 2.5 percent. According to the findings of a Yugoslav study on private peasants' households, those where the young people already left the village made up 30 percent of the total number, 40 percent did not wish their children to go into farming, and only 30 percent were encouraging their children to stay in farming.[18]

Polish peasant farms in 1970 (according to census data) held 5.4 hectares of land on average, of which 90 percent was agricultural land. The size was practically the same as ten years earlier (5.5 hectares). Since 1960, the share of farms with 2–5 hectares had declined, the percentage of those with more than 10 hectares had increased, and that of the other size classes remained roughly the same. The total number of farms with more than 0.5 hectare slightly declined between 1960 and 1970, from 3.22 to 3.1 million, while the number of those with less than 0.5 hectare rose from 348 to 365 thousand. But later on, from 1970 to 1976, the smallest holdings increased their share, and so did the larger farms with more than 10 hectares. Of all those with more than 0.5 hectares, the latter held 38.3 percent of peasants' agricultural land, and those with 0.5–2 hectares, 6.3 percent, while the bulk was in the 2–10 hectares group (55.4 percent of the land, and 57.0 of the number of farms). The average size in 1976 was 4.9 hectares of agricultural land for all farms above 0.5 hectares, and 6.5 for those above 2 hectares.[19] According to the 1970 census, peasants of 60 and more years owned roughly 30 percent of all farms, and of these, every third had no heir. Because of widespread part-time farming, mainly of males, 37 percent of the farms over 0.5 hectares were managed by women. Under these circumstances, many such holdings are likely to disappear in the foreseeable future as peasant farms, to reappear as part-time plots of nonagricultural workers and employees, comparable to those in countries with collectivized agriculture.

Contrary to a dogmatist notion that a decline of peasant farm sizes should be tolerated because it prepares the ground for collectivization, the Polish leaders since 1963 have taken measures to prevent such a latent process. The motive was that smaller farms, especially part-time farms of "worker-peasants," reduce the share of marketings in their output. By a law of June 1963, splitting farms through sale or inheritance was made difficult, and the former restrictions on private peasant land purchases to enlarge existing farms were lifted under certain conditions. Three more laws on peasant farming were enacted in January 1968. These concerned: pensions for those peasants (mainly the aged and those without heirs to the farm) who gave their land to the State Land Fund; land consolidation and exchange of fields among peasants to amalgamate the many scattered small strips; and compulsory sale of land to the State Land Fund if peasants did not achieve the minimum standards of farming techniques. An important, but not the only, intent of this legislation was to expand the area of state and collective farms.

Continuation of collectivization in one form or another was discussed almost permanently in Poland, but not made official policy. Only in 1973 did a semiofficial tendency toward gradual socialization become discernible. It was discontinued, however, three years later when strikes and social disturbances followed the ill-founded attempt of summer 1976 to curb the increasing demand for quality food by higher consumer prices. Since then, the government again has emphasized the important role of individual peasants in more rapid growth of agricultural production.

It has often been argued that private plot activities in socialized agriculture will drastically shrink and finally disappear as soon as remuneration from collective work markedly exceeds income (per labor input) from the private plot. This is only partly true. It might become wholly true when rural cash income can without great difficulty buy those indispensable food items that are privately produced (meat, milk, eggs, vegetables, potatoes, fruit), but is unlikely to happen in the near future. In addition, there would have to be full employment, agricultural and otherwise, for everybody in the agricultural sector who wanted to earn some additional income. In countries like Rumania and Albania, and in regions like Slovakia and the southern and southeastern parts of the Soviet Union, much labor still finds only part-time employment in the socialized sector and prefers productive and income-earning employment in the private sector to nonemployment, even if the rewards per hour worked are less than on the huge collective and state farms.

Hungary, with its unorthodox policy toward the private sector since 1968 (and before), has demonstrated how a certain liberalization can mobilize such labor resources for the individual's as well as society's benefit. But even there, such possibilities within the framework of socialized agriculture have by now reached a labor ceiling. There, as well as in most parts of Eastern Europe and the Soviet Union, the labor resources of the private sector are fully mobilized at present, and will soon be shrinking for demographic reasons. Only a fundamental change of socio-economic parameters, instead of cosmetic measures as in some CMEA countries at present, can bring about an additional, sizeable contribution of this sector to an overall increase in agricultural production. In parallel, the income contribution of the private sector in agriculture will continue to diminish in relative terms, and at best will hold (with rising prices) its absolute volume.

Notes

1. R. G. Vartanov in *Problemy izmenenia sotsial'noi struktury sovetskogo obshchestva* (Moscow, 1968), pp. 114–117.
2. *Rocznik statystyczny, 1978,* p. 200.
3. *Concise statistical yearbook of Poland, 1973* (Warsaw, 1973), p. 133, and *1975* (Warsaw, 1975), pp. 146–147.
4. *Bujqësia socialiste,* no. 5 (1975), pp. 1–4 (according to *ABSEES* 6:4 (October 1975), p. 127).
5. For Hungarian doubts on the usefulness of "so-called agro-chemical centers," see *IIIrd National Conference on Agricultural Economics* (Research Institute for Agricultural Economics, Bulletin no. 42) (Budapest, 1978), pp. 31–32.
6. G. I. Shmelev and V. N. Starodubrovskaya, *Sotsial'no-ekonomicheskie problemy razvitia sel'skogo khozyaistva evropeiskikh sotsialisticheskikh stran* (Moscow, 1977), p. 85.
7. *Ibid.,* p. 122.
8. *CMEA statistical annual, 1979,* pp. 260–262, 264. The end-of-year sow numbers of 1975–77 are compared to pork output during 1976–78.
9. J. F. Karcz, "Agricultural Reform in Eastern Europe," in M. Bornstein, ed., *Plan and Market* (New Haven, 1973), p. 220; for the Soviet Union, see Wädekin, *The Private Sector in Soviet Agriculture* (Berkeley/Los Angeles/London, 1973), p. 61.
10. *Statisticheski godishnik na NR Bulgaria, 1971,* p. 191; *Statistická ročenka ČSSR, 1977,* p. 258; *Statistical Yearbook, 1974* (Budapest, 1974), p. 225.
11. I. F. Suslov, *Agrarnii sektor ekonomiki stran sotsializma* (Moscow, 1978), p. 248.
12. *The situation of small-scale farming in Hungary and its development* (Research Institute for Agricultural Economics, Bulletin no. 43) (Budapest, 1978), pp. 40–43.
13. Jan Baryl, *Otázky míru a socialismu,* no. 9 (1975), p. 19.
14. Decree of 15 July 1971, according to M. Kaser, "Albania," in H.-H. Höhmann, M. Kaser, K. C. Thalheim, eds., *The New Economic Systems of Eastern Europe* (London, 1975), p. 261.
15. T. Tonev, *Dunavska pravda,* 5 May 1975, p. 4 (according to *ABSEES* 6:4 (October 1975), p. 148).
16. *Bashkimi,* 22 February 1976, pp. 2–3 (according to *ABSEES* 7:3 (July 1976), p. 109).
17. *Rocznik Statystyczny, 1978,* p. 194; *Statistički godišnjak SFRJ, 1978,* p. 160.
18. S. Jelacić, *Rad,* 19–25 September 1975, p. 3 (according to *ABSEES* 7:1 (January 1976), p. 235).
19. Augustin Woś and Zdzisław Grochowski, *L'agriculture polonaise—les dernières mutations* (Warsaw, 1979), pp. 62–63.

7. Shifting Policy Goals and the Comparative Place of Agriculture in the Economy

EVERYWHERE IN EASTERN EUROPE and the Soviet Union, agricultural output during the 1950s lagged behind the ambitious quantitative plan targets, and its growth even slowed down between 1958 and 1965. At the same time, rapid industrialization, rising incomes, repressed demand, and generally rising expectations of consumers increasingly showed their effect. Fast output growth became an objective necessity. Population growth also had an influence, although not a dominating one, as it was becoming slower. By the mid-1970s, it was still more than two percent only in Albania, was roughly one percent per year in Poland, Yugoslavia, Rumania and the USSR, while it was lower in the rest of the countries, with even an absolute population decrease in the GDR. Overcoming the factors impeding a more rapid growth of agricultural production became a goal of high priority, once the overriding socio-political goal of the socialization of agriculture was attained. In Poland and Yugoslavia, such a change of priority had come about earlier. There the necessity of output growth, along with the internal political difficulties, had been the main reason to renounce collectivization for the foreseeable future. In those countries where peasant agriculture was already collectivized, the transfer of labor from agriculture became less desirable and feasible. A Hungarian author stated that "the main objective of our agricultural modernization is the increase of crop yields, the expansion of production, rather than the release of manpower or retention of the actual live labour at any price."[1]

It will be shown in chapter 8 that nutrition standards as such were not unsatisfactory in the area, with the possible exception of Albania. If, all the same, the demand for food has not been adequately met, this is not a consequence of an absolute shortage of food, but mainly of the fact that the supply of nonfood, especially industrially produced consumer commodities, is not adequate in quantity and/or in quality. It cannot absorb the rising cash incomes of the population, and therefore the purchasing power is often diverted to high-quality or semiluxury food.

Under such conditions, calculating the usual coefficients of income and price elasticities of the demand for food (even where the data would allow) does not yield meaningful results. In markets—abstracting here from the "parallel," illegal, half-legal, or legal free markets in all countries concerned—where supply is not directly influenced by demand and free prices, the consumer cannot choose among the commodities he wants most, but only among those that are offered. This is why population and real income growth have a very direct, almost proportionate, bearing on the necessity to increase agricultural output, especially as the latter's growth in most of the countries was more than doubled by that of real incomes. Recalculated as three-year averages of 1976–78 over 1969–71, according to official CMEA and semi-official FAO and West-

Table 7.1 Average Annual Growth of Real Incomes (1970-77) and of Agricultural Production (average 1976-78 over 1969-71[a])

	Real income of population (CMEA official and Yugoslav statistics)	Agricultural production		
		CMEA official[b] (gross of feed and seed)	FAO[c] (net of feed and seed)	USDA[d] (net of feed)
Soviet Union	5.1	1.6	2.3	2.4
Albania	5-6 (estimate)	6-7	3.3	...
Yugoslavia	4.8	...	3.4	3.4
Rumania	7.7	5.8	6.6	5.9
Bulgaria	5.3	2.5	2.3	1.7
Poland	8.9	2.4	1.6	1.9
Hungary	4.5	3.2	3.5	3.5
Czechoslovakia	6.9	1.9	3.0	2.3
GDR	5.4	2.1	3.1	1.8

[a] Recalculated from growth indices of the individual years, unweighted.
[b] In national, "comparable" (over time) prices, which do not all have the same year as a basis. For Albania, national statistics.
[c] In national prices, base period 1969-71.
[d] In West European average prices of 1961-65.
Source: *Production Yearbook* (FAO), vol. 32 (1978) (Rome, 1979), p. 78; *Indices of agricultural and food production for Europe and the U.S.S.R.*, USDA, Statistical Bulletin no. 620, Washington, D.C. (June 1979), p. 3; *CMEA statistical annual, 1972*, p. 166, and *1979*, pp. 8-9, 27-36, 221. The growth of real incomes was calculated by combining population growth and the indices of real income per head of population. Albanian official gross output growth rate estimated on the basis of that for the 1971-75 plan fulfilment of overall 33 percent, that of the fulfilled 1977 plan (planned: 8 percent), and the announcement that 1976 was a successful year for agriculture: for 1971-75, see P. Dode, *Probleme ekonomike*, no. 1 (1977), pp. 3-23; official plan for 1977, *ibid.*, p. 125; and 1976 success, *Zeri i popullit*, 14 July 1977, p. 1 (according to *ABSEES*, no. 53 (September 1977), pp. 26-27, and no. 54 (January 1978), p. 15).

ern USDA indices, the resulting agricultural growth rates are shown in Table 7.1 For Albania, no such data are available, but her average annual population growth during 1970–77 alone (2.8 percent) came close to that of her FAO-estimated agricultural output growth, while the per capita real income increase was officially indicated at 2.7 percent achieved during 1971–75 and 2.1–2.7 percent planned for 1976–78.[2]

All underlying indicators are gross in the sense of not subtracting nonagricultural inputs (the CMEA indicators do not subtract feed and seed either). Because of the two record or excellent harvests in the USSR and Czechoslovakia during 1976–78, the results of these two countries look better than they otherwise would. The weighted average for the whole area is heavily influenced by the great quantitative weight of Soviet agriculture, and therefore amounts to only 2.5 percent annual production growth according to the USDA and FAO data.

The diverging developments of incomes and food supply, combined with rapid urbanization and the insufficient diversion of income increments to nonfood commodities, demonstrate that the situation was and is not merely a problem of food production. It is even more one of the economy at large, and of the role attributed to agriculture. The high share of expenditures for food and beverages in East European and Soviet private household budgets bears witness to this. In the early 1960s, this share was distinctly above 40 percent, in some cases even above 50 percent. Since then it has decreased, but except for Czechoslovakia and the GDR, it has not yet approached the level of Western industrialized countries, where today it amounts to 15–30 percent (excluding alcoholic beverages, tobacco, coffee, and comparable items) for average wage-earners' households. For low income groups in the West, such as pensioners and the majority of farmers, the figure is usually 5 percentage points more; for high income groups, about 5 percentage points less. The corresponding shares (in some cases for the whole population) in recent years in European Communist countries, according to their statistical annuals, are listed in Table 7.2. As distinct from the other countries, Rumania and Albania have not published such data. However, the director of the Institute of Agricultural Economics Research (Bucharest) in 1976 quoted internal estimates that the share in Rumania was "some 50 per cent" in 1970 and would be brought down to 40 percent by 1980.[3]

According to Soviet calculations which exclude the "social consumption funds" (i.e., "consumption" out of nonmonetary transfer income),[4] the shares of income spent on "food" (possibly including beverages) in the consumption expenditures of the population of five East European countries in the early 1970s were considerably higher than indicated in Table 7.2. The figures ranged from 44.9 percent in the GDR to 51.9 percent in Hungary, with Czechoslovakia, Poland and Yugoslavia in

Table 7.2   Percentage of Expenditures for Food and Beverage in European Communist Countries

Soviet Union (1978, industrial workers' households only, recalculated to exclude non-monetary transfer incomes)	35.7
Yugoslavia (1976, total population)	38.9
Yugoslavia (1977, workers' households, 4 members)[a]	29.7
Bulgaria (1977, total population)	39.4
Bulgaria (1977, workers' households only)	39.8
Hungary (1977, total population)	30.3
Poland (1977, total population, from personal incomes)	35.1
Poland (1976, workers' households, married couples with 2 children)	36.4
Czechoslovakia (1977, workers' households)	25.7
GDR (1978, blue- and white-collar employees' households)	26.4

[a] *Statisticki bilten,* no. 1085, p. 10.

between. It was clearly an economic as well as a political necessity to increase the share of expenditures for nonfood, and at the same time to enable agriculture to better meet the remaining demand for food.

The concrete measures for achieving a "comprehensive strengthening" and "new forms" of agriculture during 1971–75 in all countries of the CMEA have been summarized by a Soviet author in the following terms: (1) more rapid growth of agricultural investment, supply of machinery and fertilizer, irrigation and drainage works, and industrial and other non-agricultural subsidiary production within collective farms; (2) a comprehensive program for "agro-industrial complexes"; (3) improved organization and productivity of labor; (4) cooperation and concentration in agricultural production; (5) socialist (international) integration of agriculture; and (6) "coming closer to" [apparently, not yet quite attaining] a full satisfaction of the growing demand of the population for food and of industry for raw materials [of agricultural origin] in all countries of the CMEA.[5]

This catalog remained valid beyond 1975. It outlines the short, as well as the medium-term, agrarian policy goals, and indirectly also reveals the still persisting restraints under which such policy has to be carried out. Long-term socio-political demands continued to determine the ways and forms which seemed acceptable to the political leaders for achieving the goals, and those which were a priori excluded. They did not impair, however, the medium-term planning for output growth as such. The ambitious figures the leaders felt compelled to set as medium-term goals for agriculture were determined in part by plans for overall economic growth and the resulting rise in consumer demand. Moreover, not just the rate of population growth as such, but also, to a degree exceeding this, the increasing urbanization of the growing population, were other factors encouraging the setting of high output goals. On the other hand, the growing dependence on imports for feeding the population pressed on the increasingly negative foreign trade balances. This was considered a deficiency of the productive performance of agriculture. As a Soviet author stated, "imports from non-socialist countries underline the necessity to increase the growth rates and the efficiency of agricultural production in the socialist countries."[6] A goal of producing 1000 kilograms of grain per head of the population is sometimes mentioned in CMEA publications, but it was also pointed out that in "economically developed countries," 800–900 kilograms per head are consumed.[7]

Also, an ideological and political factor motivated the desire to attain self-sufficiency. For Communist leaders and functionaries, it was hard to bear, and impaired their perception of their socio-political legitimacy, that the agrarian system created by them and praised as being the most progressive, showed such severe weaknesses, which were obvious to their own population as well as to foreign observers.

All this accounts for the great attention paid to agriculture and the sharply increasing inputs accorded to it, as implied in the above enumeration of measures. Significantly, these measures did not amount to reforms of the existing system, but only allowed for its organizational (cooperation, concentration, integration) and technical improvement.

Besides all other problems, the success of the new measures was jeopardized by a development in agriculture which apparently had not been considered a threat earlier, and the economic consequences of which had not been fully realized immediately: agricultural labor shortage in many parts of Eastern Europe and the Soviet Union. The problem was compounded by, or even resulted from, the fact that the capital inputs accorded to agriculture up to the 1960s were insufficient to turn the new large-scale farms into truly modern, mechanized enterprises, and to substitute for labor. The outflow of labor was caused by the industrialization and collectivization drive and very probably also by a psychological reaction of practically expropriated private farmers. Agricultural labor shortage made itself felt most in the GDR, Czechoslovakia, in certain Soviet regions, and in the northwestern parts of Yugoslavia, that is, where a country or region had previously (before 1950 or so) had only small labor reserves.

A goal not specially mentioned in the preceding enumeration, but often found in Soviet and East European publications, is that of lowering production costs in agriculture. It seems that initially, the opinion prevailed that such cost reduction would take place by itself in the course of forming large-scale farms, and with the mechanization and concentration of agricultural production. Yet exactly such a process requires more investment than output growth in itself would demand. The

availability of machines, adequate buildings, and other industrially produced inputs was far from what was needed for modern large-scale
farms. On average, there were fewer of these items per land or product
unit than were available to small and medium-sized Western farms of the
time. That is why costs of expanding investments were high and remained so on the collective and state farms. At the same time, labor costs
rose quickly because the gulf between rural and urban incomes had to be
bridged not only to slow down the labor outflow, but also generally for
social reasons. Whether or not one considers cost efficiency as a policy
goal in itself, it proved to be an important aspect, and contributed to the
shift in emphasis of Communist agrarian policies as outlined above.

The new situation was taken into account by, or perhaps was the
paramount motive for, the reforms of the socialist economic system as a
whole and of its agricultural sector specifically. By the mid-1960s,
reforms were inaugurated in the agrarian sectors of all CMEA countries
and also Yugoslavia. They changed many aspects of agricultural production, but not the system as such. New forms of planning and
organization, more investment, higher producer prices and remuneration, and a general improvement of the rural infrastructure were part of
these reform measures. What attention was devoted to more fundamental changes was replaced towards the end of the decade and the
beginning of the 1970s by emphasis on the concepts of "agro-industrial"
cooperation and integration, or else of the "industrialization" of agriculture, which have remained the order of the day up to the present
(1980). Officially it was considered that the reforms were on the whole
successfully completed and had been adequate, and that now the opportunities made possible by them had to be realized. Accordingly, the
new tasks were more of a technical and current nature, and further
reforms in the agrarian sector, or even their desirability, were no longer
mentioned in official statements or in relevant publications (see chapters
11 and 12).

It is generally considered that the predominance of the industrial and
service sectors characterizes the degree of development in the national
economies of our time.[8] Vice versa, a still considerable, if not dominant,
role of agriculture, and a large share of labor employed in agriculture, is
considered a sign of deficient economic development. Using such
criteria, a ranking of the European Communist countries is noticeable,
which is similar for most indicators. It groups together the Soviet Union
and the Balkan countries (Albania, Yugoslavia, Rumania, Bulgaria) at
the lower end, Czechoslovakia and East Germany at the upper, and
Poland and Hungary in the middle. As to Albania, statistical data are not
available for all indicators chosen, but there are enough of them to
define that country as being the most agrarian of all. For the Soviet
Union, its vast territory comprises regions of such great natural and

economic differences that the ranking among the least developed CMEA
countries does not really characterize certain European parts of it, nor
some islands of concentrated industrial development east of the Urals.

For the purpose of defining the importance of agriculture in the
overall economy, its net material product (NMP) is a more appropriate
measure than its gross product. First of all, agriculture's gross social
product (in Communist definition), like socialist national income, more
or less (although not to quite the same degree in the individual countries) excludes nonproductive services.[9] Even more important, it is gross
not only in the sense of not making an allowance for depreciation and
expenditure on material inputs from other branches of the economy,
but also because it double-counts crop produce consumed in animal
production.

A special case is that of land as an input. Land is of paramount
importance for agriculture, and plays a minor, though not wholly
insignificant, role in other sectors of material production. If it were given
a value in money terms in Eastern Europe, one might lump it together
with capital, and the resulting picture for capital inputs would deviate
greatly from that given below. Although work on a cadastral survey is
going on, socialist countries, except for the GDR, as yet do not have a
comprehensive evaluation of land which could be used for assessing
agriculture's comparative efficiency. In any case, the attributed value
would be questionable, and the emerging picture doubtful, because land
prices are often distorted, hardly less in the dirigist planned economies
of socialist countries than in the West.

Measured in terms of its contribution to the NMP as a percentage of
the national total, the importance of agriculture clearly follows the
ranking by countries as explained above. This is demonstrated in Table
7.3 which also includes data on the relative share of agriculture and
forestry in total employment.

Examining the size of the agricultural labor force in the total domestic
labor force for Poland, the dominant role of private agriculture and its
greater labor force means that this country no longer fits in the above
grouping, though the share is still clearly less than in Yugoslavia, where
the private sector also dominates in agriculture. Generally, it is true for
all East European countries and the Soviet Union that employment in
agriculture as a share of total employment has declined (see chapter 8).

The reduction in the relative contribution of agriculture to the net
material product since 1960 would appear even more marked for most
countries—except for the GDR, where comparable prices are used (see
note to Table 7.3)—were it not for the fact that in recent years, the prices
paid to agricultural producers have risen more rapidly than the purchase prices of other products; or, expressed another way, for the fact
that the 1960 (artificially determined) producer prices gave too low a

Table 7.3 Contribution of Agriculture and Forestry to the Production of National Income (Net Material Product) and Their Relative Share in Average Annual Employment (A= contribution to the production of the net material product, in percent.[a] B = numbers of those employed in agriculture and forestry (average per year) as a percentage of the total labor force.[b])

	1960		1970		Average 1975-77	
	A	B	A	B	A	B
USSR	20.7[c]	38.8	22.0[c]	25.4	17.2[c]	22.3
Albania	44.4	ca.57[d]	34.5	ca.54[d]	...	...
Yugoslavia	25.0[e]	56.9[f]	18.3[e]	44.6[f, g]	15[e]	...
Rumania	34.9	65.6[h]	19.1	49.3[h]	17.5	36.2[h]
Bulgaria	32.2	55.5	22.6	35.7	20.5	26.9
Poland	25.8	44.1[h]	17.3	34.6[h]	15.3[h]	30.8[h]
Hungary	30.8[i]	38.9[h]	17.8[i]	26.4[h]	16.1[i]	22.3[h]
Czechoslovakia	15.2	25.9	10.5	18.6	8.5	15.3
GDR	18.0	17.2[h]	12.9	13.0[h]	10.2[j]	11.1[h]

[a] The sources name current prices as the basis of calculation, except for Hungary in 1960 (prices of 1968) and for the GDR, where they are "comparable prices" (without closer specification, possibly constant prices of an unspecified year). For Albania, no price basis is indicated.

[b] Workers and employees can be regarded as fully employed, whereas collective farmers and private producers, because of the way of calculating in the underlying statistics, are seen as less than fully employed, but they do additional work on their private plots (cf. chapter 8).

[c] Agriculture's contribution is understated because of the Soviet statistical method of attributing turnover tax revenue.

[d] For Albania (estimate), see Table 8.8.

[e] Contribution to the net social product, not according to CMEA definition and including in particular "non-material" production; instead of forestry, fishery is included.

[f] The Yugoslav data include those employed primarily or seasonally in agriculture, and therefore appear to be high in comparison to the other countries. For this reason, the percentages of a Polish source on Yugoslavia (*Rocznik Statystyczny, 1976* (Warsaw, 1976), p. 554) have been used, which are lower and apparently were made comparable (see chap. 8).

[g] In 1971.

[h] Share in the work force for Rumania as at 31 December, for Hungary (including water management) as at 1 January of a given year; in both cases, the share is lower than it would be with a yearly average. Thus, the Rumanian *Revista de statistika*, no. 7 (1973), p. 4 (according to *ABSEES* 5:2 (April 1974), p. 231) gave much higher percentages, apparently not on the basis of December counts, for the total population engaged in agricultural and forestry production: 72.0 for 1960, 56.0 for 1970, and 47.8 for 1975 (expected at the time of publication). For Poland, too, the percentages are presumed to be based on winter counts. For the GDR, they are likely to be as at 30 September, and close to an annual average.

[i] Excluding forestry.

[j] Because of changed methods of calculation, not comparable to the data for 1960 and 1970.

Source: For Yugoslavia: *Statisticki godisnjak Jugoslavije, 1970*, p. 87, *1974*, p. 141, and *1978*, pp. 119, 153; *Statisticki bilten*, no. 720, p. 17, no. 866, p. 11, and no. 889, p. 11f; for Albania: *30 Vjet Shqiperi Socialiste*, p. 188; share in employment according to FAO estimates of population economically active in agriculture *(Production Yearbook, 1974*, vol. 28-1, Table 5) (The FAO percentages deviate from those given by the CMEA countries). For all other countries: *CMEA statistical annual, 1978*, p. 41 (net product), and pp. 393-95 (share in employment).

valuation to the contribution of agriculture. This also partly, but not completely, explains why the share of agriculture in the NMP in the Soviet Union appears to have gone against the general trend, increasing between 1960 and 1970.

The share of agriculture and forestry in the production of the NMP and the numbers of those employed would have fallen even more markedly if this branch of the economy had not had to catch up a great deal, and if not all of the countries had undertaken great efforts to raise agricultural production. If, on the one hand, one compares the agricultural and forestry industries' share in the production of the NMP (national income) to their share in the number of employed persons (in material production) and, on the other hand, also the shares of all the other sectors of material production in the NMP and in the numbers employed, one can construct a rather crude scale for agricultural labor productivity compared to that of the major part of the rest of the economy. For this purpose, the proportions of those employed in the "non-productive sphere", that is, nonmaterial production, have been eliminated, unlike in the previous table, except in the case of Yugoslavia. In Table 7.3, the agricultural labor productivity so derived is overstated for Hungary, Poland, and Rumania because of the winter counts of employment (see footnote h, Table 7.3).

It should be stressed that in Table 7.4, one is comparing labor productivity in agriculture and forestry not to that in the economy as a whole, but to that in the whole economy minus agriculture and forestry and (excepting the Yugoslav case) minus the "non-productive sphere." This means that the figures for agriculture appear relatively low. Compared with industry alone, they would appear even lower. In comparison with labor productivity in the whole economy, labor productivity in agriculture would compare better, the larger its share in the NMP. The greater weight of agriculture would force down the average for all sectors and considerably influence the comparison in favor of agriculture.

Nevertheless, these are relative levels within each country which make agricultural labor productivity in that particular country appear higher or lower in relation to nonagricultural production. Even if there were more detailed data for drawing comparisons among socialist countries, it is well known that international comparisons of absolute levels of labor productivity are a particularly thorny problem. Apparently for the early 1970s, the East German Statistical Administration, taking labor productivity in GDR agriculture as 100, calculated it for Czechoslovakia at 65, for Hungary at slightly over 50, and for Poland at 35 percent. From another calculation, quoted in the same Soviet source, it emerges that Bulgarian and Soviet agricultural labor productivity was of the Polish (small peasant) order of magnitude.[10] According to another Soviet

Table 7.4 Net Labor Productivity in Agriculture and Forestry as Percentages of the Net Labor Productivity of All Other Materially "Productive" Sectors of the Economy: 1960, 1970, and 1975-77 (labor productivity in the other sectors = 100)

	1960	1970	Average 1975-77
USSR	31	61	52
Albania	38-43[a]	27-31[a]	...
Yugoslavia[b]	25	28[c]	...
Rumania	22	19	31
Bulgaria	30	42	55
Poland	36	32	33
Hungary	54	48	52
Czechoslovakia	41	40	40
GDR	86	78	70

[a] Approximate estimations of a share of 12-16% in the "non-productive spheres" of em-employed in 1960 and 14-18% in 1970; around 1974, this share was 18% (*Zeri i popullit*, 22 June 1975, p. 2) or else 20% (*Rruga e partise*, no. 5 (1975), pp. 24-35) (both cited here according to *ABSEES* 6:4 [October 1975]), p. 137).

[b] Related to the net social product, including the "non-productive sphere"; therefore slightly too low compared to the other countries.

[c] Labor for 1971.

Source: Calculated from proportional employment figures cited in Table 7.3 and from the contribution to production of the "national income" (net material product) or, in the case of Yugoslavia, the net social product, and from the relevant numbers of employees in other sectors of the economy (same sources), eliminating those in "non-productive spheres" (except for Yugoslavia).

computation, based on the national statistics of five countries, the labor productivity gap was widening rather than narrowing during 1961–75. The GDR moved ahead fastest, followed by Czechoslovakia and Hungary, while Poland and the USSR made very slow progress.[11]

One of the reasons that labor productivity in agriculture in Eastern Europe and the Soviet Union is poor when compared with other sectors of the respective economies is the low level of capital per worker in agriculture. This becomes apparent when one compares agriculture's share in the civilian labor force (Table 7.3) with its share in fixed assets (Table 7.5). It is only in the GDR that the share in capital has reached a level roughly equal to that in the labor force. This is the main reason why the disparity in labor productivity was smallest in East Germany. The GDR is, in fact, approaching the situation that has characterized the highly industrialized Western countries for some two decades: thanks to a high level of capital inputs (on a per capita basis, sometimes exceeding

Table 7.5   Percentage Share of Agriculture in Total Fixed Productive Assets and (including forestry) in Productive Gross Investment in the National Economy, 1960, 1970, and 1975-77

	Fixed productive assets[a]			Investment[b]		
	1960	1970	1977	1960	1970	Average 1975-77
USSR	24	20	21	20	26	28
Albania	...	...	...	15[c]	16[c]	...
Yugoslavia	...	...	...	22.8	11.7[d]	5.8[g]
Rumania	26	18	15	26	20	17
Bulgaria	25[e]	21	18	40	21	19[f]
Poland	32	27	25	19	22	20
Hungary	14	22	22	20	29	23
Czechoslovakia	11	12	15	23	15	17
GDR	12	13	13	15	16	14

[a] The term *fixed productive assets* refers to the "basic production funds" (osnovnye proizvodstvennye fondy), not including those of the "non-productive sphere," nor the land value. They are on the books at their original prices and are periodically revalued in accordance with more recent prices. In the Soviet case, the prices are those of 1973; in the GDR, they are "comparable prices" (without closer specification, possibly constant prices of an unspecified year); in Poland, the price basis seems to be an estimate at 1971 prices (and the basic animal stock not counted in with the assets); for Czechoslovakia, possibly prices of 1967. The sources give no prices for the latter two and the other countries.

[b] Agricultural gross investments as a percentage of those of all "branches of material production," that is, excluding the "non-productive sphere." The price bases of evaluation are: for the USSR, 1969 prices with certain adjustments; for Rumania in 1960, those of 1959; afterwards, those of 1963; for Bulgaria and Hungary, current prices; for Poland, 1971 prices; for Czechoslovakia, those of 1 January 1967; and for the GDR, those of 1975.

[c] Share in total investment, including that of the "non-productive sphere." Averages of 1961-65 (instead of only 1960), and of 1966-70 (instead of only 1970). One can surmise that if the "non-productive sphere" were eliminated, investment shares of approximately 17 percent and 18 percent would result. According to A.D. Stupov and V.I. Storozhev, *Razvitie sel'skogo khozyaistva i sotrudnichestvo stran SEVa* (Moscow, 1965), p. 27 (table), they were 18 percent in 1960, and 13.8 percent in 1961.

[d] The private investments contained herein are those of 1968 according to the estimate of M. Jevdjovic in *Ekonomika poljoprivrede*, no. 1-2 (1970), p.4; without these, they would only be 6.5%.

[e] Including forestry.

[f] Excluding forestry (planting of trees).

[g] In 1978, the private sector accounted for 3 percent.

Source: Investment for Albania, 1961-65 and 1966-70 from *Edonomia popullore*, no. 5 (1971) (according to *ABSEES* 2:4 (April 1972), p. 133); investment for Yugoslavka, 1960 and 1970, according to *Statisticki godisnjak Jugoslavije, 1965*, p. 273, *1972*, p. 256, and *1979*, p. 225; all others, according to *CMEA statistical annual, 1978*, pp. 45, 135, 139 (shares in gross investment calculated).

that in other, non-industrial sectors of the economy), labor productivity in agriculture comes close to the average for the whole economy.

Another reason for low agricultural labor productivity in Eastern Europe and the Soviet Union is that the transport sector is so deficient, in terms of both roads and also truck numbers. In some cases, the losses from bad country roads are said to amount to more, over a number of years, than it would cost to reconstruct and enlarge the road network.[12]

Table 7.4 shows that between 1960 and 1970, the gap between labor productivity in agriculture and in nonagricultural branches of the economy was closing quickly in the Soviet Union and Bulgaria, and more slowly in Yugoslavia. These improvements came partly because of the increase in the value of agriculture's net product due to price increases. However, after 1970, there was a tendency for net labor productivity in agriculture to fall behind that in the other sectors in the Soviet Union and the GDR (and perhaps also in Albania, for which such details are unavailable). In Poland, the deterioration between 1960 and 1970, and relative stagnation since 1970, is at least partly related to what through the 1960s was an increasing surplus of agricultural labor. In the other countries since 1970, net labor productivity has for the most part risen more quickly in agriculture than in the other sectors. In viewing these developments, one must note that agriculture's share in total gross investment during the whole period 1960–75 underwent a sustained increase only in the Soviet Union.

Relative net agricultural labor productivity seems not as low as one would expect in the Soviet Union and Albania compared to the relative level in the other East European countries. In the case of the Soviet Union, in comparison with other East European countries, its abundance of land appears to account in part for the seemingly "high" level of labor productivity in agriculture (see Table 7.4). A Soviet author has estimated the total value of the country's agricultural land at 180–270 billion rubles, which is more than the total value of fixed capital stock in Soviet agriculture in 1975.[13] In addition, the relatively high level of labor productivity in Soviet agriculture can be explained by the great rises in agricultural producers' prices after 1964 and also by the increasing importance since 1960 of the state farm sector, whose statistics record only fully employed labor and where labor productivity is above average. But one should note that by international standards, Soviet labor productivity in industry and trade, and in the services directly serving industry and trade, that is, the measure against which comparisons are being made, is low in itself. This point applies even more to Albania. It is not, however, true for East Germany where the state agricultural sector is small and industry is characterized by what could be considered a high level of labor productivity, at least by East European standards.

The bad showing of Czechoslovak agriculture is to be explained partly by the relatively low level of agricultural development in Slovakia, partly by the relatively high level of labor productivity in industry and the exclusion of forestry output (but not of forestry labor) from the data on agriculture, and partly by the price changes since 1966. Compared to Rumania and Hungary, the Czechoslovak data contain less of a statistical bias towards too low figures for labor inputs.

In Yugoslavia, the position is exactly the opposite. Here partly employed laborers are also counted, so that labor productivity appears particularly low. Yet one should also note that in the Yugoslav case, the production share is taken relative to the share in the net social product, a concept which excludes double-counting of seed, fodder, farm-processed products, etc. Moreover, the "non-productive sphere" is not eliminated in the Yugoslav national income and social product data, and this probably also acts towards making agricultural labor productivity look comparatively worse. If corrections were made for these three points, the Yugoslav figures would probably approximate to, though not fully attain, the Polish ones.

In the case of Bulgaria, one suspects that the figure of labor productivity outside agriculture is relatively low, thus allowing the comparison to appear favorable to agriculture. To this one must add that the surplus labor in Bulgarian agriculture is being rapidly reduced.

We have measured labor productivity relative to the NMP. That is why the reduction in the relative net productivity of Soviet agriculture after 1970 can be explained as a result of increased capital inputs from other sectors of the Soviet economy. Measured in gross investment, capital inputs increased markedly from 1970 to 1975–77, and this was also true for the share of agriculture in fixed capital stock (the so-called basic productive funds). Among all the other countries, the latter was the case only in Czechoslovakia. Table 7.5 illustrates the situation for fixed capital and investment, excluding the "non-productive sphere."

Up to the early 1960s, the share of agriculture in the gross investment of all productive sectors was generally smaller than its contribution to the net material product, but by that time had increased in most countries compared to earlier years. A comparison of Tables 7.3 and 7.5 shows that already in 1960, agriculture's share of gross investments in Bulgaria and Czechoslovakia had become larger than its contribution to the NMP. Yet in Bulgaria, one was witnessing a temporary and unusually large increase of investment, and to a lesser extent this was also true in Czechoslovakia. In the other East European countries, the contribution of agriculture in the NMP in 1960 still exceeded the share of agriculture in investment, and in the Soviet Union, it was roughly the same. The relative value of agriculture's fixed capital stock in most countries for which such data are available was less than its relative contribution to production, except for Poland and the Soviet Union.

Ten years later, the ratios were completely reversed in Czechoslovakia and the GDR, where agriculture contributed less than it owned (capital stock) and received (gross investment). In Rumania, the situation was now similar to the USSR, with the give and take of the agricultural sector roughly equal. Apart from Rumania and Bulgaria, and possibly (because recent data are not available) also Yugoslavia and Albania, that is, with the exception of the Balkans, agriculture's share of investment and capital stock in 1975–77 exceeded its share in generating the NMP in East Central Europe and the Soviet Union. The fact that in recent years agriculture has received a greater share of gross investment than it held in fixed capital is indicative also of an overdue renewal of its capital stock. The striking discrepancy in Polish agriculture between the share in fixed capital stock and in gross investment presumably is due to an insufficient renewal of capital stock in most small peasant farms. These in 1970 contributed 85.7 percent to agriculture's gross production, held 45.6 percent of its productive capital stock (prices of 1961), but accounted for only 36.4 percent of agricultural investment; by 1977, the proportions became yet more extreme: 77.9 percent of gross production, 54.6 of productive capital stock (current prices), and 22.1 of total agricultural investment.[14] The notably high share of the agricultural capital stock in Poland, even exclusive of livestock herds, can obviously be traced to the large number of old buildings still on the small peasant farms, and one should not infer the existence of large modern livestock premises and agricultural machinery on a great scale. For Yugoslavia, for which no data exist on the peasant farmers' capital stock, one can assume the same to be true. In any case, such comparisons ignore the value of land. Its inclusion as a capital asset would make the share of agriculture in fixed capital in all the countries appear larger than its contribution to NMP but would not, of course, change its relative share in investment.

Comparing agriculture's share in gross investment and in the capital stock with its relative contribution to the NMP, one may conclude that since the latter is not larger than the former two, agriculture has ceased to be used as a source of capital for the other sectors of the economy (and especially for industry). Correspondingly, industry and construction in the CMEA countries, with the exception of the GDR, have in 1975–77 shown a greater relative contribution to the NMP than was their share in total gross investment in the branches of material production, whereas earlier they received more than they contributed, or at least an equal share.

If the relative productivity of agricultural labor (see Table 7.4) were considerably above the level of its comparative remuneration, there would still be a capital transfer to other sectors of the economy. But neither for recent times nor for the 1960s was the wage or income disparity as large as the difference of labor productivity. The picture is one of not only wage disparity being smaller than the disparity of labor productivity, but also of the wage disparity diminishing over time, in some cases disappearing altogether (see chapter 9).

It may be assumed that agricultural labor productivity, as shown in 1960 official statistics, was artificially low because of the comparatively low producer prices fixed for agriculture by the state. This also applies to the private peasant sector, which at that time still carried some weight even in collectivizing countries like Hungary and Rumania. By 1970, such a distortion was no longer, or at least not to a sizable degree, the case.

Data on fixed capital stock and on investment do not tell the whole story about the capital supply for agriculture, as they do not yield information on the turnover capital, which is impossible to assess on the basis of East European statistics. Yet it seems that in the collective and state farms of the CMEA countries, the turnover capital in monetary terms on the whole was proportionate to fixed capital and investment. The utilization of turnover capital, though, is likely to have been less than efficient, and moreover it may too often have been used for long-term investment instead of for current requirements.

Up to the early 1960s, a transfer of capital out of agriculture obviously was taking place (with the short-term exceptions of Bulgaria and Czechoslovakia around 1960). However, by 1970 and later, this was no longer the case in East Central Europe and the Soviet Union, where in part the direction of the transfer was reversed. The Balkan countries remained an exception to this. In Bulgaria, though, the remuneration of agricultural labor has become relatively high, exceeding the relative labor productivity of agriculture, while there is no sizable underemployment any more.

The freeing of agriculture from its earlier task of accumulating capital brings into relief the change in agricultural policy which has taken place in the Soviet Union and in East Central Europe and which will probably also occur in Rumania and Bulgaria in the future. Its cause is principally the decrease in the size of the agricultural work force which has taken place everywhere, and which, without injections of capital, would make the production levels of this branch of the economy sink. Even when the outflow of labor from agriculture is still relatively slow or delayed by countermeasures, this outflow would, without proportionate and even overproportionate capital investment, still be a disrupting factor in carrying out the economic plans, as these aim at providing for a continued and speedy increase in agricultural production.

Deagrarization, or the decreasing weight of agricultural population and employment in the overall economy, has been seen as a typical phenomenon accompanying industrialization. If one takes as a yardstick for deagrarization the (roughly estimated) share of agriculture in producing the national income according to Western definition (that is, somewhat smaller than in Table 7.3) and in labor inputs, then the GDR and Czechoslovakia would rank not much behind the highly industrialized countries of Northwest and Central Europe, as well as North America. The remaining East European countries (including the Soviet Union) would then appear to be a kind of transitional type, still with strongly differing degrees of the agrarian elements in each country. But such all-inclusive characterizations are questionable. Another common characteristic of Eastern Europe and the Soviet Union is more important. On the one hand, the weight of heavy industries in the industrial sector is greater than in other countries experiencing a comparable transition period, and on the other, in the development of its production potential, agriculture has remained farther behind the development of the industrial sector than is the case in Western industrialized countries of today. In this, East European and Soviet agriculture has much in common with the service industries of these Western countries. Obviously, an industrializing process with planned priority for the heavy industries, along with struggles for autarky and temporary neglect of consumer interests, lead to less stimulation in the long term for agriculture. East European and Soviet economic policy-makers have, however, recently been trying to overcome the disproportions which have arisen in this way, by means of an improved supply of capital for the food production sector.

Notes

1. B. Nagy, *Hungarian Agricultural Review*, no. 2 (1973), p. 1 (as rendered in English from *Mezögazdasági Szemle*, no. 5 [1972]).
2. K. Reimer, *Wissenschaftlicher Dienst Südosteuropa*, no. 8 (1976), pp. 158–160.
3. *Symposium on forms of horizontal and vertical integration in agriculture* (UN/FAO: ECE/ Agri/29), vol. II (New York, 1977), p. 369.
4. N. Rimashevskaya and S. Shatalin, *Voprosy ekonomiki*, no. 12 (1975), p. 101. The Soviet figure in Table 7.2 has been recalculated by the present author to eliminate this factor, on the basis of the breakdown in *Narodnoe khozyaistvo SSSR, 1978* (Moscow, 1979), p. 390.
5. V. Simchera, *Vestnik statistiki*, no. 12 (1971), p. 35.
6. L. P. Zlomanov, *Mezhdunarodnye ekonomicheskie sopostavlenia* (Moscow, 1971), p. 155.
7. M. E. Bukh, *Problemy effektivnosti sel'skogo khozyaistva v evropeiskikh stranakh SEV* (Moscow, 1977), p. 64.
8. The following section is in part a condensed and updated revision of an article by the present writer on "The Place of Agriculture in the European Communist Countries: A Statistical Essay," *Soviet Studies* 29:2 (April 1977), pp. 238–254 (by kind permission of *Soviet Studies*). Some data have been changed, following amendments in the most recent CMEA statistical annuals, but data for 1978 have not been included, because the record harvests in the USSR, Hungary, and Czechoslovakia in 1978, followed by disappointing results in 1979, tend to distort rather than clarify the general picture.
9. On the problems of comparison, see Benedykt Askanas, *Zur Berechnung des Brutto-Nationalproduktes in Osteuropa nach westlichen Methoden am Beispiel Polens*, Österreichisches Institut für Wirtschaftsforschung, Forschungsberichte, no. 1, Vienna (July 1972) (hecto-graphed).
10. M. E. Bukh, *Problemy, op. cit.*, p. 77.
11. M. E. Bukh, *Voprosy ekonomiki*, no. 5 (1976), p. 96. Whereas four of the country indices in this source are based on net output, the one for the Soviet Union is in terms of gross output, apparently in order to camouflage the lagging behind.
12. M. E. Bukh, *Problemy, op. cit.*, p. 41.
13. R. M. Gumerov, *Sovershenstvovanie tsenoobrazovania i razvitie khozraschetnykh otnoshenii v sel'skom khozyaistve* (Moscow, 1976), p. 57 (referring to *Voprosy ekonomiki*, no. 4 [1973], p. 62). The total fixed assets (osnovnye proizvodstvennye fondy) in 1975 were officially given as 168 billion rubles in *Narodnoe khozyaistvo SSSR v 1975 g.* (Moscow, 1976), p. 58.
14. *Rocznik statystyczny, 1971*, pp. 151, 163, 267, and *1978*, pp. 101, 106, 194.

8. *Performance and Factors of Agricultural Production*

Feeding the Population

SEVERE SHORTAGES OF basic foods were observed after the mid-1960s only in individual countries and regions of Eastern Europe and the Soviet Union, and only at certain times, but have become more frequent again in recent years. As shown in the preceding chapter, these shortages are mainly a consequence of rising incomes and demand. Apart from such irregularities, the growth of agricultural output has on the whole been continuous. Yet it was less than soils permit, and less than earlier aspirations aimed at.

It is still of some interest to remember the production goals of the early 1960s. For the Soviet Union, they were broadly outlined in the (so far not abrogated) Party Program of 1961, and a Moscow economic weekly subsequently projected them more specifically and added those for the other CMEA countries.[1] Combined with the output figures for 1978—an excellent harvest year with better results than in 1979—they yield the data for the CMEA countries (excluding non-European members) contained in Table 8.1.

There was not the slightest prospect for attaining these goals in the remaining year, except for sugar. Yet in terms of quantitative consumption, all the countries concerned, with the exception of Albania, do not essentially differ from Western industrialized countries. Calorie consumption per head and per day, according to data of the UN's Food and Agricultural Organization (FAO), attained 3,000 calories by the mid-

Table 8.1   Agricultural Production, CMEA Countries

Product	Projection for 1980	Achievement, 1978
Grain (million tons)	467	320
Unginned cotton (million tons)	17.0	8.5
Meat (million tons, slaughter weight)	35.9	25.6
Milk (million tons)	198.5	134.6
Eggs (billions)	155	95
Fruit and berries (million tons)	56.6	24.1
Sugar (white, million tons)	18.7	17.8 (excluding Cuba)

1960s. Things looked different, though, for protein, especially animal protein, and for oils and fats, where four of the countries belonged to the same category as Spain, while Poland, Hungary, Czechoslovakia and the GDR exceeded that level, and Albania in this respect ranged even lower. In all of these countries, however, one cannot speak of insufficient nutrition, as is the case, for example, in South Asia. In fact, in contrast with some less developed regions of the world, the nutrition levels have clearly and continually risen.

The hectare yields of the main crops and the meat and milk yields per animal regained their prewar levels during the 1950s, and for grain they surpassed them by 1960. For the two most important row crops (potatoes and sugar beets), however, they did not quite achieve that level in Poland, Czechoslovakia, and the GDR. The most marked progress during 1950–60 was with the milk yields per cow and the meat yields per livestock unit. This applied to the collectivized countries as well as to Poland and Yugoslavia. The yield increases have continued since then, although at a reduced rate during the early 1960s for row crops and milk (see Table 8.2).

In the whole of Eastern Europe, excluding the Soviet Union, total agricultural production, according to the 1975 survey of the UN's Economic Commission for Europe, achieved and surpassed the prewar level from 1957 on. The same is valid for the gross product in G. Lazarcik's definition (i.e., the end product minus nonagricultural inputs), and the net product (gross product so defined minus depreciation) per head of the population.[2] For the Soviet Union, the prewar gross product (Communist definition, i.e., social gross product, not deducting nonagricultural inputs and the farm's own output of feed, seed, etc.) per head of population was fully attained by the middle of the 1950s (population losses and annexation of territories prevent an exact statement as to the year). Since then, the growth of agricultural production in Eastern Europe and the Soviet Union has been more rapid than that in Western Europe, where food demand pressure levelled off and restraining, rather than speeding up, growth became a major concern of politicians.

According to Lazarcik's calculation, and deviating from official indices, those countries where collectivization was pushed through had a markedly slower agricultural output growth during 1950–61 than Poland and Yugoslavia with their predominantly peasant farming; for the net product, the growth was only half. Afterwards, during 1961–73, the Yugoslav and Polish growth rates lagged somewhat behind those of the countries with fully socialized agriculture, but not in terms of net product, as the inputs were less and labor amply available. However, it would be premature to draw far-reaching conclusions from these contrasts, especially as one might not get quite the same results if the periods

Table 8.2 Hectare Yields of Three Basic Crops (5-year averages)[a] and Milk and Pork Yields (100 kilograms)

	Grain, including leguminous grain		Sugar beet		Potatoes		Milk per cow		Pork per breeding sow[b]	
	1961-65	1974-78	1961-65	1974-78	1961-65	1974-78	1966	1976-78[a] 3-year averages	1966	1976-78 3-year averages
USSR	10.2	15.5	165	232	94	115	1880	2244	10.8	12.6
Albania	10.1	19.3	169	244	76	71	680	1482	2.4c	4.5c
Yugoslavia	19.3	33.0	279	415	87	91	1243	1406	3.8c, d	3.9c, d
Rumania	15.9	27.3	149	240	85	134	1686	1951	8.1	8.7
Bulgaria	19.0	33.8	205	268	85.5	110	1912	2312	14.5	12.1
Poland	17.0	26.1	267	293	154	186	2365	2803	7.9	8.3
Hungary	20.3	39.3	246	333	79	134	2101	3031	13.3	15.6
Czechoslovakia	21.8	36.3	270	341	114	164	2146	2964	13.1	15.3
GDR	25.3	35.7	243	267	166	164	3038	3424	9.7e	10.3e

a Unweighted averages of annual yields.

b Pork(slaughter weight) includes bacon and fat. The figures relate to the numbers of littering sows at the end of each preceding year, because this number fluctuates less between the usual end-year counts than the total number of pigs and because in this way the number of piglets per sow is indirectly taken into account. The same kind of indicator was chosen by K. Kalinin and J. Nagy, *Internationale Zeitschrift der Landwirtschaft*, no. 6 (1974), p. 621. Pork output per sow will, of course, in this way seem higher, the more other sows will be used, e.g., for one-time farrowing. In Albanian statistics, only numbers for all kinds of sows are indicated, and therefore the pork yield is more below the figures of the other countries than would otherwise be the case. For the GDR, see footnote e.

c Pork production and sow data for Albania from *Vjetari Statistikor i RPSh, 1971-1972* (Tirana, 1973), pp. 107-108, and for Yugoslavia, from *Statisticki godisnjak Jugoslavije, 1978*, pp. 229, 231.

d Meat output divided into number of littering sows, including young ones after farrowing, annual average. If they were counted at the end of the year, the number of sows would be somewhat lower, that is, output per sow would be higher, especially if the number of young sows could be eliminated.

e In 1966: socialist sector only. The surprisingly low output per sow in the otherwise highly productive GDR agriculture is hard to explain. The author assumes that practically no other than genuine breeding sows are used, which is not the case in most other countries concerned, and that therefore the meat output per breeding sow, as given in the above table, is overstated for the other countries. In any case, FAO data indicate the highest average slaughter weight of pigs in the GDR, compared to the other East European countries (*Production Yearbook*, vol. 32 [Rome, 1979], p. 220).

Source: Pork in all cases was derived from pork output and sow numbers in the given sources: for Albania and Yugoslavia, FAO *Production Yearbook*, vols. 29 and 31 (Rome, 1976 and 1978), and as in footnote c; for all other countries, *CMEA statistical annuals*, various years.

were chosen in another way and the differing characteristics of labor application in peasant and in large-scale socialized farms were taken into account.

The picture during 1950–60 is somewhat more favorable for the countries of the Soviet bloc if one looks at the official indices. These reflect gross output (in Communist definition, see above) and are not wholly comparable among the countries because the prices underlying the index calculation (although said to be comparable over time) are not based on the same years. Yet these indices, too, reveal slower growth for the second half of the decade, when collectivization was being pushed through, with Bulgaria as an exception. The rapid output increase in the Soviet Union in 1954–56 was due mainly to tolerant policy toward the private sector and to the expansion of the arable area by the Virgin Lands campaign. The overall growth remained unimpressive during the 1961–65 Five-Year Plans. In Czechoslovakia, the absolute volume of agricultural output even decreased, while Bulgaria again performed best, with Poland in second place. During the following quinquennium, the growth became more rapid again, except in Poland and Rumania, and for the same period, the plan targets were more realistic than before.

The faster growth that began in Yugoslavia after the end of the collectivization drive, ebbed by 1959. Only in 1963 was it resumed, and then continued at a more moderate rate. According to Yugoslav statistics, it amounted to an annual average of not quite 5 percent during 1950–60, of 2–3 percent during the following decade, and of 3–4 percent since 1970. The annual variations were great, though.

Albania's available official statistics seem to exaggerate the increase. For the ten years 1949/50–1959/60, the growth rate of agricultural gross production was indicated to have reached an annual average of 3–4 percent, and from 1959/60 to 1970/71, and again during 1971–75, even 5.9 percent. Yet such rates still remained below the unrealistically high plan figures, which provided for an annual average of 10–12 percent during 1950–65 and 1971–75, and of "only" 7–8 percent during 1966–70 and 6.7–7.1 in 1976–80.[3] There clearly were intervals of stagnation and even setbacks in agricultural production following the two main periods of Albanian collectivization, 1955/56 and 1966/67. The FAO indices indicate virtual stagnation from 1967 to 1972: less than one percent annual average increase. Albanian plans for 1978 and 1979 stipulated fantastic rates of 28 and, after nonfulfilment, even 30 percent.[4]

Obviously, more investment and higher producers' prices (see chapters 8 and 10) were the main reasons for the new growth momentum after the early 1960s. However, this growth slowed considerably in the CMEA countries after 1974, and over the whole period 1970–78 (see

Table 7.1) it was, with the one exception of Rumania, less than planned toward 1980.[5] The growth rates envisaged in the five-year plans of the CMEA countries for 1971–75 and 1976–80 were supposed to be attained mainly by higher yields per land and livestock unit. Defined in gross product (Communist definition), the planned aggregate growth up to 1975 was achieved or surpassed only in Czechoslovakia and Rumania. For 1976–80, the planned rates were obviously based on those actually achieved during the preceding quinquennium, and were sizably higher than the latter only in the Soviet Union and the Balkan countries. On the whole, the growth targets were less ambitious than those planned for 1971–75 and not achieved, in the main ranging between 2.7 percent in Czechoslovakia and 3.9 percent in Yugoslavia (with targets for Rumania and Albania exceeding that range). Most plans, except for those of Rumania and Bulgaria, unmistakably now implied greater emphasis on crop production (fodder and feed grain in the first place) and temporarily a slower growth for animal production compared to the 1971–75 plans, apparently in order to achieve a more stable feed basis.

The agricultural area available per head of the population, and the great differences among countries as to this indicator, are demonstrated in Table 8.3. Deducting the arable and perennial crop areas from the total agricultural land (Table 8.4), one is able to find the area of cultivated and natural meadows and pasture and to determine where this extensive kind of land utilization is widespread (Soviet Union, Albania) or very minor (Hungary). The degree of urbanization is shown by the percentage of people living in settlements categorized as urban (an administrative and at the same time economic definition).

Table 8.3   Tilled Land under Arable and Perennial Crops per Head of Total Population (hectares)

	1960	1970	1978
USSR	1.04	0.94	0.88
Albania	0.29	0.28	0.25 (1977)
Yugoslavia	0.45	0.40	0.37 (1977)
Rumania	0.56	0.51 (1971)[a]	0.48
Bulgaria	0.55	0.47	0.47
Poland	0.52	0.46	0.42
Hungary	0.57	0.54	0.50
Czechoslovakia	0.40	0.37	0.35
GDR	0.29	0.28	0.31

[a] For 1971, because the decrease in arable use in 1970 was abnormally strong.
Source: FAO *Production Yearbook* for Yugoslavia and Albania; CMEA statistical annuals for the other countries.

Table 8.4 Population and Agricultural Area in 1970

	Population at end of year		Agricultural land		Land under arable and perennial crops	Green land (natural meadows and pasture, excluding sown grass and similar)	
	Total (1,000)	Proportion urban	1,000 hectares	Hectares per head of total population	1,000 hectares	1,000 hectares	As a percentage of all agricultural land
USSR	243,873	57%	545,778	2.24	227,940	317,838	58
Albania	2,136	34%	1,229	0.58	599	630	51
Yugoslavia	20,523a	62%b-	12,437c	0.61c	8,137	4,300c	35c
Rumania	20,361	41%	14,930	0.73	9,954	4,976	33
Bulgaria	8,515	53%	6,010	0.71	4,017	1,993	33
Poland	32,658	52%	19,543	0.60	15,121	4,422	23
Hungary	10,354	47%	6,875	0.66	5,594	1,281	19
Czechoslovakia	14,362	62%	7,093	0.49	5,334	1,759	25
GDR	17,668	75%	6,286	0.36	4,817	1,469	23

a According to the census of January 1-10, 1971.
b Deviating from the definition in the other countries, this percentage represents total minus agricultural population. The minus sign after the percentage is meant to indicate that the figure is overstated in comparison to the other countries.
c Actually utilized. If the areas categorized as agricultural, but not actually utilized are included, the figures are: 14,595,000 hectares in total, 0.71 hectare per head, and 44% green land.

Source: For Yugoslavia and Albania, national statistics; for all other countries, *CMEA statistical annual, 1971*.

The per head consumption of selected food items, as given in CMEA statistics (Table 8.5), illustrates a more or less steady improvement for meat, which is a most important qualitative indicator. Meat consumption in 1960 may be considered unsatisfactory in physiological terms, but during subsequent years, most countries achieved a more satisfactory level, with the exception of Rumania, Yugoslavia, and Albania. Yet it has to be pointed out that the meat consumption data for the Soviet Union, Rumania, Bulgaria, and Poland are overstatements insofar as they include not only inferior subproducts, but also animal fats. The national statistics for Poland (as distinct from the Polish figures in the CMEA statistics, which seem to have been made comparable) indicate a meat consumption, excluding animal fats, which is only 42.5 kilograms in 1960, and 69.1 in 1977 (compared with the figure of 77.4 kilograms for 1977 in CMEA statistics). Thus, a deduction of at least 10–15 percent, possibly more, should be made for some countries, but less so for Yugoslavia, Hungary, Czechoslovakia, Rumania, and the GDR. Generally, and in spite of the physiologically satisfactory standard attained, the demand for meat continues to rise as its income elasticity is very high.

Eastern Europe and the Soviet Union as a whole are net importers of food products. The negative net balance of Soviet foreign trade in such goods roughly equals that of the other East European countries, taken together. Calculated per head of population, though, it is much less. On such a basis, the GDR is the biggest single importer, followed by Czechoslovakia. In absolute volume, agrarian foreign trade of the Soviet Union and Eastern Europe has increased during the last 10–15 years, mainly because of more imports.

Due to two record harvests in the Soviet Union (1976 and 1978), the annual average output of grain in the CMEA countries, taken together, during 1976–78 was exactly the quantity planned for the average 1976–80 (298.6 million tons, as against 255 million during 1971–75). Yet Soviet grain exports to the European partner countries remained much smaller than previously, and Soviet imports from the world market stayed on a high level. The 1979 harvest was disappointing in all the countries, and led the Soviet Union to envisage record grain imports, which in the event were scaled down by the US part-embargo on grain purchases.

Hungary was most successful in expanding her exports of agricultural products; in her total exports to convertible currency countries, they accounted for 47.8 percent by 1965. As a share of total foreign trade turnover, agricultural products are generally losing in weight because of the progressing industrialization of the East European and Soviet economies and the corresponding changes in foreign trade. Even in Hungary, the share in total exports to the West, although still re-markably great, declined to 29.0 percent by 1978.[6] In Rumania and

Table 8.5   Per Head Consumption (kilograms) of Selected Foods

	Year	Grain products[a]	Potatoes	Sugar[b]	Meat[c]	Fish	Milk[d]	Eggs (units)	Vegetable fats and oils
USSR	1960	164	143	28.0	40	9.9	240	118	5.3
	1978	140	120	43.0	57	16.9	321	230	8.2
Albania (output per head)	1975	256	...	8.1	29	..	...	...	...
Yugo-slavia	1960	185	70	14.9	30[e]	0.7	80[f]	66	3.7
	1976	181	67	32.8	48.4[e]	2.6	101[f]	164	10.5
Rumania	1965[g]	208	76	15	26.6[i]	5(1970	105[h]	115	8(1970)
	1978	189 (1975)	96(1975)	20	51.0[i]	...	143[h]	254	...
Bulgaria	1960	190	35	17.7	32.7	2.0	126	84	9.6
	1978	159	30	35.0	64.0	6.6	217	197	14.6
Hungary	1960	136	98	26.6	47.6[e]	1.5	114[h]	160	0.8
	1978	118	63	34.0	72.0[e]	2.5	155[h]	320	6.0[j]
Poland	1960	145[k]	223[k]	27.9	49.9	4.5	363	143	1.0
	1978	120[k]	166[k]	42.7	78.8	7.3	450	219	2.7
Czecho-slovakia	1960	126	100	36.6	56.8[e]	4.7	173[h]	179	4.1
	1978	107	88	37.5	83.4[e]	5.5	220[h]	315	6.9[l]
GDR	1960	102	174	29.3	55.0[e]	...	94.5[h,m]	197	2.2
	1978	94	139	38.9	86.2[e]	6.9	99.5[h,m]	284	1.9

[a] Recalculated on a flour basis.

[b] Recalculated on a white sugar basis, including sugar products.

[c] Slaughter weight, incl. fats, bones, and "edible subproducts," except where otherwise stated.

[d] Except for Yugoslavia: including milk products, recalculated.

[e] Excluding fats.

[f] For Yugoslavia in liters.

[g] Rumanian data available only for 1950 and 1965.

[h] Excluding consumption of butter and butter fats; in the GDR such consumption amounted to 13.5 kgs. in 1960, and 15.2 in 1978.

[i] In addition, 17.0 kilograms of animal fats were consumed in 1978.

[j] Including margarine.

[k] During the economic year.

[l] Including certain kinds of butter or butter fats.

[m] Recalculated at 2.5% fat content; source: *Statistisches Jahrbuch 1979 der Deutschen Demokratischen Republik*, p. 33.

Source: For Rumania, *Neuer Weg* (Bucharest), 14 June 1978, 7 July 1979, 23 September 1979 [fish in 1970: H. Trend, *Agriculture in Eastern Europe* (Radio Free Europe Research, 15 May 1974] ; for Yugoslavia, *Statisticki godisnjak Jugoslavije, 1963*, p. 287, and *1978*, p. 165, for Albania, *Economic Survey of Europe in 1976, Part II* (New York, 1977), p. 61; all other data according to CMEA statistical annuals (and see note m).

Bulgaria, too, agriculture contributed much to the exports, which exceeded agricultural imports, but to a diminishing degree in recent years. The great Soviet, Polish, and Yugoslav agricultural output potential as yet has not come fully to bear. That makes these countries net importers of food.

The Soviet and East European trade balances in agricultural products reflect to a high degree the grain trade balance, which is its most important single part. Excepting Yugoslavia, grain imports have increased in leaps and bounds, from an average total of a few million tons in individual years (1963/64 being the first high point), to an annual average of 7.1 million during 1970–72, and 16.4 million tons during 1973–75. For 1976 and 1977, grain imports remained at that high level, most of them being accounted for by the Soviet grain purchases, while Poland's imports increased greatly. In the 1970s, Bulgaria's previous grain surplus turned into a deficit in individual years, Rumania remained a net exporter of varying quantities, while Hungary, which during the 1960s had an almost even grain balance, became a major grain exporter. The Soviet Union up to the late 1960s was the main grain supplier to Eastern Europe, above all to Poland, Czechoslovakia, and the GDR, but since then, it has been unable to play that role, in spite of its own increasing output.

Calculated per head of the population, the GDR, Czechoslovakia, and Poland until recently were heavily dependent on grain imports. The first two succeeded in reducing that dependence since 1971 and 1973, respectively. In October 1975, the Czechoslovak Party Central Committee explicitly set the goal of not only decreasing the dependence on food and raw material imports, but also specifically achieving self-sufficiency in grain within a few years. The Polish 1976–80 plan provided for an expansion of the area sown to grain, which in the event shrank from 8.1 million hectares in 1975 to 7.9 million in 1979. Per head of the Soviet population, grain imports up to 1977 did not exceed 40 kilograms on average, although in individual years it was more, for example, almost 100 kilograms in 1973. Yet for the most recent three-year average, 1977–79, the net balance of grain imports reached 60 kilograms per head and was likely to exceed that level for 1980 and 1981. On the whole, however, this did not imply a crucial dependence in view of the 700–800 kilograms of domestic output per head.

The grain imports are destined to close the feed gap in animal production. The considerable share of bread grain in the imports only seemingly contradicts such a statement, because these in the main serve to save some low quality bread grain, which then may be used for feed grain. A most significant case of the role of grain imports for the animal sector is Poland. Only with their help could Warsaw hope to satisfy the internal demand for meat. In addition, some meat so produced was exported, thereby paying for part of the grain imports. This made sense in utilizing the peasant labor surplus for more intensive production. However, this policy became doubtful in view of the slanted prices and the labor numbers not only decreasing, but also more and more consisting of an over-aged and predominantly female peasantry. Czechoslovakia and the GDR achieved self-sufficiency in meat on a feed basis which in part consisted of grain and other feed imports. During the 1970s, except for these two, as well as for the Soviet Union and Albania, the other countries (and primarily Hungary) were net exporters of meat (Poland only up to and including 1976). Taken together, they would have been able to cover the Soviet meat import needs during 1971–75 of close to 250,000 tons per year on average, had they not exported a great part to Western countries. This does not apply, however, to more recent years, when Soviet meat imports ranged at 0.4–0.5 million tons during 1976–78. Yet their meat trade surplus did not mean that internal demand was wholly met. Rather, the meat exports contributed to shortages, at given prices and rising incomes, in the countries themselves.

Self-sufficiency in grain is one of the major goals of the current five-year plan of Albania. No exact information can be given on this country's foreign trade in food. Per head, feed grain imports must be sizable in individual years. Agricultural products in general accounted for 5–10 percent of Albania's total imports, and more than 15 percent of total exports, during the ten years 1955–64. According to FAO data, her agricultural imports increased in value more than the exports during 1969–73, whereas during 1973–78, Albania became an agricultural net exporter again. In 1977, agriculture and its processed products were said to account for 35–40 percent of total exports.[7]

One of the major problems, for which foreign trade represents an alleviation, is the supply of digestible protein for the livestock sector. The "protein gap" has been felt in almost all of Eastern Europe and in the Soviet Union, and since the early 1960s, serious efforts have been made to improve the population's diet. All the countries increased their imports of soybean and other oil cakes, fish meal, etc., beginning in the second half of that decade. According to FAO data, Yugoslavia during 1976–78 imported 291,883 tons of soybean (including cake and meal) on average annually, and the CMEA countries, 4.7 million tons. Exports of sunflower seed cakes and meal, still sizable in 1969, practically ceased subsequently. Although the Soviet Union is the biggest cotton producer of the world, cottonseed cakes were exported recently in only insignificant amounts. Apart from the Soviet Union, Poland, the GDR, Czechoslovakia, Hungary, and Yugoslavia are major net importers of all kinds of oilseeds, oil cakes, and fish meal. In this respect, it must be noted that the Soviet Union, Rumania, Bulgaria, and Hungary are making efforts to increase their domestic soybean production.

Other items that deserve mention are sugar, fruit, and vegetables. Since Cuba proved unable to fulfil the hopes for meeting the excess sugar demand of her European CMEA partners, new efforts have been directed at increasing the somewhat neglected beet sugar production. Czechoslovakia set the goal of becoming a main sugar exporter, as she was before World War II. The Balkan countries supply part of the fruit and finer vegetables to their northern partners. Especially Bulgaria and Hungary are active in this field, the Soviet Union being the main buyer. However, efforts have also been directed towards exporting as much as possible to the West.

It has been pointed out above that Hungary, as an important food exporter, contrasts with the rest of the countries. Even so, more than half of the products of her agriculture and food industry went, except for the years 1978 and 1979, to CMEA countries, in part because trade restrictions and marketing difficulties since the early 1970s prevented an expansion of her food exports to Western markets. The growth of Hungary's agricultural production also resulted in the necessity to import more industrially produced inputs for agriculture and the food industry. Not all of them could be bought from CMEA partners, and therefore part of the hard currency earned by agricultural exports had to be spent on such imports.

In order to save hard currency for investment and other industrial goods, the governments of the Soviet Union and Eastern Europe feel impelled not to import more food than is absolutely indispensable. Yet for the reasons outlined at the beginning of the preceding chapter, the notion of what is indispensable has changed. Tropical and subtropical agricultural products have to be imported anyway for obvious reasons, if one disregards a few small producing areas on the southern Balkans and in Soviet Asia. For such goods, too, the notion has changed. Bananas, coffee, etc., although in CMEA countries still either very expensive or hard to obtain, were an almost unimaginable luxury thirty years ago. Many of the food imports which presently are considered a burden on the foreign trade balance, would not be indispensable by sheer nutritional standards. They have become so by the demand standards of urbanizing societies. The possible way out was not to cut down any longer on living standards, but rather to assign to agriculture the task of making at least part of the food imports unnecessary.

The Development of the Crop and Animal Sectors

As in most industrialized or industrializing countries with few or no reserves of unused arable land, nonagricultural users in Eastern Europe and the Soviet Union have been continually taking away more land from agriculture than could be regained by putting new land under cultivation. The Soviet Union and Albania were exceptions at times, insofar as vast stretches of low productivity pasture were available and turned into arable land. Similarly, Bulgaria after 1973 and the GDR after 1970 increased their arable area by soil improvement at the expense of natural meadows and pasture. Per head of the increasing population, however, the decrease of arable land was noticeable in all the countries except for the GDR, with her declining (up to 1978) population numbers. Soil erosion and the "unprofitable utilization of small-sized lots in hilly and forest areas"[8] also contributed to the losses of tilled land, and this Soviet evaluation, referring to the whole of Eastern Europe, is likely most of all to apply to the large-scale socialist farms.

In recent times, efforts have been made to increase the areas of intensively used land (including high yielding meadows and pasture) by soil improvement. Such improvement relies on irrigation or on drainage, depending on a country's climate and relief. With drainage, the regaining of previously deserted and/or neglected tilled land is also an important aspect, for example, in northwestern and western Russia, or in Lithuania, Belorussia and the Kaliningrad *oblast'* (former Eastern Prussia). The formerly modest expectations of land gains by improvement have been greatly stepped up since the second half of the 1960s. A semiofficial overview prognosticated in 1973 that by 1985, the irrigated land of the European CMEA countries would expand to 31.5 million hectares (from 14.3 million in 1970), and that the drained land would reach 44.1 million hectares (from 20.4 million).[9] Albania expanded her irrigated arable area from 29,000 hectares prewar to somewhere around 400,000 hectares, or close to half the country's total arable land, by the end of 1974, and planned to increase it by 8 percent over the 1975 level during 1976–80.[10] Nevertheless, it has been reported that many new irrigation systems in Albania were never put into operation or utilized only at half their capacity.[11]

During 1971–75, irrigated land as a share of total arable land expanded most of all in the GDR and Rumania, increasing until 1977 from 5.6 and 7.3 to 16.9 and 20.0 percent, respectively.[12] The great absolute increase in the Soviet Union produced a relatively modest proportional rise, increasing the percentage from 4.9 to 7.1. Bulgaria had 21.9 percent of her arable land under irrigation by 1965 and increased this to 31.1 percent by 1978, one of the highest shares in the world, and still rising.[13] Yet on the whole it does not seem likely that the CMEA overall goal set for 1985 will be attained. In Yugoslavia, the irrigated area expanded very slowly, from 118,000 hectares in 1965 to only 124,000 in 1976.[14] Compared to total arable land, this amounted to 1.7 percent.

The change of the cropping pattern during 1960–70 generally consisted in the expansion of the areas under cash crops for industrial use and for processing (except in Czechoslovakia and the GDR, where such an expansion was resumed only in 1970 and 1971), and under perennial crops such as fruit and wine. Over the period, the grain area contracted,

Table 8.6   Cropping Patterns in 1960 and 1978 (as percentages of groups of crops in total sown areas)

	Grain and leguminous grain (including feed grains)		Cash crops for industrial use and processing		Potatoes		Fodder crops (excluding feed grains)	
	1960	1978	1960	1978	1960	1978	1960	1978
USSR	56.9	58.9	6.4	6.8	4.5	3.2	31.2	30.1
Albania, harvested: as a percentage of total sown area	77.0a	59.9 (1973)a	13.1	14.7 (1973)	1.0	2.6 (1973)	5.1	15.1 (1973)
as a percentage of total arable land	61.8 (1973)	69 (1977)b	...	...	...	2.8 (1976)	...	...
Yugoslavia	81.2c	69 (1977)	4.6	1.4 (1977)	4.1	4.8 (1977)	10.1	13.5 (1977)
Rumania	73.9	66.2	9.2	13.1	3.0	3.0	11.3	13.8
Bulgaria	65.3	59.1	13.1	14.7	1.1	1.0	18.0	21.4
Poland	62.5	55.0	4.7	7.1	18.8	16.3	11.5	17.8
Hungary	67.8	63.3	5.3	9.5	5.4	2.3	17.9	20.6
Czechoslovakia	50.3	57.9	7.6	7.6	11.0	4.5	29.1	28.5
GDR	50.0	54.5	8.9d	8.0	16.0	12.1	23.1d	22.5

a Excluding leguminous grain.
b FAO estimate.
c Including all kinds of maize (which in practice is mostly maize for grain).
d For 1960, sugar beet for livestock is included in cash crops and not in fodder crops.

Source: For Albania, 30 Vjet Shqiperi Socialiste (Tirana, 1974), p. 109-10 (sown area), and FAO Production Yearbook (arable area); for Yugoslavia, Statisticki godisnjak Jugoslavije, 1978, p. 224; for all other countries, CMEA statistical annual, 1979.

except in the USSR, GDR, and Czechoslovakia, and continued to do so in Poland, Yugoslavia and Hungary. It began to expand again in the USSR from 1971 until 1975, and again in 1977, and in Rumania and Bulgaria during 1974–76. Grain's relative shares of agricultural and sown areas were slightly different, depending on whether the total agricultural and sown areas decreased or increased in the individual countries. As a percentage of the total sown areas, grain's share after 1970 continued to be greater than in 1960 only in the USSR, GDR, and Czechoslovakia. If feed grain is added to the other fodder crops, these clearly expanded in absolute as well as in relative terms. (Feed grains are not separated from other grains in Table 8.6 because the statistics do not reveal the exact share of other grains, such as wheat, actually fed to livestock.) The share of cash crops, which had decreased for brief periods during the late 1950s or (in Hungary, Czechoslovakia, and the GDR) the 1960s, resumed its earlier expansion. For reasons of space, no more than a few main crops can be dealt with here and in Table 8.6 (for hectare yields, see Table 8.2).

Grain (including leguminous grain) is the single most important crop in all of Eastern Europe and the Soviet Union. Its output increased considerably faster than total crop production during 1970–75, except in Rumania. This development is due mainly to the increasing grain yields per hectare, on which emphasis is put for the future as well. The yield increase is to be achieved mainly by more fertilizer application, and also by irrigation, the introduction of new varieties, improving seed quality, changing crop patterns, and more careful and faster execution of field work. A goal of 435–450 million tons a year for the CMEA bloc (possibly including the negligible Cuban and Mongolian grain quantities) has been mentioned several times. It is hoped thus finally to overcome the permanent "grain problem," which was a slogan of the CMEA as early as in 1956. Even starting from a relatively low level, the five-year grain output average (for the European CMEA countries) in 1971–75 was only 41 percent more than 10 years earlier in 1961–65. Judging by past performance, the above-mentioned goal, which is 45–50 percent above the achieved 1976–78 average with its two Soviet record harvests, is unlikely to be attained before the late 1980s, if by then. In 1979, the total for the CMEA countries was again down to 255 million tons, after the 1978 record of 320 million tons.

Potatoes are an important crop for human consumption as well as for feed in the GDR, Czechoslovakia, and especially in Poland. The European CMEA countries produce more than half of the world's potato output. The share of potatoes in the arable area has generally declined since 1960, except for Yugoslavia, Albania, and Bulgaria, where potatoes are of minor importance. The declining share of private plots in overall tilled land has much to do with this development, as potatoes are the most important crop on private plots in terms of the area occupied.

For the production of vegetable oils, sunflower seed is the main crop in Eastern Europe and the Soviet Union, and is concentrated in their warmer regions. However, a growing sunflower output in Eastern Europe could not fully compensate for a decrease in the Soviet Union.

For cotton, the Soviet Union is overwhelmingly the main producer, with the quantities grown in the Balkan countries (including Albania) negligible in comparison. Soviet production has risen rapidly, from an average of 5.0 million tons a year in 1961–65, to 8.8 million tons in 1977–79. These figures relate to raw, unginned cotton, and to make them comparable to Western data on lint cotton, about two-thirds would have to be deducted. As United States cotton output has been on the decline for more than ten years, the USSR nowadays ranks first in the world in cotton production, with the USA in second place and China third.

Total East European and Soviet output of sugar (exclusively beet sugar) during 1971–75 was somewhat less than during 1966–70, which was contrary to the worldwide trend. Moreover, the sugar content in beets went down, in the USSR from 14.05 percent in 1950 to 10.94 percent in 1974. Thus, the rise of 36 percent in the Soviet Union's average hectare yield of beets resulted in an increase of sugar per hectare of only 6 percent over those 24 years.[15] During 1972–76, the Soviet area under sugar beet expanded again, in accordance with the 1976–80 plans, possibly under the influence of Cuba not attaining her ambitious 1970 target and switching back to a more diversified agriculture. The two-year output average of 1977–78 exceeded the 1967–68 record by 17 percent.

The supply of mineral fertilizer, although much increasing in recent times (see Table 8.13), does not yet satisfy the demand of agriculture, and the available quantities were and are often not put to the optimal use. What fertilizer supplies were available were formerly applied almost exclusively to cash crops such as sugar beet and cotton. However, in recent years, more and more fertilizer has been applied also to grain, potatoes, and green fodder crops.

Arable farming has to supply most of the feed for animal production, but has done so only inadequately in Eastern Europe and the Soviet Union. The share of grain in the feed balance has been growing in recent times, and presently about two-thirds of the total grain output is fed to animals.[16] As a percentage, human consumption and industrial uses of grain are to decline in an overall growing grain output. Reflecting the intention to step up animal production, it was planned, at least until very recently, to increase the share of grain for feed, which, of course, implied a still more rapid increase of the absolute quantities of grain fed to animals.[17] It is doubtful, though, whether all this feed grain is really needed. More balanced feed rations and a better utilization would result in less feed consumption per animal output unit. In such a case, either animal production could increase more, or less sown area would be needed for feed grain.

It has often been said, and correctly so, that the animal sector is the weaker part of the two main subdivisions in the overall weak agricultural sectors of the Soviet Union and Eastern Europe. The share of animal production in total gross production has been low. Whereas in Western industrialized countries this share usually amounts to 60–80 percent, it has continued to hover around half of total gross production in the countries to be dealt with here. The difference is only in part due to differing statistical methods.[18]

In comparison to prewar times, the share of animal production in total gross agricultural production had scarcely grown by 1960. As can be seen from Table 8.7, subsequent progress in this sphere has been slow, in spite of the considerable capital invested in premises and machinery for large-scale animal farming since the mid-1960s. The rapid expansion of Hungary's animal production by an annual average of 3.9 percent during 1965–78 (by CMEA indices), surpassed only by Rumania, was to a great extent due to the private sector's contribution and the tolerant official attitude toward it.

The number of animals per land unit differs considerably between the countries. According to a Polish calculation of livestock units per hundred hectares of agricultural land, there were only 20 in the USSR and as many as 100 in the GDR; Poland had 70 (exclusive of goats), Czechoslovakia 67, Rumania 49, Bulgaria 47, Yugoslavia 43.5, and Hungary 40.[19]

With mechanical draught power increasingly being substituted for animals, horse numbers went down in all the countries, most of all in the highly industrialized GDR and Czechoslovakia. The decline was less in Rumania, Yugoslavia, and the USSR, and least of all in Poland and Albania. Sheep herds declined in Yugoslavia, Albania, and (until 1973) in Bulgaria, Hungary, and the GDR. This was compensated for by their increase in Rumania (up to 1972) and the Soviet Union (up to 1974), and even (up to 1970) in Czechoslovakia. Beginning in 1974 or 1975, government efforts made sheep numbers go up again in all CMEA countries. Wool production was and is a sensitive issue, since demand in those countries still exceeds supply, especially for the finer qualities.

Poultry farming has shown a remarkable upsurge in Rumania, Bulgaria (up to 1977), Czechoslovakia, and the USSR, and has also expanded in the other countries. At the same time, it has partly shifted from private plots to large-scale specialized farms or farm sections. This

Table 8.7   The Share of Animal Production in Total Gross Production of Agriculture
(percentages according to Communist statistics)[a]

| | Three-year averages[b] | | |
	1960	1970-72	1976-78
USSR[c]	52.2	50.9	55.4
Albania	43.8[d]	28.8 (1970 only)	29.0 (1975 only)
Yugoslavia[e]	28.0	37.7 (1971 only)	41.3 (1975-77)
Rumania	35	39.3	42.8
Bulgaria[c]	32.7	37.2	44.4
Poland[d]	41.8	45.6[g]	42.8
Hungary[f]	39.2	43.4	47.8
Czechoslovakia[f]	49.4	54.8	55.4
GDR[c]	51.7	56.0[g]	62.5

[a] For the impact of double-counting of feed, see footnote 18, above).
[b] Unweighted, arithmetical averages of the indices of the three years.
[c] In constant prices of an unspecified year.
[d] In prices of 1971.
[e] For 1975-77, derived from annual growth indices. The Yugoslav statistics for the final
product of agricultural do not double-count farm produce used on the farm, and therefore,
in view of the low share of animal production, tend to overstate its share (see footnote 18, above). Moreover, products processed in the farm household (plums for liquor,
grapes, olives, etc.) are accounted for separately and are not part of the division applied
here, thereby also making the share of animal output look larger.
[f] The price bases for Hungary and Czechoslovakia are not indicated in the sources. Presumably current prices.
[g] Probably overstated because of an unspecified change in the price basis.
Source: For Yugoslavia, *Statisticki godisnjak, Jugoslavije, 1974*, p. 163, and *1978*, p. 223,
and *Statisticki prikaz razvoja poleprivrede 1950-1971* (Belgrade, 1973), p. 46; for Albania:
in 1960, *30 Vjet shqiperi Socialiste* (Tirana, 1974), p. 112; in 1970 and 1975, *Economic
Survey of Europe in 1976, Part II* (New York, 1977), p. 61; for the other countries, *CMEA
statistical annual, 1973*, and *1979*.

relates to egg as well as to broiler production. However, the organization
of modernized large poultry farms has often been impeded by shortage
of capital, skilled labor, and infrastructural facilities.

Overall, livestock herds during the 1960s grew only very slowly, and in
Czechoslovakia, Yugoslavia, and Albania, they even declined. A more
rapid growth set in during the early 1970s, although there was a setback
after the 1975 drought. East European authors have admitted that this
sluggish development was a consequence not only of the great annual
variations in crop yields and the resulting unreliable feed basis, but also
of the marked decline in private livestock holdings. In the Soviet Union,
however, the growth was about even over the 1960s and 1970s, with the
exception of 1975/76 and 1979/80.

After 1970, the producers' price increases and the more favorable attitude toward the peasants in Poland, as well as the systematic expansion of the livestock sector in Rumania (both in the socialized sector and, with a more tolerant policy, in the private sector), brought about a sizable increase of herds. The Yugoslav development after 1973 was similar for a few years. It seems there that the increased demand (through higher incomes), the raised producer prices, and the intensified cooperation of peasants and socialized enterprises had a combined positive effect on the livestock sector for a short interlude. However, in more recent years, only pig numbers have continued to increase in Yugoslavia, while cattle, sheep, and goat herds have declined. Little information is available on Albanian animal production during the early 1970s, although it seems that livestock numbers continued to stagnate, and plans were far from being fulfilled. Already in 1973, the meat supply situation was said to worry the authorities, and since then, no sizable successes have been reported.[20] In the CMEA countries, too, the upswing of the early 1970s soon levelled off, with the notable exception of Hungary.

In countries with collectivized agriculture, the less restrictive policy (except in Albania) towards the private sector after Khrushchev's removal from power in October 1964 apparently brought less than the expected output growth of animal production. The only exceptions here were in Bulgaria, during 1974–76, and again in 1978, and in Hungary, for private pork production. It is the present writer's opinion that the disappointing development of livestock production was not due only to higher wages in the public sector and a resulting decline of interest in private animal farming. Rather, social and demographic change (diminishing rural population and labor) and a changed attitude towards work seem to have had fundamental effects.

Yields per animal (see Table 8.2) are generally low by Western standards, but in the GDR are approaching these standards. Especially for pork, the yields seem surprisingly high in Hungary and Czechoslovakia (but see note e, Table 8.2). Yet it also must be pointed out that the increased yield per animal over the 1960 level was remarkable in most cases, so that output grew more than the numbers of livestock, especially during the 1970s.

A main weakness in Soviet and East European animal farming is the feed conversion ratio—feed consumption per output unit is too great. In the Soviet Union, for example, 9.22 feed units (= 9.22 kilograms of oats, recalculated) were needed in 1971 to produce a weight gain of one kilogram for pigs (average for all kolkhozes and sovkhozes); for beef and veal, the ratio was 10.32; for poultry meat, 4.5; and per kilogram of milk, 1.44. For pigs, 6.93 of the feed units consisted of concentrates; for beef and veal, 1.78; for poultry meat, 3.7; and for milk, 0.32.[21] For pigs and poultry, this was roughly 50–100 percent more than on average

Western farms. Inadequate livestock premises under a severe climate and a winter shortage of feed, allowing the animals to survive but considerably curtailing their production, is part of the explanation. Yet above all, it is the lack of digestible protein in animal feed that causes this low feeding efficiency. As long as these factors, together with inadequate care for the animals, persist, even improved animal breeds cannot develop their full productive potential. Moreover, overall replacement of the low productivity herds takes a long time to bring about, especially in cattle production. After a period of putting the main emphasis on increasing feed grain quantities, since 1978 it has been "increasingly acknowledged that the protein problem which exists in most countries of the region can best be solved by improving pasture production rather than by relying exclusively on grain and synthetic protein additives."[22]

For Poland in 1965–67, a Soviet author estimated that 15–30 percent (1.5–2 million tons) of the total grain fed to animals was overfed.[23] Since then, the overfeeding is said to have increased.[24] Two Soviet authors in 1975 estimated that in the USSR, the protein deficit in the feed ratios alone caused overfeeding of 14 million tons of oats units annually, which in 1974/75 amounted to roughly 20 percent of total concentrate grain feed. However, even these figures may be too low, since a leading article a few months later implied overfeeding of the level of 30 or 40 percent.[25]

Slow growth of herds and low animal productivity put a brake on the animal sector's growth. To the degree that large-scale animal production under "industrial" methods is being introduced, such a development becomes anachronistic and very costly. It is the paramount and indeed indispensable goal of agrarian policy in Eastern Europe and the Soviet Union to overcome such deficiencies in the modernization of animal farming.

Resources and Inputs of Labor

A precise and internationally comparable quantitative assessment of the agricultural labor available, and of that actually employed, is not possible for Eastern Europe and the Soviet Union. Indeed, it is hard to achieve this in Western countries as well. Apart from a general lack of precision in the data and the differences in statistical definitions of the individual countries, one has to deal with the problem of assessing family labor which is not fully employed in agriculture, especially on the small peasant farms of Poland and Yugoslavia and on the private plots in collectivized agriculture. The statistical reporting on labor on collective, as distinct from state, farms is also different by countries and sectors. There are socio-political motives in some countries to distort the agricultural labor data. "In almost all socialist countries, a direct accounting of the working time put in is lacking", said a Soviet author in 1971.[26]

There are no recalculations on a full-time worker's basis in official published statistics of most of the countries concerned.

In Eastern Europe and the Soviet Union as elsewhere, rural-urban migration and outflow of labor from agriculture are not identical. In industrialized and industrializing countries, a sizable part of the work-force continues to live in the countryside, but is employed in nonagricultural branches, either in the village or in a nearby town or city. In the GDR, Czechoslovakia, and Hungary in 1975, those making their living in agriculture were already a minority of the rural population (40.5, 43.1, and 45.6 percent, respectively). In the other countries, they still formed a majority of 55–65 percent, and for Albania one might estimate it at approximately 80–90 percent.[27] Already in the early 1970s, 47 percent of people living on Yugoslav farms had permanent employment outside their farms.[28]

Generally, the farm population derived a considerable part of their income from nonagricultural work outside the farm, even if still working in agriculture. An extreme situation was revealed by a Rumanian time budget survey of 1973 in Argeş district: of the male collective farm members' time (inclusive of private plot work, housework, and childcare) in the plain area, 46.8 percent was spent on work outside the collective farm, and in the hilly-plain area even 59.6 percent, but only 14.4 and 5.5 percent, respectively, on collective farm work. For female members, the shares attributed to the collective farm were about the same, and most of the rest was expended on housework and childcare.[29]

On the other hand, 20 percent of the permanent collective farm work-force in the GDR in 1978 were nonmembers (12 percent in 1970), and of these, every third had a nonagricultural occupation within the farm. In Hungary, the corresponding percentage was 22 as of 1 January 1978.[30] The number of workers in the socialized large-scale farms doing nonagricultural jobs serving agricultural production is growing. It is among these that the younger, male, and skilled labor is predominant.

The average annual numbers of workers in agriculture, as given in Soviet and some East European statistics, are not those of physical persons employed in agriculture. Underemployment is a generally known phenomenon in small peasant agriculture (because only during sowing and harvesting time is the available labor fully employed on the farm), and a similar phenomenon exists for collectivized agriculture. Thus, in the Soviet Union, the number of physical persons employed in agriculture declined by one-third according to the 1959 and the 1970 censuses, but according to current labor statistics of the same eleven years, the average annual number of workers declined by only 10 percent. While the pension scheme for kolkhoz peasants, as introduced beginning in 1965, may be considered to have increased this numerical disparity, one of the reasons obviously is a higher degree of employment of previously underemployed labor resources. In all Communist countries, members of collective farms who are invalids, pensioners, housewives, or juveniles are counted as part of the "labor collective" to the degree of their participation in collective work. "The collective farm is obliged to provide work for all its members, but does not guarantee them full employment and does not pay for days without work. . . . For the majority of farms, this is a problem not yet solved."[31]

The amount of labor spent on the private plots poses special problems of statistical assessment, which have been extensively elaborated upon for the Soviet case. The best available Western estimates are those of Rapawy, who gives full-time equivalents for total agricultural employment, including the private sector, of 40.0 million in 1960, 37.6 million ten years later, and 37.0 million by 1974, i.e., sizably more (for good reasons) than the published official estimates of 32, 29 and "about" 28 million, respectively.[32] The Soviet figures account only for those in the private sector who are working there exclusively. In favor of the Soviet estimates, it may be taken as a reasonable assumption that the kolkhoz workers spend free time on the plot, and that both types of work, taken together, make them full-time equivalents. Yet this still leaves unaccounted for the agricultural work privately put in by non-kolkhoz wage-earners.

On average, the number of days worked on the collective farm by an able-bodied member has come close to that of a state farm worker. Since the five-day working week has become the rule in industry, the annual number of days worked in agriculture's six-day week is similar to that in nonagriculture. These facts diminish the discrepancies in statistical reporting of labor inputs. Within agriculture, however, and especially in the collective farm or kolkhoz sector, the differences between occupations have remained great. An enquiry on Bulgarian collective farms near Sofia during 1965–72 revelaed an annual average of more than 300 days worked per worker in animal production (as in Soviet kolkhoz animal farming), in contrast to 181 days in crop production, against an average of 237–239 days per year in industry. In Czechoslovakia, where shortage of agricultural labor predominates, industrial workers worked an average of 228 days in 1973, against 253 days for collective farmers and 257 days for state farm workers. At the same time, female pig tenders on Czechoslovak collective farms worked 25 percent more than the average (and others correspondingly less).[33] These relations are more or less typical for all European countries with socialized agriculture, although the annual average employment on collective farms is less in countries like Rumania and Albania or in some regions of the Soviet Union. Such figures also explain the seeming contradiction that in Soviet agriculture, the minority of workers is employed in animal and fodder production, but that, according to one Soviet author, two-thirds of the total agricultural labor inputs are spent in this branch.[34]

Hungarian statistics apply relatively refined methods for assessing the numbers of labor in agriculture, and the results may be assumed to have parallels in other countries with collectivized agriculture. In connection with the census, it was found in Hungary that by 1969, there were almost 1.8 million people working in agriculture, although 700,000 of them worked less than 90 days per year in agriculture (many only 30–40 days). Within the last category, more than 200,000 had their main employment in other branches of the economy, another 130,000 were pensioners, and 300,000 were family members without independent income. Of the manual workers on farms, 24 percent were employed in nonagricultural occupations,whereas in 1960 this figure was only 7 percent. Because of two changes in statistical methods, during 1960–62 (upon full collectivization) and in 1970, the decrease in the numbers of agricultural labor in Hungary looks greater than it actually was, although this has in part been taken into account statistically by recalculating the older data. The Hungarian author who reports these facts states that the actual decline in agricultural employment numbers during the last twenty years was not very rapid.[35]

Because of all the pitfalls in finding comparable data of labor actually put in, or of full-time worker equivalents, the more comprehensive and less quantifiable category of "economically active" persons in agriculture, as applied by FAO, is also used, together with official CMEA data and one Eastern and one Western estimate in Table 8.8.[36] According to the FAO statistics, the share of those in agriculture, compared to the total economically active population, dwindled rapidly in Eastern Europe and the Soviet Union. In total for the CMEA member countries and Albania, it was down to 23.9 percent in 1975, compared to 42.4 percent fifteen years earlier. Most of the decline occurred during the 1960s. It makes sense that in the table the numerical discrepancy between the average annual workforce and the economically active population is greatest in those countries where the agricultural labor surplus still plays a great role, such as Albania, Rumania, and Yugoslavia. However, the similar discrepancy in Bulgaria, and the considerably smaller shares of economically active persons (as against average annual labor) in the USSR, Hungary, and Czechoslovakia, are hard to explain, and raise doubts as to the FAO data basis.

The decline of the labor resources and of their actual employment over time is of special interest. Where the output is constant, such a decline implies a corresponding increase of labor productivity. Where output is growing, as was the case, labor productivity grows by a percentage greater than the proportional decline in labor input num-

Table 8.8   Labor Employed in Agriculture as a Percentage of All Labor in the Economy, According to Various Sources and Statistical Calculations

	Year	CMEA annual averages[a]	Doleshel	Elias and Rapawy	FAO: economically active population
USSR	1960	38.7	38.3	41.4	41.9
	1974	23.2	23.0	29.2	20.5 (1975)
	1978	20.9	. .	. .	18.1
Albania	1960	ca. 57[b]	. .	. .	71.4
	1975	ca. 50[b]	. .	. .	63.3
	1978	. .	. .	. .	61.5
Yugoslavia	1960	56.9[c] (1961)	. .	. .	63.7 (71.5[d])
	1970	44.6[c] (1971)	. .	. .	49.8 (64.5[d])
	1978	. .	. .	. .	39.8
Rumania	1960	65.6[e]	65.4	68.1	64.5
	1974	40.0[e]	39.8[f]	. .	51.6 (1975)
	1978	32.8[e]	. .	. .	49.0
Bulgaria	1960	55.5	54.7	57.1	56.5
	1974	30.1	29.6	. .	39.7 (1975)
	1978	25.2	. .	. .	35.7
Poland	1960	44.2	43.0	44.5	48.2 (48.0[c])
	1974	31.1	30.5	. .	34.6 (1975)
	1978	30.8	. .	. .	32.0
Hungary	1960	38.9[g]	37.9	38.9	38.1
	1974	23.3[g]	22.2	. .	19.9 (1975)
	1978	21.7[g]	. .	. .	17.3
Czechoslovakia	1960	25.9	24.2	25.9	25.7
	1974	15.7	14.4	. .	13.2 (1975)
	1978	14.5	. .	. .	11.4
GDR	1960	17.2	16.6	17.6	17.6
	1974	11.5	10.9	. .	11.2 (1975)
	1978	10.6	. .	. .	10.2

[a] Including forestry. The CMEA data are not uniform as to the methods of statistical accounting in individual countries, and therefore are not wholly comparable among each other.

[b] Author's own estimate, to make the Albanian data comparable to those for Soviet agriculture. It is based on the Albanian share of rural in total population (69.1 percent in 1961, and 66.3 in 1970), on the assumptions that this share contains some non-agricultural employees among rural inhabitants and that the degree of agricultural underemployment was sizable. The Albanian newspaper, Zeri i popullit, on 25 May 1975, p. 2 (according to ABSEES 6:4 [October 1975], p. 141) indicated that the "peasantry," presumably excluding state farm workers and other state employees in agriculture, made up 49.4 percent of the active population.

[c] According to the Polish statistical yearbook Rocznik Statystyczny 1976 (Warsaw, 1976), p. 554.

[d] Yugoslav census results of 1960 and 1969 (instead of 1970). Current Yugoslav data (67.5 percent for 1960, 60.0 for 1970, 53.0 on the average for 1975-77) are also higher than those of FAO and the Polish source because they include those only seasonally employed.

[e] As of 31 December.

[f] This percentage does not fit with Doleshel's own absolute figure of 6.233 million.

[g] At beginning of year, including apprentices, and water management workforce.

Source: CMEA statistical annual, 1976 and 1978; O. Doleshel, op.cit., p. 531; FAO Production Yearbook, 1975 (Rome, 1976), Table 6 and 1978, Table 3; Andrew Elias, "Magnitude and Distribution of the Labor Force in Eastern Europe", Economic Development in Countries of Eastern Europe, Joint Economic Committee, U.S. Congress (Washington, D.C., 1970) pp. 216, 224-225; Stephen Rapawy (for USSR only), Estimates and Projections of the Labor Force and Civilian Employment in the U.S.S.R. 1950 to 1990, Foreign Economic Report no. 10, U.S. Department of Commerce, Bureau of Economic Analysis, Washington, D.C. (September 1976), pp. 40-41 (Table 13); for Yugoslavia: Statisticki godisnjak, 1970, p. 87, 1974, p. 141, and 1978, p. 119; Statisticki bilten, no. 720, p. 17, and no. 866, pp. 11-11-12.

bers. In the countries concerned, the rates of decline in the agricultural labor force, in terms of annual averages, were quite diverse (Table 8.8). Over the fourteen-year period 1960–1974, they ranged from less than 1 percent in Poland, Yugoslavia (for nine years only), and the USSR, to 2–3 percent in Czechoslovakia and the GDR, and 3–4 percent in Rumania, Hungary, and Bulgaria.[37] The rate of decrease in the USSR, GDR, Rumania, and Czechoslovakia accelerated after 1967. On the whole, the decline was slower during the second half of the 1960s than in the first two or three years of the decade or again in the early 1970s.

By the mid-1970s, it was expected that the decline in the agricultural labor force would become more rapid up to 1990 in the countries under study taken as a whole, especially if one excludes Yugoslavia and Poland (where the decline will be slower).[38] In Yugoslavia, the reduction seems to have continued after the 1969 census. In Poland, the decline was minimal during 1960–74, and in full-time equivalents there was no decline at all up to 1977.[39] It may have set in towards the end of the decade, but in any case, the rate is expected to accelerate in the future. The planned actual 3 percent annual rate of decrease up to 1980 in Bulgaria is somewhat slower than the 1960–74 average, which was the highest of the countries (3.7 percent annually). This still fast rate does not necessarily contradict the recent Bulgarian efforts to prevent people from migrating to urban areas, as it may include many who have changed over to non-agricultural employment but continue to live in the countryside within the new agro-industrial complexes. The estimates for Albania indicate a small decline between 1960 and 1975 in the absolute number of people agriculturally employed. Apparently the government did not try to encourage such an outflow.

If for the Soviet Union one disregards the private sector and includes seasonal labor coming from outside agriculture, no decrease at all of the agricultural labor force was noticeable from 1971 to 1978. The number of days worked in agriculture in fact even increased slightly, mainly due to increases in workers' numbers as well as more intensive production in the southern and Central Asian regions. For most of the country (excluding the Baltic Republics), the rate of decrease would be similar to that of Rumania (see Table 8.8). Similarly, the development in the Slovak part of Czechoslovakia, together with the relatively high labor remuneration in agriculture, account for the fact that the overall diminution was not as rapid as might have been expected in a highly industrialized country.

In sum, only in Bulgaria did the rate of decrease of the agricultural work force exceed that in the Western industrialized countries in the 1950s and 1960s, when the decline in the latter was most pronounced. It came close to the Western rate in Czechoslovakia, Rumania, and the GDR, but was considerably slower in Poland, Yugoslavia, and the USSR.

As these last three countries account for three-quarters of the total agricultural work force, the decline for the whole of Eastern Europe and the Soviet Union was slower than in the West. Yet a decline was clearly visible, except in Poland and possibly also in Albania. Its expected further acceleration is partly a result of the great percentage of collective farmers of pension age who will leave the work force in the near future (over one-fourth in Czechoslovak collective farms, and one-fifth in total Polish agriculture during the 1970s).[40] Hungarian agriculture (including the private sector and forestry) on 1 January 1978 had a staggering 51.4 percent of people of pension age among the "active earners," and 74.1 percent of the "active earners" were 40 years and older. [41] Complaints about aging and feminization of the agricultural work force have been sounded in all the countries, although at times it was said that the process had been reversed, for example, in Czechoslovakia. [42]

Albania witnessed heavy rural-urban migration, which was said to have levelled out since 1965, as the towns' needs for labor were satisfied.[43] The high rural birth rate apparently prevented a depopulation of the Albanian countryside, and since 1965, the rural has grown as fast as the urban population. Even so, a "return to the village" campaign was launched by the mid-1970s.[44] The feminization of collective farm labor obviously was in part caused by male out-migration. In 1975, 53 percent of Albania's economically active village population, and 55 percent on collective farms, were said to be women (although only 30 percent of the total, including men, were over 40 years old), and for some districts, percentages up to 80 were registered.[45] These percentages might even exclude those who were only seasonally employed. In 1978, the number of female labor on Albanian collective farms was said to have increased twenty-fold since 1973.[46] On the whole, however, Albanian collective agriculture still seems to be overmanned, judging by the recurrent complaints about absenteeism and lack of discipline and control over the labor force.[47]

The availability of land per worker is a telling indicator for the labor intensity of agricultural production, as reciprocal to capital intensity. According to the 1969 census, agricultural labor intensity measured thus was greatest in Yugoslavia (no data for Albania available), with only 1.7 hectares, or somewhat more of compound land units, [48] per worker. By 1974, there were less than 2.8 compound land units per worker in Rumania, 2.9 in Poland, 3.7 in Bulgaria, less than 5.1 in Hungary, and less than 5.3 in Czechoslovakia, while in the GDR, the figure was 5.7.[49] Unexpectedly, the Soviet Union comes out with the greatest area (more than 10.5 compound land units) per worker. This is not due to a greater capital intensity but to enormous areas with extensive dry farming and natural pasture utilization. One has to bear in mind that in Canada, a Western country with comparable soil and climatic conditions, there are

50–60 hectares of compound land units per average annual worker. The land/labor ratio in Hungary, Czechoslovakia, and the GDR was of about the same order of magnitude as in Western Europe with her small- and medium-farm agriculture, although in socialized large-scale farming, one would have expected less labor application per land unit.

As the capital per worker available to collective farms gradually catches up with that in state farms, the increase of the labor productivity in the collective sector on average is faster than in the state sector. For the private sector, one has to assume that its possibilities for capital inputs are small and therefore the increase in labor productivity on plots is very slow. For this reason, the overall increases tend to be lower than those in the official East European and Soviet statistics, which account for the socialist sector only. Exceptions from this general statement are periods when, after a time of restrictions, the private sector is given more scope and therefore increases its output without a corresponding increase in labor inputs because it now merely has better possibilities for utilizing the same resources. This happened, for example, in 1964/65 in the Soviet Union, after 1968 in Hungary, after 1970 in Poland, after 1973 in Bulgaria, and may have occurred again in the USSR during 1976–78.

The net productivity of labor, that is, with material inputs deducted from output and then recalculated per average annual worker, in most cases more than compensated for the decline in the labor forces, so that net output grew all the same. There are a few exceptions to this, the most important being that of the USSR during 1966–68, when the growth of material inputs exceeded that of the gross output. Gross labor productivity did, of course, grow faster than net labor productivity. Yet whether gross or net, it surpassed the growth of labor productivity in the nonagricultural branches only in the GDR, whereas such more rapid growth has been generally characteristic for Western agriculture during the period of its rapid substitution of capital for labor. Thus, the gap between agriculture and the rest of the economy is not yet being narrowed in most of Eastern Europe and the Soviet Union.

Generally, it is correct to say that compared to Western countries, the agriculture of Eastern Europe and the Soviet Union was and still is overmanned. Yet this statement needs three qualifications. First, to make the comparison with Western farming, ten percent or so should be subtracted from Soviet (and some other East European) labor data, because Western owner-occupier farmers usually expend more working time on their farms. The Soviet full-time farm worker by definition, if not in practice, is one who spends 275–280 days of seven hours on his job, or, with a five-day week, 225–235 days of 8.2 hours—roughly 1,900 hours per year. By contrast, some Western statistics assume 2,000 and more hours (United Kingdom, Federal Republic of Germany, Netherlands, Austria), or 1,800 as a minimum (USA) for full-time agricultural labor. As to Yugoslavia, the average number of days worked was only 140, similar to French statistics, where a person working 120–140 days in agriculture is defined as a full-time farm worker.[50]

The second qualification is that soil and climate in most of Western Europe and North America lend themselves more easily to highly productive crop and animal farming, and thereby to higher labor productivity. Last, the need for great labor inputs is a consequence not only of structural and organizational deficiencies, but also of lack of capital. For the first fifteen or so years after the Second World War, agriculture in Eastern Europe and the Soviet Union received very little capital, and this deficiency has not yet been made up in spite of increasing supplies and investment from the 1960s on.

Overmanning does not exclude regional and seasonal shortages of labor in agriculture, especially as the fastest possible output growth is constantly being urged. The reserve of female labor has been almost wholly tapped, and women now constitute about half of the agricultural workforce. Under these conditions, the mobilization of the nonagricultural population to help agriculture in peak seasons of labor application has become widespread in the countries under discussion, especially during harvest time and even in highly mechanized grain production. This phenomenon is probably best known (and most fully documented) in Soviet agriculture. According to official statistics, the number of *privlechennye* for agricultural work, which for the most part are such seasonally mobilized workers, grew from 0.5 million in 1965 to 1.3 million by 1978. These figures are recalculated on a full-time workers' basis, and in physical persons would amount to 2.5–3 and 6–8 million, respectively, or, roughly one-fourth of the total workforce in socialized agriculture in 1978.

In the GDR and Czechoslovakia, the agricultural labor shortage is already considered critical, and measures have been directed at slowing the outflow, as yet with little success. The same applies for individual regions in Hungary and Rumania. The Bulgarian planners up to the early 1970s reckoned on more labor from agriculture being available for the expanding industry. However, in 1974 the Bulgarian government issued a decree against continuing rural-urban migration, presumably in connection both with the inadequate suppy of living space in the towns and cities and with the intended expansion of nonagricultural activities in the countryside in the context of "agro-industrial integration". Two other decrees, of August 1973 and April 1975, concerned the mobilization of temporary labor for agriculture from urban places and nonagricultural enterprises.

In Yugoslavia, where the unemployment rate outside peasant agriculture reached 13.5 percent by 1979, the problems are of a different kind. As not enough nonagricultural jobs are available, the government so far has not encouraged labor outflow from peasant agriculture. Anyway, more than half of all Yugoslavs employed abroad in the early 1970s had been peasants at home. [51] The demographic and economic situation in Poland has been similar, at least until recently, but may change in the near future. For the time being, it applies to both countries that small peasant farming helps local underemployment not to turn into open unemployment.

The substitution of capital for labor in agriculture requires technical and comparable skills of the work force. While the shortage of equipment operators ("mechanizers") is still a problem, most of all in the USSR and Bulgaria, the share of workers trained for modernized large-scale agriculture is gradually growing. By 1975, the share of the work force with such skills reached more than 80 percent in the GDR, whereas in Hungary it was 30 and in Czechoslovakia 25 percent on the socialist farms. [52] The continued increase of the percentage is also a function of the imminent retirement of the many overaged workers, who for the most part have only the traditional peasant skills.

Capital for Agriculture

According to Soviet publications, the capital per worker in Soviet agriculture in the early 1970s was half as much as that available in industry, and only one-sixth in Rumania. For 1973, a Bulgarian calculation showed a capital availability per worker in agriculture, compared to industry, of 38 percent in Bulgaria, 43 in Poland, 67 in Czechoslovakia, 71 in Hungary and 83 percent in the GDR. Except for Poland, this was an improvement over 1960 and 1965.[53] With continuing priority of capital supply for agriculture, the proportion should have further improved since 1973, but not to a degree that would make agricultural labor better endowed with capital than industrial labor, as has become the case in some Western countries.

Of the three production factors—land, labor and capital—the first two have very slowly (land) and moderately (labor) declined, whereas capital has greatly increased during the past fifteen or more years. Therefore, capital is obviously the factor that effected the progress in labor and land productivity. Improvements in organization and intensity of work and in land allocation may have contributed to an unquantifiable degree but cannot have done so decisively, as there was no fundamental change in farm organization and land use. If "agro-industrial integration" also played a role, this could have become noticeable only in the most recent years, when it started to encompass more than a small segment of agriculture. Moreover, it was closely connected with the increased capital supply.

Capital input grew faster than gross output, so that each additional capital unit brought an additional unit of output, but a proportionately smaller one than the preceding capital input unit. In other words, the marginal productivity of capital declined rapidly.[54] Especially the share of the most productive kind of capital assets, namely machinery and implements, has been too low. For the CMEA countries, it was 13–22 percent by the mid-1970s, whereas 32–35 percent was considered an optimum, not attainable in the near future.[55]

The situation above is reflected in the growing disparity between gross and net output of agriculture, that is, in the growth of the share of material cost as a part of gross output. According to a Western calculation that excludes Albania and the Soviet Union, nonagricultural inputs, including depreciation (which indirectly takes account of investment, too), doubled from 1965 to 1976.[56] Therefore, net final product slightly decreased during 1965–70 and in 1976, and during 1970–75 increased by an annual average of only 2.3 percent. As the increase of inputs was smallest in Yugoslavia, this country's agriculture registered a marked growth of 2.6 percent of its final net product on average annually over the whole 1965–76 period, while that for the other countries remained below one percent.

A growing disparity of gross and net product characterizes any modernizing agrarian sector. Apart from capital cost, it is attributable to the increasing share of products requiring on-farm inputs (especially feed for animal production) and of crops of more refined quality, which make costs increase disproportionately. During such a development, increasing cost per labor unit may be compensated for by so great a decline of labor inputs that overall cost still would be reduced. Yet this is precisely what did not, or not to a sufficient degree, happen in Eastern Europe and the Soviet Union, except in the GDR.[57]

Almost all the five-year plans for 1976–80 provided for a reduction in the material cost of agricultural production—not an absolute reduction but one per unit of product. Thus, gross output was planned to grow more than material cost. This did not seem to have worked out at the time of writing. At least during the preceding ten years, the development was to the contrary, as is evidenced by the increasing share of material costs (Table 8.9). As distinct from the mid-1960s, material cost and depreciation in the 1970s accounted for more than half and up to two-thirds of the gross product. The two exceptions were Bulgaria (possibly due to the change of the price basis), and Czechoslovakia (where the share was over 60 percent already by 1965).

Soviet authors estimated that of agriculture's production expenditures, "materialized work" (material inputs) in the first half of the 1970s accounted for 67 percent in the GDR, 62 in Czechoslovakia, 52 in Bulgaria, 50 in the USSR, and 42 in Rumania, the remaining percentages being those for "live work." In some countries (e.g., the GDR and

Table 8.9   Material Cost and Depreciation as a Percentage of the Gross Product of Agriculture, 1965-1970 and 1975 (at constant prices; percentages at current prices
added in parentheses)

	1965	1970	1975
USSR	38.5 (38.5)	44.2 (39.2)	53.4 (50.1)[a]
Yugoslavia	.. (38.2)[b]	.. (52.5)	.. (52.4)[b]
Rumania	45.9 (..)	53.2 (..)	56.4 (..)
Bulgaria	39.0[c] (35.1)	50.7[c] (45.1)	46.3[c] (44.8)
Poland	.. (60.8)	62.3 (64.5)	69.4 (65.4)
Hungary	50.5 (53.0)	56.2 (55.7)	60.1 (59.2)
Czechoslovakia	65.2 (66.0)	64.6 (64.4)	67.3 (68.1)
GDR	43.0 (..)	50.8 (..)	62.5 (..)

[a] In the drought year of 1975, the gross product was unusually small, and therefore the
share of material cost was abnormally great in proportion; in the excellent harvest year
of 1973, the share (in current prices) was only 43.9.
[b] 1967 instead of 1970, and 1973 instead of 1975.
[c] For 1965 and 1970, in prices of 1962; for 1975, in prices of 1971.
Source: For Yugoslavia, derived from data in *Statisticki godisnjak, 1969*, p. 110, *1972*,
p. 108, *1975*, p. 138; all other data derived from the Table in *Economic Survey of Europe
in 1976, Part II* (New York, 1977), p. 33, which is based on official information from the
countries concerned.

Hungary), material inputs of nonagricultural origin exceeded those
produced by agriculture itself, and the trend in the other countries went
in the same direction.[58]

The surprisingly high share of material cost and depreciation in
Poland's predominantly peasant agriculture may be ascribed at least in
part to a great share of imported and other purchased feed and to high
(in comparison to agricultural producers' prices) costs of industrially
produced inputs bought from the state and/or an overestimate of fixed
capital (old buildings on peasant farms). For Yugoslavia in 1970 and
1973, the same sources also give separate figures for private and
socialized farms. They show that the share on peasant farms was almost
as great as in Poland, while on socialized farms, it was strikingly low (35.7
and 39.9 percent in respective years), probably because of the output
mix and the more favorable terms of trade accorded by the state.

East European and Soviet sources admit that the returns on agricultural investment after 1960 became lower than before. They have
correctly argued that the switch to large-scale farming, and the formation later on of enormous production units by way of "industrializing"
agriculture, required great investments which did not at once come to a
full productive bearing. Yet this does not fully explain the situation.
Rather, general problems of management and planning, and the fact
that modernization was in part premature, in view of the development
level of the economy at large, also played a negative role. Otherwise, the
great investments of the 1960s should by now have yielded much better

Table 8.10   Total Gross Investment in Agriculture and Forestry of CMEA Countries (value in million units of national currencies), as Given in CMEA Statistical Annuals

	1950 (1971 ed.)	1960		Annual average 1966-70 (1971 ed.)	1970		Annual average 1971-75 (1976 ed.)	Annual average 1976-78 (1979 ed.)
		1979 ed.	1971 ed.		1979 ed.	1971 ed.		
USSR	1,796	5,473	(5,695)	12,531	14,401	(20,261)c	20,152	25,133
Rumania	742	5,431	(5,431)	10,613	13,102	(9,782)	15,836	24,226
Bulgaria	40.3a	405a	(405)a	519a	559a	(560)a	68a	820a
Poland	5,068	14,007	(12,381)	29,097	37,141	(36,247)	53,867	105,405
Hungary	1,778	6,171a	(7,577)	13,031b	21,970a	(22,707)a	19,525a	30,009a
Czechoslovakia	1,597	9,754	(9,754)	9,500	9,757	(17,963)	13,204	17,084
GDR	421	1,898	(1,929)	3,820	4,281	(2,379)	4,615	5,066

a In current prices.
b For 1968-70, in current prices. The index numbers of the same source, as given in constant prices, reveal that in such prices, the 1966-70 average would be 10-11 percent lower.
c Including buildings for health care, social insurance, and sport.
Source: CMEA Statistical Annual, 1971, 1976, and 1979.

returns. In any case, the investments during and immediately after collectivization had to replace the former peasant farms' fixed capital stock, which could no longer be used in the new large-scale farms. To that degree, these investments did not create additional production capacities.

The aggregate sums of investments in monetary terms (Table 8.10) do not give more than a very crude approximation in countries where the exchange rates of the currencies are fixed by the governments and only to a very limited degree by real purchasing power. Price rises, unequal for the various commodities, occur in centrally planned economies as well as in the West, although not to the same extent. In the Yugoslav decentralized, and to a certain degree, liberalized, economy, inflation even exceeds that of most Western countries. Hungarian and Polish sources, too, admit a certain inflation, and generally in Communist countries, cost rises, hidden or open, for investment goods and labor are an often-mentioned phenomenon. Czechoslovakia, Hungary, and Poland have informed the FAO of the development of their prices for nonagricultural inputs, and these have been published in the form of indices in the FAO *Production Yearbook*. They show modest price rises of investment goods for agriculture of 2–6 percent (Czechoslovakia) and up to 30–50 percent (Hungary and Poland) during 1967–76. One may assume that they reflect the actual development of production costs only in part, especially since they do not comprise construction costs. CMEA data on investment in constant prices (except for Bulgaria and, since 1968, for Hungary), mirror the development over time in a way which is not wholly comparable among those countries, because the base years for the price data differ and are changed from time to time. In Table 8.10, the discrepancy of data for 1960 and 1970 (according to the 1971 and 1979 CMEA statistical annuals) reveals sizable investment price changes for Poland, USSR, Rumania, Czechoslovakia, and the GDR.

Albania published indices of agricultural investment only up to 1969. They are for the socialized sector only and do not account for peasant investment of precollectivization times. Thus, their rapid increase between 1950 and 1960 mainly reflects the expansion of the socialized sector. Their stagnation, and in part decline, between 1960 and 1966 was followed during 1967–69 (i.e., after the completion of collectivization) by a new rapid rise by 45 percent in 1969 as compared to 1966. The present writer has no information beyond 1969, but it seems that a great part of investment goes into land improvement (mainly irrigation), as during earlier years.[59]

The investment growth rates planned for 1976–80 were smaller than those of the early 1970s in the USSR, Poland, Hungary, and Czechoslovakia, but greater for Bulgaria. In the GDR, the planned growth rates were also greater, but this was due to the fact that the rates during

1971–75 were the lowest of the CMEA countries, even lower than the 1966–70 GDR rates. The slowdown in most CMEA countries is hardly due to deliberate moderation, rather to a general shortage of investment finances and capital goods. Compared to total investment in the economy, agriculture continued to fare well, with its share in the generally slowing investment growth keeping up or being reduced only slightly.

The Communist data are for gross, not net, investment. However, the simultaneous growth of the so-called basic production funds (roughly, fixed assets), including livestock herds, reveals approximately how much went into re-investment and how much was left as net investment. A Soviet calculation, based on the statistical yearbooks of the countries, yielded the data in Table 8.11.

The Soviet author explains the fluctuations of the ratio over time by unfinished capital construction and also by a greater or smaller volume of such construction started in the preceding period. He offers no explanation for the striking differences among countries, and does not mention possibly differing retirement rates. A Western source, based on official information from the countries concerned, shows somewhat different rates, among them one for Rumania which is similar to those for Poland, Czechoslovakia, and the USSR.[60]

In countries under a Communist regime, especially in those where agriculture is socialized, the state provides considerable means both for infrastructure and for the branches supplying inputs to agriculture and processing and distributing its output. Part of these capital inputs may be considered to benefit agriculture, although this share cannot be defined in quantitative terms. Some of such investments and current expenses are financed from public budgets in Western countries also, for exam-

---

Table 8.11   Growth of Fixed Assets as a Percentage of Gross Investment (five-year averages)

	1961-65	1966-70	1971-75
Soviet Union	49	41	61
Bulgaria	62	53	46
Poland	49	80	65
Hungary	56	62	73
Czechoslovakia	56	60	77
GDR	68	48	36

Source: M.E. Bukh, *Problemy effektivnosti sel'skogo khozyaistva v evropeiskikh stranakh SEV* (Moscow, 1977), p. 85.

ple, research, plant and animal selection, some veterinary services, soil conservation and improvement, and educational and cultural facilities in the countryside. On the other hand, some costs, which in the West would be part of public budgets, are financed by socialized agriculture, for example, some of the road construction, school building, and energy supply systems in Soviet kolkhozes. On the whole, the Communist state takes over more of agriculture's infrastructural burden than its Western counterpart. As a result, agriculture's investment shares look somewhat bigger in Communist countries than in the West, and also bigger if related to productive investment only (as was done in Table 8.11). Agriculture proper, plus forestry, received the shares of productive investment expressed in Table 8.12 as a percentage of total (including nonproductive) investment during 1966–78 (three-year averages, except for Yugoslavia).[61]

Apart from capital stock and investment, the availability of turnover capital is an important factor in the productive performance. There are no comparable official data on turnover capital in East European and Soviet farms, and those available are less comprehensive than those on capital stock and investment. Various Communist authors complain that turnover capital is too small in proportion to fixed capital stock, on collective farms more so than on state farms. Yet seemingly, what data are available do not confirm this opinion. Thus, a Soviet book, basing the statement on a Hungarian and a Bulgarian estimate, indicates an average of 30–40 percent of the value of fixed assets as the volume of agriculture's turnover capital for the whole of the CMEA countries.[62] For Soviet kolkhozes in 1973, a percentage of 50.8 (apparently including the wage funds) was indicated, and the authors pointed out that a higher

---

Table 8.12   Agricultural Share of Productive Investment (percent of total)

	1966-68	1972-74	1976-78
USSR	16.8	20.2	20.4
Yugoslavia	7.2 (1970)	10.4 (1974)	. . .
Rumania	15.8	14.3	14.1
Bulgaria	17.8	16.2	13.9
Poland	15.9	13.9	16.3
Hungary	16.0	15.0	15.3
Czechoslovakia	12.4	11.4	11.9
GDR	14.3	12.4	10.7

Source: For all countries except Yugoslavia, *CMEA statistical annual, 1972,* pp. 136, 140, *1976,* pp. 139, 143, and *1979,* pp. 169, 173. The figures for all countries except Yugoslavia are not strictly comparable for the three time periods because of changed prices. The data for Yugoslavia, given in a Polish source (*Rocznik statystyczny rolnictwa i gospodarka zywnosciowej, 1978* (Warsaw, 1978), p. 507), are in constant prices.

percentage was desirable. For the whole of Soviet agriculture, they indicated 51 billion rubles of "material turnover funds" (excluding wages), and this must have been equal to 37 percent of the total "basic funds" in agriculture as of 1 January 1974.[63] For 1966, two American authors derived the share of agriculture's current purchases from all sectors, including intrasectoral, at a sum amounting to 42 percent of agricultural capital stock, including productive livestock.[64] Current purchases of a year, of course, exceed available financial means at a given point in time.

In Yugoslavia, socialized enterprises have for a long time generally been suffering from inadequate financial liquidity, in agriculture still more so than in the rest of the economy. Availability of turnover capital of a value of fifteen days' output in liquid money was considered necessary for normal farm activities, but in 1972 only four days' worth was available to socialized farms.[65] The great investments, and the concomitant need for greater current means for their utilization, have contributed to shortages of turnover capital.

A Soviet author criticized that kolkhozes often use short-term credits for financing investment, and thereby aggravate their shortages of turnover capital.[66] Kolkhozes have assigned a growing share of their net income to capital stock increases, and declining shares to turnover capital. This has become possible because they now receive more state credits than formerly. About one-fourth of their investment and one-third of their current production cost are financed through credits from the state.[67] In absolute terms, their assignments to turnover capital decreased by more than one-third during 1970–74.[68] Such a behavior is not irrational in view of the State Bank's low interest rates, but it puts an additional burden on the state budget.

Under these circumstances, the complaints about shortage of turnover capital seem understandable, although the available turnover capital by Western standards is sizable in comparison to capital stock and to the wage funds. In addition, the potential productive effect of turnover capital is jeopardized by its slow rate of turnover. Such inputs as fertilizer, plant protection chemicals, spare parts, etc. generally are in short supply and therefore tend to be bought as reserve before they are really needed. Or else the farms do not buy exactly what they need, but what is to be had at a given moment and which then cannot be put to optimal use. Such behavior freezes part of the financial means, or makes material current inputs less efficient.

International comparisons of capital inputs in agriculture are even more problematical than comparisons of labor inputs. The relation of agricultural to industrial and other producer prices, the state subsidies, and the differing demands for and shortages of the various material means of production prevent monetary capital data from telling the whole story. However, East and West German agriculture present fewer difficulties in this respect, and a comparison will be risked at the end of the present chapter in spite of remaining uncertainties.

The proportion of agriculture's share in total investment or in fixed capital and its contribution to a country's material net product, as shown in Chapter 7, yields a meaningful picture only in relative terms. It demonstrates that quite generally, the capital supply for East European and Soviet agriculture has very much increased in recent times, but this knowledge does not help to assess and to compare internationally the capital inputs per worker, or per land unit, or per unit of output.

More comparable than monetary data are the quantities of physical units of capital goods in agriculture. Some of them are summarized in Table 8.13. Yet a number of qualifications have to be added to each of these data.

Concerning tractors, a relative saturation point has recently been reached in some of the East European countries, where therefore tractors no longer characterize the overall availability of capital. The demand shifts to other production means. Yet the data for other machines which have come into greater demand usually are less easily comparable, and because of the increasing specialization, such diverse machinery has gained more importance.

The average horsepower per tractor has increased everywhere, except for Yugoslavia. Measured in motor h.p., it ranged between 37 in Poland and 60–70 h.p. in the rest of the CMEA countries in 1978. This was, on the average, about one-third more than ten years earlier. The data for Yugoslavia were not quite 40 h.p. in 1970, against slightly more than 40 in 1960. The stagnation is probably due to a growing number of tractors on small peasant farms. It is envisaged that by 1985, the average horsepower per tractor in CMEA countries will have increased by 40–80 percent, to 84 h.p. in Bulgaria and to more than 100 in Hungary, Czechoslovakia, the GDR, and USSR.[69] Similarly, the greater or unchanged grain area per harvester in 1978 in some countries (Rumania, Czechoslovakia, GDR) is to be explained by the increased harvesting capacity per unit. On Albanian socialist farms, part of the grain is still mown by manual workers.[70]

In a worldwide comparison, Adolf Weber has demonstrated that "the quantity of mechanical technologies can be measured with the concept of tractorization and electrification, because power is the common denominator for using them."[71] Yet in view of the statistical uncertainties, far-reaching conclusions shall not be drawn here from the data on tractor horse-power and energy consumption per worker. Instead, a Soviet author shall be quoted who for the Soviet Union considers 35–40 h.p. availability of total energy (apparently including stationary electrical motors) per agricultural worker the optimum to be attained. O. Doleshel

Table 8.13   Available Grain Harvesters per 100 Hectares of Grain Sowings; and Tractor Power, Annual Supply of Fertilizer, and Consumption of Electric Energy in Agriculture per 100 Hectares of Compound Land Units[a] and per Average Annual Worker

	Mineral fertilizer supply per 100 hectares[a] (tons of effective nutrient)			Tractor power (total motor h.p.) Per 100 hectares[a]			Per worker[b]	Hectares of grain sowings per one grain combine harvester (physical unit)			Consumption of electric energy, thsd kWh, in agriculture (including non-productive uses) Per 100 hectares[a]			Per worker[b]
	1960	1970	1978	1960	1970	1978	1974	1960	1970	1978	1960	1970	1978	1974
USSR	0.9	3.6	6.2	17	38	61	5.2+	232e	191	184	3.5	13.2	32.4	2350+
Albania[c]	1.7	7.5	8.4 (1976)	15d	27d	33d (1973)	...	702e	300e	291e (1973)	...	...	...	...
Yugoslavia[c]	3.3	7.5	10.0 (1977)	16	36	117 (1976)	0.5+ (1969)	1827	419	412 (1977)	0.5f	2.0f	7.3f (1974)	26+f (1969)
Rumania	0.7	6.4	13.0	16	57	76	1.8+	409	134	136	9.3g	6.5g	22.8	500-g
Bulgaria	3.4	14.5	16.3	20	49	85	2.45	372	245	210	3.6g	15.3g	21.8	770g
Poland	4.5	15.1	23.0	12	49	123	2.5	3087	617	271	3.7	9.1	27.5	480
Hungary	2.8	14.3	27.2	23	54	69	3.1-	857	263	208	1.4g	14.8g	38.9g	1400-g
Czechoslovakia	8.8	21.6	31.3	40	102	150	6.9-	412	163	152	7.9g	27.7g	49.2	2110g
GDR	17.9	30.0	31.2	43	126	158	7.7	377	131	197	25.4g	36.6g	70.0	3330g

a One hectare of land under arable and perennial crops taken at unity; one hectare of other agricultural land (mainly natural meadows and pasture) at one-fifth of its physical area. E.g., 100 hectares of compound land units may consist of 60 hectares of arable plus 200 hectares of other agricultural land, or of 85 plus 75, etc.

b The plus (+) and the minus (-) signs behind the data indicate that in comparison to other countries, the number of workers probably is overstated and therefore the tractor power and energy consumption per worker understated (a minus sign) or, vice versa, overstated (a plus sign).

c For Albania, cultivated area (arable and perennial crops) instead of compound land units; for Yugoslavia, agricultural and arable land actually cultivated.

d h.p. at drawbar, that is, for motor h.p., it should be roughly doubled.

e Grain for human consumption, which accounts for 85-90 percent of total grain sowings.

f Only agricultural machinery and melioration works.

g Including consumption in forestry, but not including forestry workers; for Hungary in 1977: including forestry and water management.

Sources: Number of workers, according to O. Doleshel, *op. cit.*, p. 531 (where it should read 1974 instead of 1970, obviously because of a misprint); for Albania, *30 Viet Shqiperi Socialiste*, pp. 105, 109, 129, 133 (mineral fertilizer in 1976 according to *Landerkurzbericht: Albanien 1978*, Statistisches Bundesamt Wiesbaden [Stuttgart, Mainz, 1978], p. 13); for Yugoslavia, number of workers according to 1969 census, and other data from *Statisticki godisnajak, 1970*, p. 172; *1974*, p. 173; *1977*, p. 275; *Jugoslavija 1945-1964. Statisticki prikaz* (Belgrade, 1965), pp. 99, 110, 111, 119; *Statisticki bilten*, no. 434, p. 58; no. 439, p. 13; no. 692, p. 7; no. 988, p. 8; no. 1049, pp. 13, 50; no. 1065, p. 11. All other figures are derived from data in *CMEA Statistical Annual, 1979*.

more generally speaks even of 50–55 h.p. for the future.[72] (Note that for electrical energy, these are data in terms of h.p. and not of kWh, as in Table 8.13.) The actual availability in the mid-1960s was 2–5 h.p. in Bulgaria, Hungary, Poland, and Rumania, and 8–10 h.p. in the GDR, Czechoslovakia, and the Soviet Union. By the first half of the following decade, these indicators rose to 3–6.5 h.p. in the first group of countries and to 12–15 in the second, still far below the optimum.[73] At a Hungarian conference in 1977, it was pointed out that "a number of important operations—besides harvesting—cannot be carried out at an optimum time for lack of machine capacity (e.g., soil preparation, sowing). This may also cause a failure of crop and reduce the efficiency of other factors (e.g., fertilizer) as well."[74]

The application of mineral fertilizer in the GDR and Czechoslovakia seems to approach a degree of saturation, where diminishing returns on additional inputs may be expected. All the same, the GDR planned for 1976–80 to increase it further, while Czechoslovakia already achieved her stated goal for 1980. Generally, state farms have been more generously supplied with fertilizer in Eastern Europe and the Soviet Union, so that the reserves of yield increases through more fertilizer application seem to be greater in the collective farm and kolkhoz sector. It must be noted though, that the low Soviet doses of fertilizer are of about the same magnitude as those of Canada, a Western country with roughly comparable climatic conditions. For the Canadian farmer, the cost/price ratio is an important aspect in determining how much fertilizer to apply, whereas for the Soviet farm head, the goal of output increase is paramount and makes heavier fertilizing desirable for him—or for the planners. One has to add that the available data refer to supply, not actual application of fertilizer, and that a deduction of 10–15 percent for losses and inadequate handling is rather a conservative estimate, at least for the less developed socialist countries.[75] The Albanian plan target of applying 128,600 tons of fertilizer by 1980, or roughly 16 tons per 100 hectares, is not likely to be achieved.[76]

Consumption of electrical energy in agriculture is an indicator that becomes more meaningful the higher the level of technical modernization. Electrical energy supply is to a great degree determined by stationary machinery, especially in on-farm transport, repair, threshing and, above all, modern large-scale animal production, which in all countries concerned has not yet attained saturation. It also plays a role in irrigation (pumps, sprinklers, etc.) and nonagricultural activities of farms. In electric energy consumption, the Soviet Union and Bulgaria ranked unexpectedly high, comparatively, up to the 1970s. In Bulgaria, the great relative share of irrigated land probably accounts for a sizable part of the energy consumption, which has become relatively low by CMEA standards. The GDR by this indicator clearly comes out with the technically most advanced agriculture. Poland, in spite of her small peasant farms, ranks higher in electrical energy consumption than does Rumania. Rumania and Poland had the highest rates of growth in electrical consumption between 1970 and 1978, while Bulgaria had by far the lowest. Yugoslav electric consumption lags far behind the rest, but less so in mineral fertilizer supply and tractor power. Albania, which ranks lowest (although her indicators are per cultivated area, i.e., overstated in comparison), planned to increase tractor power by 20 percent and fertilizer supply by 65 percent during 1976–80, which implies that she will remain in the same relative position.[77]

Electrical power consumption per agricultural worker in Communist Europe, even in the GDR, by 1970 was still far below the 1967/69 Western standards of 4000 and more kWh (about 9,000 in USA).[78] Yet since then, the gap, in terms of electrical power consumption per worker, has narrowed somewhat, as power consumption has risen and the workforce has fallen in the Communist countries. With more workers per land unit in Eastern Europe (excluding the Soviet Union) than in the West, the gap in terms of electric power consumption per land unit was not quite as great in any year. The picture was similar for tractor horsepower in 1969, although for the reasons indicated above, this indicator may have lost some of its significance since then.

The data of Table 8.13 give a picture which is neither complete (because of the neglect of other capital inputs) nor an exact mirror of reality (because of its partial coverage). Machines, fertilizer, and electrical energy may be put to a more or a less efficient use. The level of efficiency achieved depends on the ability of workers and managers to handle the inputs, and also on factors like the quality of the machines, the availability of repair facilities, of special machinery for large-scale application of fertilizer, etc. In 1978 and 1979, official appeals for a less wasteful energy consumption became frequent, especially with regard to agriculture. Another important aspect is the shortage of building materials which is felt almost everywhere. It not only delays construction and in part makes it more expensive, but also impairs adequate maintenance of machines and storage of fertilizer and other chemical means. Such factors cannot be quantified, but must have negative effects on capital utilization.

For decades, lack of spare parts for agricultural machinery has been a recurrent complaint in the CMEA countries. Apparently, this problem affects Albania as well,[79] but less so Yugoslavia. Shortage of spares naturally reduces the productivity of the available machines. This phenomenon reveals what quite generally characterizes socialist farming (with reference to machinery, construction, and supply of fertilizer and other chemicals and, in part, of electrical energy): it is not only lack of financial means of the farms which impedes capital inputs, but also the inadequate capacity of the state-owned industry supplying these means. As was stated by an East German author: "To a considerable degree, the industry supplying the means of production becomes a factor which determines the level and development rates of agricultural production."[80]

From a comparison with Western industrialized countries, low efficiency of capital inputs seems to emerge for East European and Soviet agriculture. Thus, the capital channelled annually into West German agriculture in recent times amounted to only 3–4 percent of total gross investment in the economy, and in USA this percentage was 5–6 (compared with percentages in the CMEA countries of 10–20, see Table 7.5). In both Western countries, agriculture contributed a share to the material net domestic product that was roughly equal to the share it received of investment.[81] Still, the decline in the agricultural workforce was faster than in most East European countries and in the Soviet Union, and gross agricultural output grew, although less than in the East. During 1976–78, agriculture's contribution to net material product was less than its share of investment received in the Soviet Union, while the former exceeded the latter in the other CMEA countries. Yet such an East-West comparison loses part, although not all, of its meaning if one takes into account that in both Western countries, total investment in the economy—to which the percentage for agriculture refers—was much greater, but not so the agricultural area. Thus, per land unit and per worker, agricultural investment in the West was not so fundamentally different as the percentage shares seem to imply.

East and West German agriculture lend themselves to a comparison more than other countries, although here, too, the resulting figures should be taken as orders of magnitude rather than as exact indicators.[82] Assuming that by the mid-1970s the East and the West German marks were of about equal value in the production sphere,[83] agricultural investment per compound land unit (see footnote 48, preceding) was practically identical in 1974: 744 marks (in the GDR, including forestry, constant prices of 1967), as against 707 marks (Federal Republic, current prices). Almost the same applies to agricultural investment per agricultural worker, if one equates the East German labor numbers (*Berufstätige,* as of September 30) minus 5 percent (to account for seasonal overemployment) to the full-time equivalents (*Arbeitskräfte-Einheiten*) of West German statistics: 5,698 marks per worker in the GDR (including forestry), and 4,793 marks in the Federal Republic.

Things look different, though, for fixed assets. Compared to East German "basic funds" (*Grundmittel,* in prices of 1966, including forestry), they were higher in West Germany by 89 percent per land unit, and by 67 percent per worker (*Brutto-Anlagevermögen,* in prices of 1962, including forestry and fishery). However, if one converts the fixed assets into current capital cost by imputing on them an interest charge of, say, 8 percent p.a., the similarity appears again. Then the current capital cost (gross investment plus interest on fixed assets) in 1974 was 1,301 marks per land unit and 9,962 per worker in the GDR, as against 1,757 and 11,915 marks in the FRG. Most of the remaining cost differential must be seen against the higher labor incomes it generates in modern Western agriculture. It turns out that in West German agriculture, previous investment (as consolidated in the fixed assets) required great outlays for reinvestment (as expressed in amortization), and left only 6 percent of gross for net investment. In contrast, the growth of East German "basic funds" in 1974 over those of 1973, which may be roughly taken for net investment, made up 38 percent of gross investment.

East German agriculture, as shown, is the most capital intensive among all the Communist countries, and a comparison between West Germany and any of the others would naturally be less favorable. But then, agriculture in Italy and Spain, even in France, also is less capital-intensive than that of West Germany.

In Communist European agriculture, relatively high rates of investment (compared with the West) represent an effort to make up for previous underinvestment and/or the effect of socialization, which made worthless most of the earlier peasant investment. The resulting capital cost of modern giant-scale agriculture is excessive only in proportion to the weakness of the overall nonagricultural economy, and makes the percentage accorded to agriculture look so large. As Adolf Weber termed it:

> Nations that create big farms according to a timetable must pay the price of high agricultural investment levels for several decades. Countries with a relatively small industrial base do not have the investment capacity to mechanize agriculture. The establishment of large farms requires heavy investments that can be obtained only from other, more productive branches of the economy. To reach this goal in two decades can mean only a pyrrhic victory. . . .[84]

One has to add that obviously the low labor productivity that prevails in all branches of Communist economic systems, not only in agriculture, aggravates the consequences of their costly tribute to ideology and power. Even the GDR, being the most industrialized country in Communist Europe, feels agriculture's capital needs as a heavy burden on the rest of the economy.[85] However, the GDR is in a better position to supply that capital than the other countries of the area, except Czechoslovakia. For the others, the percentage share of total investment as an indicator of capital supply is even more misleading than for the GDR.

It has been shown (Chapter 7) that a transfer of capital out of agriculture took place until the early 1960s (with the short-term exceptions of Bulgaria and Czechoslovakia around 1960), and that by 1970, in Central East Europe and the Soviet Union this was no longer the case. In part, the direction of transfer was even reversed. As to the special case of Bulgaria, it strikes this observer that there were no signs of capital transfer in favor of agriculture, except if one makes allowance for a high value of land as an additional part of fixed agricultural assets. One would have expected large capital injections into agriculture (and a greater increase of fertilizer supplies) in connection with the creation of huge agro-industrial complexes since 1969 (see chapter 12). Yet Bulgarian agriculture's share in fixed capital (neglecting land value) and in gross investments has in fact been somewhat less than its contribution to material net product in recent times (up to and including 1978). Bulgarian agro-industrial complexes in practice seem to have been more a form of organization for stricter employment of the still inadequately mobilized labor force than a vehicle for accelerated technical modernization of agricultural production processes. Perhaps this is why in Bulgaria, from 1966–68 up to 1972–74 (three-year-averages), the increase in the gross as well as in the net agricultural product has been less than in all other East European countries.

On the whole, the picture is still one of relatively small productive effects of agricultural investment in Eastern Europe and the Soviet Union. However, the possibility should not be excluded that the greatly increased investment of the past decade will come to bear in the future, then to result in higher capital productivity.

Notes

1. *Ekonomicheskaya gazeta*, 19 January 1963.
2. Gregor Lazarcik, "Agricultural Output and Productivity in Eastern Europe, and Some Comparisons with the U.S.S.R. and U.S.A.," *Reorientation and Commercial Relations of the Economies of Eastern Europe*, Joint Economic Committee, Congress of the United States (Washington, D.C., 1974), pp. 335, 346.
3. Plan data for 1976–80 in *Probleme ekonomike*, no. 1 (1977), pp. 123–124 (according to *ABSEES*, no. 53 [September 1977], p. 26).
4. *Zëri i popullit*, 22 February 1978, 21 February 1979 (according to *ABSEES*, no. 56 [September 1978], p. 21 and no. 59 [September 1979], p. 23).
5. See *Economic Survey of Europe in 1977, Part I* (New York, 1978), pp. 71, 83, 87.
6. *Külkereskedelmi statisztikai évkönyv, 1978* (Budapest, 1979), p. 35; for 1965, B. Kádár, *Acta oeconomica* (Budapest), no. 1/2 (1978), p. 150 (Table).
7. T. Thomaj, *Probleme ekonomike*, no. 2 (1977), pp. 15–29 (according to *ABSEES*, no. 54 [January 1978], p. 16.
8. I. F. Suslov, *Agrarnii sektor ekonomiki stran sotsializma* (Moscow, 1978), p. 50.
9. *Internationale Zeitschrift der Landwirtschaft*, no. 2 (1973), pp. 218–233.
10. *Bashkimi*, 24 February 1974, p. 2 (according to *ABSEES* 5:3 [July 1974], p. 114), and *Economic Survey of Europe in 1976, part II* (New York, 1977), p. 42.
11. *Zëri i popullit*, 25 May 1975, p. 2 (according to *ABSEES* 4:2 [October 1973], p. 123). See *Zëri i popullit*, 26 April 1978, p. 3, and 10 October 1978, p. 2, on investment plans for irrigation and amelioration not being fulfilled (according to *ABSEES*, no. 56 [September 1978], p. 21, and no. 57 [January 1979], p. 22).

12. Not strictly comparable, though, because a minor area of irrigated meadows and pasture is included.

13. *Economic Survey of Europe in 1976, op. cit.,* pp. 34–35, and *1978, Part I,* p. 97, Table.

14. *Statistički bilten,* no. 434, p. 60, and no. 1049, p. 51.

15. R. M. Gumerov, *op. cit.,* p. 117.

16. Sixty to 65 percent already in the early 1960s, according to W. Gurtschenko and A. Nagy, *Internationale Zeitschrift der Landwirtschaft,* no. 4 (1974), p. 378.

17. Even before the American part-embargo on grain, in his speech of 4 July 1978, Brezhnev pointed out the need to utilize more fully the potential of non-grain feed.

18. Communist statistics count gross crop and livestock production on farms separately and then add them up, thereby accounting for feed twice. If the shares are 50 percent each, this does not result in a distortion of their weights, because both sectors profit equally from the double-counting. Otherwise, the smaller sector profits more in the resulting statistical proportion. For the special Yugoslav case, see note e, Table 8.7. G. Lazarcik ("Agricultural Output," *op. cit.,* p. 341) arrived at higher percentages for the livestock sectors during 1968–72, namely from 43.0 (Bulgaria) up to 71.1 percent (GDR). Taking into account the final products only, Lazarcik (p. 334) apparently did not count the feed used in animal production as output of the crop sector, and thus followed the practice of most Western statistics. Even so, the share is lower in Eastern Europe and the USSR. The data and conclusions in the present work are based on CMEA statistics, that is, on the Communist methods of calculation.

19. *Rocznik statystyczny rolnictwa i gospodarka żywnościowej 1978* (Warsaw, 1978), p. 501.

20. *Bashkimi,* 4 May 1974, pp. 1–2 (according to ABSEES 5:4 [October 1974], p. 119).

21. *Ekonomika sel'skogo khozyaistva,* no. 5 (1972), pp. 36–42.

22. *Economic Survey of Europe in 1978, Part I* (New York, 1979), p. 105.

23. V. I. Storozhev, *Ekonomicheskaya reforma v sel'skom khozyaistve sotsialisticheskikh stran,* E. V. Rudakov, S. A. Mellin, V. I. Storozhev, eds. (Moscow 1968), p. 21. The figures are as stated in the source.

24. Z. Grochowski, *Życie gospodarcze,* no. 9 (1979), p. 6.

25. F. Chapurin and A. Terekhov, *Ekonomika sel'skogo khozyaistva,* no. 4 (1975), p. 40, and *ibid.,* no. 9 (1975), p. 5.

26. L. P. Zlomanov, *Mezhdunarodnye ekonomicheskie sopostavlenia* (Moscow, 1971), p. 115.

27. For Albania, author's own crude estimate; for the other countries, R. J. Fuchs, G. J. Demko, *The ACES Bulletin,* no. 1 (1977), p. 22.

28. *Gledišta,* no. 11 (1972), pp. 1388–99 (according to *ABSEES* 3:4 [April 1973], p. 259).

29. Michael Cernea, *Sociologia Ruralis,* no. 2/3 (1978), p. 116.

30. *Statistisches Jahrbuch 1979 der DDR,* pp. 85, 87, 99, 101; *Statistical Yearbook, 1977* (Budapest, 1979), p. 120.

31. I. F. Suslov, *op. cit.,* p. 110.

32. Stephen Rapawy, *Estimates and projections of the labor force and civilian employment in the U.S.S.R. 1950 to 1990,* U.S. Department of Commerce, Bureau of Economic Analysis, FER no. 10, Washington, D.C. (September 1976). For the official Soviet estimates, see *Narodnoe khozyaistvo SSSR, 1960,* p. 521, *1970,* p. 404, and *1974,* p. 448 (in each case, the bottom footnote). For a penetrating discussion of the categories and methods of Soviet labor statistics and their not being "false", but misleading in some respects, see Philipp Grossman, *Soviet Studies* 14:3 (January 1968), pp. 398–404.

33. G. I. Shmelev and V. N. Starodubrovskaya, *op. cit.,* pp. 196–198.

34. M. E. Bukh, *Problemy, op. cit.,* p. 49.

35. János Timár, *Die Erwerbstätigkeit in Ungarn* (Forschungsberichte, no. 50, Wiener Institut für Internationale Wirtschaftsvergleiche), Vienna, January 1979, pp. 18–19.

36. In the table, most of the CMEA data and those of the Czech author Doleshel are annual averages and are closer to a recalculation on a full-time basis than those of the FAO, but include forestry. In view of seasonal underemployment existing in part of the collective farms, they are still greater than a full-employment recalculation would make them appear. As to the Yugoslav data, a Polish official source (*Rocznik Statystyczny, 1976,* p. 554) apparently tried to make them comparable to those of the CMEA statistics, thereby making them deviate by a wide margin from the FAO as well as from the national Yugoslav data. For Poland herself, the percentages in the Polish source are very similar to those of the FAO and seem to comprise also those people who are not fully employed in agriculture.

The CMEA, as well as the Doleshel, percentages for Poland are considerably lower. The data for Albania, as estimated by the present writer for comparability with CMEA data, leave room for a wide margin of error.

37. For the underlying absolute figures, see O. Doleshel, *Internationale Zeitschrift der Landwirtschaft*, no. 5 (1976), p. 531, where the year 1970 is a misprint for 1974.

38. *Ibid.*

39. *Rocznik statystyczny rolnictwa i gospodarka żywnościowej 1978* (Warsaw, 1978), p. 28.

40. M. E. Bukh, *Problemy, op. cit.*, p. 134.

41. *Statistical Yearbook, 1977* (Budapest, 1979), p. 34.

42. *Zemĕdĕlské noviny*, 4 July 1974, p. 3 (according to *ABSEES* 5:4 [October 1974], p. 155).

43. *Rruga e partisë*, no. 9 (1972), pp. 53–67 (according to *ABSEES* 6:4 [October 1975], p. 138).

44. *Bashkimi*, 22 August 1975 (according to *ABSEES* 7:1 [January 1976], p. 134.).

45. *Zëri i popullit*, 7 September 1975, p. 2, and *Bashkimi*, 23 January 1975, p. 2 (according to *ABSEES* 7:1 [January 1976], p. 126, and 6:2 [April 1975], p. 144).

46. *Zëri i popullit*, 1 June 1978, p. 2 (according to *ABSEES*, no. 56 [September 1978], p. 24).

47. For example, still in 1979, see *Zëri i popullit*, 12 April 1979, p. 2 (according to *ABSEES*, no. 59 [September 1979], p. 24).

48. By *compound unit*, a hypothetical land unit is understood, where arable and perennial crop area is taken at unity, and other areas (because of their lower productivity) at one-fifth of their physical extent.

49. Except for Yugoslavia, all underlying data are for 1974. "More than" or "less than" means that the data, as taken from O. Doleshel, *op. cit.*, and from the Yugoslav census, seem to contain an overstatement or an understatement of the agricultural labor force as compared to that of the other countries.

50. K. Fazilija, *Economia*, no. 1 (1975) (according to *ABSEES* 6:4 [October 1975], p. 251).

51. *Jugoslovenski pregled*, no. 2 (1973), pp. 39–40.

52. M. E. Bukh, *Problemy, op. cit.*, pp. 136–138.

53. The Bulgarian source (*Ikonomicheska misul*, no. 4 (1975), p. 90) as quoted by G. I. Shmelev and V. N. Starodubrovskaya, *op. cit.*, p. 136; cf. Yu. Shintyapin and B. Frumkin, *Voprosy ekonomiki*, no. 2 (1975), p. 85, and G. Bychkov, *Voprosy ekonomiki*, no. 1 (1976), p. 105. M. E. Bukh (*Problemy, op. cit.*, p. 25) gave somewhat lower percentages for Bulgaria and Hungary.

54. This is admitted also in Communist publications, e.g., by M. E. Bukh, *Problemy, op. cit.*, pp. 87–89, and see especially pp. 89–90, n. 40.

55. *Agrarno-promyshlennaya integratsia stran SEV* (Moscow, 1976), pp. 17–18.

56. G. Lazarcik, "Comparative Growth and Levels of Agricultural Output and Productivity in Eastern Europe, 1965–76," *East European Economies Post-Helsinki*, Joint Economic Committee, U.S. Congress (Washington, D.C., 1977), p. 295.

57. *Kooperation* (East Berlin), no. 3 (1976), p. 140.

58. M. E. Bukh, *Problemy, op. cit.*, pp. 81–82.

59. No attempt was made to fit these incomplete Albanian data into Table 8.12. These statements are derived from the official Albanian statistical yearbooks for 1966, 1967/68, and 1969/70.

60. *Economic Survey of Europe in 1976, Part II*, pp. 40–42.

61. For all countries except Yugoslavia, *CMEA statistical annual, 1972*, pp. 136, 140, *1976*, pp. 139, 143, and *1979*, pp. 169, 173. The figures for all countries except Yugoslavia are not strictly comparable for the three time periods because of changed prices. The data for Yugoslavia, given in a Polish source (*Rocznik statystyczny rolnictwa i gospodarka żywnościowej, 1978* [Warsaw, 1978], p. 507) are in constant prices.

62. M. E. Bukh, *Problemy, op. cit.*, p. 86.

63. V. P. Efimov and V. I. Manyakin, *Effektivnost' sel'skogo proizvodstva v SSSR* (Moscow, 1977), pp. 117–118, 126.

64. G. B. Diamond and C. B. Krueger, "Recent Developments in Output and Productivity in Soviet Agriculture," *Soviet Economic Perspectives for the Seventies*, Joint Economic Committee, U.S. Congress (Washington, D.C., 1973), pp. 330, 332.

65. I. Lončarević and D. J. I. Matko, "The Organization of Yugoslav Agriculture,"

forthcoming in vol. II (*The Organization of Agriculture in the Soviet Union and Eastern Europe,* ed. by E. M. Jacobs) of the present series.

66. M. P. Kazakov, *Finansy SSSR,* no. 9 (1977), pp. 41–46. Similar statements are not infrequent in Soviet publications.

67. I. F. Suslov, *op. cit.,* p. 196.

68. R. M. Gumerov, *op. cit.,* pp. 239, 243, 244.

69. *Agrarno-promyshlennaya integratsia stran SEV* (Moscow, 1976), p. 67.

70. *Zëri i popullit,* 14 June 1978, p. 1 (according to *ABSEES,* no. 57 [January 1979], p. 22).

71. *Contributed Papers Read at the 15th International Conference of Agricultural Economists* (Supplement to *International Journal of Agrarian Affairs,* Double Number, 1974–1975), Oxford, 1975, p. 220.

72. G. Bychkov, *Voprosy ekonomiki,* no. 1 (1976), p. 105, and O. Doleshel, *op. cit.,* p. 532.

73. M. E. Bukh, *Problemy, op. cit.,* p. 27.

74. *IIIrd National Conference on Agricultural Economics,* Research Institute for Agricultural Economics, Bulletin no. 42 (Budapest, 1978), p. 31.

75. Application data for Estonia, one of the most developed parts of the USSR, in 1969 imply a disparity towards supply of roughly 25 percent; see E. Vint, *Intensiivse pollumajanduse majanduslik efektivsus Eesti NSV-s* (Tallinn, 1971), p. 200. For Hungary, as late as in 1978, losses of 15–20 percent for unsatisfactory storing and handling were mentioned as being typical; see J. Váncsa, *Társadalmi Szemle,* no. 3 (1978), pp. 24–31 (according to *Hungarian Agricultural Review,* no. 4 [1978], p. 33).

76. T. Thomaj, *Probleme ekonomike,* no. 2, 1977, pp. 15–29 (according to *ABSEES,* no. 54 [January 1978], p. 16).

77. *Economic Survey of Europe in 1976, Part II,* p. 42.

78. See the graph of A. Weber in *Contributed Papers, op. cit.,* p. 219.

79. The production plans for spare parts often are not fulfilled (*Zëri i popullit,* 4 March 1977, p. 3), and agricultural machinery is not being used to full capacity (*Zëri i popullit,* 6 September 1977) (according to *ABSEES,* no. 53 [September 1977], p. 25, and no. 54 [January 1978], p. 15, respectively).

80. G. Reimann, in *Integratsia sel'skogo khozyaistva s drugimi otraslyami narodnogo khozyaistva v stranakh-chlenakh SEV* (hectographed) (Budapest, 1976), p. 102.

81. Since in CMEA statistics only material production is part of the net product, one has to make a rough adjustment to that extent for Western countries. The contribution to gross (including nonmaterial) production by 1977 was 2.8 percent in both the USA and West Germany.

82. The data are from *Statistisches Jahrbuch 1976 der DDR,* pp. 41, 44, 49, 173, and *Statistisches Jahrbuch 1976 für die Bundesrepublik Deutschland,* pp. 199, 209, 311, 528.

83. A. Weber, *Studies in Comparative Communist* 6:3 (Autumn 1973), p. 290.

84. *Ibid.,* p. 295.

85. See the quotations from East German publications in *ibid.,* p. 291.

9. Labor Incomes in Agriculture

ONE CLEARLY FINDS a common tendency and economic reasoning behind the development and reforms of the remuneration system in socialist agriculture in Eastern Europe and the Soviet Union. The common policy goal in these countries, even including Hungary, since the mid-1960s has been to raise agricultural labor incomes and, at the same time, to ensure a relationship between incomes and productive performance.

This is not to say that today everything is being ordered from Moscow, as under Stalin. Some developments and proposals apparently have originated in other CMEA countries (e.g., social insurance, or guaranteed minimum wages for collective farmers), and seemingly are merely being slowed down or coordinated by Moscow. It is a number of basic exigencies held in common that make for a certain apparent conformity, with minor national variations.[1]

With agricultural output lagging behind rising urban consumer demand, and with agricultural labor surpluses dwindling, it became imperative as an incentive to close the income gap between agriculture and the rest of the economy. This was less a question of a systemic reform of labor remuneration than of the necessarily changing priorities of economic and social policies. In this way, economic factors have been at work independent of the will of political leaders, much as in the West. Apart from ideology, however, the one fundamental difference is that in Western industrialized countries, economics has been working against agriculture, but the politicians, often prompted by a more or less articulate "farm lobby," have kept farm incomes above the equilibrium level. Under Communist regimes, farmers wield no such power. It is the economics of unbalanced growth and the resulting unsatisfactory productivity of agriculture (as developed under Communism with its strict political control) which nowadays tend to raise farm incomes. Although they are a most important factor, higher wages of course are not the only means of diminishing the outflow of skilled labor from agriculture.

The remuneration of labor in socialized agriculture in Eastern Europe and the Soviet Union has changed considerably from what it was before the 1960s. As for state farms, their wages and salaries have always been modelled after those of nonagricultural state enterprises, although at a generally lower level. In 1955, the overall average was 40 percent below the level of industrial wages in the USSR, and 20–30 percent below in Eastern Europe.[2]

With respect to collective farms, one must understand the older system, based on the Soviet trudoden' (labor day unit, cf. p. 20, above) in order to assess the changes that have occurred during the past two decades. On 9 April 1948, after the currency reform of late 1947, the USSR Council of Ministers issued detailed instructions to Soviet kolkhozes on how to apply the rules of the Model Charter for the remuneration of labor. The system became more rigid and more uniform throughout the country, while the incomes of the kolkhozes remained depressed and the means available for labor remuneration did not markedly increase.

It was a fundamental shortcoming of Stalin's kolkhoz system that although it was geared to income differentiation and to material incentives for good performance, the remuneration level was much too low to serve these purposes. If an average collective farmer in 1952 (a very good year) earned 200 labor day units at 75 kopecks each (7.50 pre-1961 rubles) for the whole year, he did not really care whether in this way he received 10 rubles per month or 14. The amount was negligible anyway, and he was keen to earn additional income elsewhere. It was different for premia in kind. These were highly valued, either for own human consumption or for feed to be turned into high-quality food otherwise unattainable, or in great demand on the free market. But such premia in kind were frowned upon and, together with the whole private sector, were restricted as much as feasible.

The possibility of advance payments for labor had been provided for in the 1935 Model Charter (Section 16). However, in practice this provision was not implemented because the overwhelming majority of kolkhozes were not in a position to make such payments. Only after the mid-1950s were advance payments given more emphasis. Advance payments did not change the final remuneration, nor did they alter the labor day system and its concomitants as such. Nevertheless, such payments made during the year could no longer be a residual of the farm's income. Nor could they be taken back when it turned out (as after an unexpected harvest failure) that they exceeded the available financial resources. (Payment in kind continued to be made only after the harvest, anyway.) To the present writer's knowledge, the possibility of reclaiming "overpayments" made in advance has never been elaborated upon in Soviet publications, although it was nowhere explicitly excluded. Under these circumstances, the practical impact depended on how often, how evenly over the year, and to what extent such payments were made. In that sense, much was left to be desired. For all practical purposes, advance payments thus became a guaranteed minimum remuneration, although a very small one.

A link between guaranteed minimum remuneration and stable cash wages has often been established in Soviet as well as in Western writings on Soviet agriculture. Such a connection is beside the point so long as the wage is not fixed, in the sense of being due to the worker irrespective of a given year's economic results (which even today has not been fully achieved). But advance payments may become a substitute for a guaranteed remuneration if two conditions are met.

The first condition is that the remuneration at the end of the year must not be less than the advance payments already made. That is, a negative residual must not be claimed back by the collective farm in the same or in a subsequent year. This condition seems to have been met in practice, although it did not become part of the theory. Advance payments thus in effect become part of the production cost, before net income is calculated. Yet in an indirect way, they may then still be taken back: if the farm has to take short-term loans in a given year because it spent more on advance payments than in the end it could afford, the repayment of the credit will squeeze its wage fund of the next year, and then the labor remuneration will be correspondingly less. The second condition is that advance payments must be made regularly, monthly if possible; they must be fixed in advance and adhered to; and they must constitute the major part of the final payments. For this, sufficient reserves must be available on the farm, for example, through a special wage reserve fund, the formation of which was recommended in Bulgaria and the Soviet Union in the late 1950s.

The latter condition was not met in most cases until the 1960s, except in East Germany, because collective farms were not in a financial position to do so. So long as they were not, calculation of the remuneration wholly in cash, instead of in cash and kind separately (cf. below), did not really change the remuneration system.

If the value of a labor day unit is to serve as the basis of a real wage, it has to be totally fixed at the beginning of the year, not at the end, when the year's overall economic results become known. A minimum remuneration, guaranteed in advance, serves such a purpose. The then still remaining residual (farm profit), in so far as it is distributed according to work completed and exceeds the wage fund so fixed, may then be considered a premium or bonus fund.

Calculating the collective farm labor remuneration in monetary terms became the order of the day in the Soviet Union when another recommendation (after those of 1953 and 1956) had been issued by the CPSU Central Committee at its December 1958 plenary session. It implied a change in so far as payment in kind was no longer kept wholly apart from cash payments. Part payment in kind now had to be calculated in monetary value terms and to become part of a unified wage fund. However, to make this a fundamental change towards cash wages, the collective farmer would have had to be free to choose, at known prices, how much of his wages and what produce he wanted in kind, and not be assigned whatever produce was available within the farm. It would then become a sale instead of a wage. Moreover, the rural retail network would have to be such that the collective farmer could buy all he needed, and could afford such purchases with his cash wage. In other words, produce would have to cease to have a scarcity value exceeding its official monetary evaluation. In the event, these conditions were certainly not met in the early 1960s and are not generally met even today, although they may be close to fulfillment in East Germany, Czechoslovakia, and Hungary.

Communist leaders appear to believe that, with generally rising agricultural labor remuneration, and if and when food supply will exceed demand at given prices, payment in kind can be gradually dispensed with. Moreover, they seem to think that the attractiveness of private plot farming, including private animal production partly based on feed supplies by payments in kind, can be reduced almost to zero. (For more on payments in kind, see below.)

From the beginning of their existence, state farms have generally paid fixed cash wages, and these are predominantly piece-rate wages, that is, wages paid per work norm fulfilled and not per unit of time worked. But it is also true that piece rates are hard to apply meaningfully in agriculture, and that the wages paid are not so stable as they are usually pretended (and believed) to be. Indirectly (through premium payments, employment of skilled workers on less remunerative jobs during the slack season, seasonal employment of family members, and the like), the labor income on state farms, especially if calculated per family, has some elasticity built in. Payment in kind, in the form of premia, also exists on state farms, although to a smaller degree than on collective farms.

Above all, the average level of state farm wages has until recently been higher than that of collective farms, and, on the whole, state farm workers are in a privileged position. The state farm sector is better supplied with capital than the collective farm sector and, making allowance for differences in land and animal productivity, consequently produces at higher capital cost per output unit. Likewise, owing to its privileged position with labor remuneration, the state farm sector also produces at high labor cost. By 1957, when more and more Soviet kolkhozes were being converted into state farms, the cost to the state budget grew accordingly, especially as many of the converted farms had been "economically weak." By 1970, the sector of state and other institutional farms in the USSR accounted for more than half of the agricultural area. This development contrasted in degree, although not in kind, with that in the East European countries (see chapter 6, Table 6.1). The growing subsidy burden had to be reduced, or at least checked. In short, state farm production had to become more cost-effective.

A principal means toward this goal—and the one of interest here—was to make state farm wages more dependent on productive performance and to induce labor to expend more effort. The Soviet Union was the first to remodel the state farm wage system, by a decree of 15 June 1961, ahead of the general economic reforms of the 1960s. The new system became paradigmatic for state as well as for collective farms throughout Eastern Europe and therefore deserves closer inspection.

The essence of the Soviet reform of the state farm wage system was to guarantee a stable minimum wage at the official wage rates for all cases of less than 80 percent fulfillment of output plan. Beyond that figure, the minimum wage per work norm increases by 1.25 percent for every additional 1 percent of output plan fulfillment, up to 100 percent. It is paid out of the wage fund, which is increased by 1.25 percent per additional 1 percent plan fulfillment. This part may become depleted through an excess of work norms spent and to be paid at the minimum wage, but, in principle, an additional 25 percent of the minimum wages is paid when the output plan is 100 percent fulfilled. This payment is considered part of the basic wage. Beyond 100 percent output plan fulfillment, various premia are paid, and wage increases continue more slowly. At the beginning of every year there is formulated a farm plan in which the expected output value of each product is divided into the total of necessary work inputs (number of work norms), differentiated by skill and the "heaviness" of the work which is expected to be required. The earlier piece-rate system is thus combined with the new system, the "piece" now being not a work norm fulfilled but a unit of output in relation to the work required for its production. The procedure is repeated at the end of the year, on the basis of the actual figures of production and work inputs, thus determining the final wage fund and its distribution, beyond the minimum wage.

Had the new system been applied to the huge state farms as a whole, no appreciable incentive effect could have been expected from the beginning. On a farm of more than a thousand workers, with seasonal labor added, so little depends on the individual that he or she cannot see a direct connection between personal effort and reward. Therefore, the annual plan for output and work inputs is also made by farm sections. In the end, the wage fund is distributed among them accordingly.

Under such a system, much depends not only on the volume of output but also on the setting of work norms with regard to the output piece wage beyond 80 percent of plan fulfillment. The more work norms actually required during the year (as against those planned in the beginning), the less the piece rate per output unit. Because the minimum wage per work norm is fixed, this relationship bears exclusively on that part of the wage fund which is available for additional payment per output unit produced beyond 80 percent of plan fulfillment. By assessing this part for individual products as well as for total output, the additional wage can be differentiated by both farm sections and work groups, that is, for production units smaller than the whole farm.

For livestock workers, a similar system of basic wages (mostly per animal tended) and additional wages (per unit of produce) had been applied also on collective farms. It took account of the fact that on the one hand, only certain livestock products can be easily counted in relation to work norms during the year, and that on the other, chances for plan fulfillment have traditionally been slight in this sector. The older wage system for livestock farming was not essentially changed and was gradually incorporated into the new overall system.

A few words are needed on premia. When the new system, which is in part a premia system, was introduced in state farms, and increasingly also in collective farms, premia for plan fulfillment beyond 100 percent, directly related to additional output value, were not a financial problem because they were paid out of above-plan farm income resulting from additional production. But there are, and have to be, other premia not directly related to output units, since additional incentives are needed for the timely execution of urgent work, the better tending of animals, greater management efficiency, savings on fuel, labor, or other cost items, and the like. Most farms, especially the collective farms, found it financially difficult to pay the minimum remuneration for 80 percent output plan fulfillment. Therefore, funds assigned for such premia dwindled to so small a percentage of the wage fund that in many cases they lost their incentive character. Additional premia funds were instituted, separate from the wage funds, into which a fixed part of overall and of above-plan farm income had to be channelled. Thus, there are now essentially three kinds of premia: those inherent in the new wage system and considered part of the wage fund, those for output beyond 100 percent of the plan, and those not directly related to output. As to that part of the premia which was paid out of net farm income, it amounted to only 0.9 percent in the majority of Soviet kolkhozes in 1974.[3] The system has a number of flaws that impair its incentive effects:

1. The evaluation of output and work is made chiefly in quantitative terms. In principle, this can be changed, and is being changed in many cases. But the question remains whether the aspect of quality is sufficiently taken into account. As long as the state procurement system is directed more toward quantitative than toward qualitative increases—and this still is the case—the farms are hard put to apply qualitative evaluations internally which conflict with the state's predominantly quantitative evaluation, as reflected in the prices paid to the farms. Nor would this be desirable from the state's point of view.
2. The additional wage is calculated as a percentage of the minimum wage, not as an increment in absolute terms. Therefore, it may be to

the advantage of the individual or the group to earn more work norms at the minimum wage, which is the basis for the percentage calculation of the additional wage, even if thereby the overall fund for additional wages is reduced (additional output being divided into additional work inputs). This applies especially when there are no great chances to surpass 80 percent of the output plan by a sizeable margin, so that the minimum wage rates will not be exceeded anyway, or not by much.

3. Whether the farm can overfulfill the plan depends not only on its own efforts, or a given year's weather, but also on the plan imposed from above. If the plan is difficult or almost impossible to fulfill, not much thought will be given to the additional wage for more than 80 percent plan fulfillment. It is quite different if the plan is "slack," that is, easy to fulfill. But with the planners' continuing emphasis on maximum increase in output (disregarding here the Hungarian case), "slack" plans are not likely to be imposed. Yet contrary to the state's explicit interest, the wage system itself directs the farm's and its workers' interest toward receiving "slack" output plans, rather than toward actual output increase. The supply of material inputs is another aspect of the problem. Material inputs are centrally planned and are often inadequate through no fault of the farm, thereby impairing output and labor productivity and frustrating efforts made on the farm. Thus, the wage system is closely connected with the problems of centralized planning (see chapter 11).

4. It is similar with plan tasks distributed within the farm. Some section and work groups may feel that they are put at a disadvantage compared with other groups, and that their labor is not adequately assessed. Farm leaders and administrators, whose additional payments and premia depend on the whole farm's performance, may not give due attention to the less rewarding production of certain farm sections, concentrating their attention (and assigning inputs) instead on those sections where success is more sure. For example, the chief livestock specialist (as distinct from the head of a livestock section) may feel that even if his branch does not fulfill the plan, he will still profit from the successes in plant production as part of overall output. The output of farm sections also depends on their specific material inputs, the assignment of which by the central management of the farm may be neither economically justified, nor adequate for output plan fulfillment.

5. Above-plan labor inputs in a given year impinge on the value of the piece rate per additional output unit per work norm. The excess labor inputs may be caused by the workers themselves, but they may also be caused by unfavorable weather, by management mistakes, or by nonsupply of necessary nonagricultural inputs or services. So the

worker may be inclined to think that his effort is overwhelmed by other factors, and therefore is not important to him personally.

6. The various factors outlined above make the final wage unpredictable for the worker. What he knows is his minimum wage; the rest is the result of fate rather than of his, or his group's, efforts. The system is so complicated that often even the farm administration does not fully understand its mechanism. Improvements on its systemic deficiencies frequently make it still more complicated, because such improvements have to remain within the parameters of the overall agrarian system, which has not undergone fundamental reform so far.

The state farm wage system was restructured in the Soviet Union in 1970 but not basically changed. It was introduced, with only slight alterations, during the 1960s on most East European state farms, except in Yugoslavia (where a similar system was devised earlier) and in Rumania. Its overall effect was, while leaving minimum wages as stable as state farm wages had generally been before, to make the additional wage (not to speak of premia proper) very elastic with respect to both output and the quantity of work (norms) spent. For Hungarian state farms, the rule was applied that a farm's wage fund should increase by 0.4 percent for each percent of gross output growth, and decrease by 0.3 percent (but by no more than a total of 15 percent) for each percent of decline in gross output in a given year.[4] (On Polish state farms, cf. p. 216.) There was a general tendency in the 1970s to increase the share of premia in total remuneration by applying various devices for increasing the premia funds. In the Soviet Union, these amounted to 17.3 percent of the wages in 1973 (an excellent harvest year), whereas in 1964, to only 3.5 percent.[5]

Soviet kolkhozes were soon advised, and after 1966 strongly recommended, to make their own wage system similar. In fact, they were already on the way to such adaptation (or perhaps the new state farm wage system had taken over some features of the evolving kolkhoz remuneration system, with its advance payments and cash wages). The same happened in the East European countries, in some of them already earlier. Before that, kolkhoz and collective farm remuneration had been dismally output-elastic, to the disadvantage of the workforce. (For the advancement-like effect of "land rent" in East European collective farms, and the early guaranteed minimum wage in the GDR, see pp. 80 and 81.) Now, gradually and falteringly, the labor day unit was abolished in the majority of cases, and a minimum wage or minimum value of the labor day unit (as a rule, below the state farm minimum) generally was introduced in the form of advance payments of between 70 and 90 percent of the planned remuneration. At the same time, the former dependence on actual production results was retained for remuneration beyond the minimum level. The Hungarian reform of 1968 brought more flexibility into that country's system and, in addition, introduced the possibility of counting work done in private animal husbandry towards fulfilling the minimum labor requirements on condition that a contract was concluded for selling such products to the collective farm. Rumania was last to introduce a guaranteed minimum remuneration for collective farmers during 1970–1976. Albania did this for its new (since 1971) collective farms of the "higher type," putting the minimum at 90 percent of the planned remuneration.

The advance payment, which now fulfilled the two conditions enumerated at the start of this chapter acted, and increasingly was recognized, as a guaranteed minimum wage. This, and not the decreasing use of the term "labor day unit," constitutes the essential change of the collective farm remuneration system. It does not really matter whether one speaks of a wage rate or of a labor day unit, so long as the value of that part which is paid in advance, or is guaranteed, is known to the collective farmer from the beginning of the year. He then knows how much he can expect for a work norm fulfilled. This part may be defined as 70, 80, 90, or whatever percent of a wage rate, the rest of the wage depending on the year's economic success; and it might as well be called the guaranteed minimum value of a labor day unit, any payment beyond it being considered an increase in this value depending on the year's economic success.

To give a numerical example: the fulfilled norm for a certain (output effective) kind of work may be evaluated at a wage rate of, say, 4.20 rubles, of which 80 percent (=3.36 rubles) is paid during the month when the work was done. It may equally well be evaluated at one and a half labor day units, the guaranteed minimum value of which is 2.24 rubles per unit (3.36 for one and a half). In both cases, the additional payment, bringing it to the full wage rate or to the final value of the labor day unit, depends on the income residual (or the residual in the planned wage fund) of the farm. The difference is purely one of terminology. One could well say that the guaranteed (or paid-in-advance) part is the stable wage, and that the rest (on both the collective and state farms) has all the characteristics of the old labor day unit or of a premium wage, which in fact the old labor day unit was, although on so low a level that it did not serve as an incentive.

Two major differences of another kind affecting labor remuneration on collective and state farms remain. One is the differing share of payment in kind, which is greater on collective farms (on which more below), and consists basically of premia in kind on the state farms. The other is that collective farm remuneration rates (minimum as well as planned full rates, wage or labor day unit) are in theory fixed by each collective farm individually according to its economic potential, whereas state farm wage rates are fixed centrally. The collective farm rates may even vary annually; but downward revisions of the planned rates—as distinct from the levels above 70 or 90 percent actually paid—from one year to the next are rare. It is the predominant opinion in Eastern Europe and the USSR that collective farm remuneration rates should come as close as possible to, but must not exceed, comparable state farm wages. The collective farms receive strong guidance on this point from local and central authorities. On Hungarian collective farms, where there is a considerable share of wage-earning non-member labor, especially among machinery operators and specialists, income disparities cause problems. On the whole, the wages of such personnel are dependent on the collective farm's overall economic situation, but less so than the remuneration of members, which in extreme cases may turn out to be much lower or much higher than that of nonmember workers of the same farm.

A similar problem arose in all countries where collective farm members and wage earners work side by side on interfarm enterprises. In Bulgaria, Hungary, the GDR, and Czechoslovakia, uniform wage rates for collective and state farms, as well as for interfarm enterprises, are being introduced. Especially for the interfarm enterprises, associations, and complexes, this is inevitable in the long run under horizontal and vertical integration. When equal rates are applied everywhere (and this may occur in the near future, except in Rumania), then one may speak of a unified wage system in socialized agriculture, with only small local variations natural to any agricultural system. However, this statement presupposes that work norms, too, are fixed and applied equally on both kinds of farms. Unification in this field in fact is in progress, and voluminous and authoritative manuals defining and evaluating the various kinds of agricultural work are being issued in great numbers in CMEA countries.

Perhaps more important than any changes in the remuneration system as such, and in fact making these changes possible and meaningful, has been the general rise in farm incomes through the state's raising agricultural producer prices and through the increased supply of nonagricultural inputs since 1966.

Available data on collective farm wages lack comprehensiveness and are not comparable among the individual Communist countries. Until recently, wages on collective farms have generally been lower than those on state farms; consequently, they have had to rise faster than the latter. Scattered data on collective farm average monthly remuneration, excluding off-farm and private plot incomes, are given below for various countries,[6] in order to throw into relief the data on state farm wages which follow in Table 9.1.

Bulgaria: A permanently, but not always fully, employed collective farm member in 1960 earned 39 leva per month on the average; by

Table 9.1 Nominal Monthly Gross Wages and Salaries, Including Premia as Actually Paid on the Average, of Blue and White Collar Workers in State-Owned Agriculture and in the Economy at Large, 1960 and 1978

		Economy at large[a]		Agriculture		
	Year	Wages (in national currency)	Percentage increase, 1978 over 1960	Wages (in national currency)	Percentage increase, 1978 over 1960	Percentage of industry wage
USSR (rubles)	1960	80.6		55.2		60
	1978	159.9	+ 98	143	+ 159	81
Yugoslavia (dinars)	1963	272		218[b]		78[b]
	1977	4,198	+1,443[g]	3,921[b]	+1099[b,g]	100[b]
Bulgaria (leva)	1960	78.3		74.4		93
	1978	157	+ 101	155	+ 108	96
Hungary[c] (forint)	1960	1,553		1416		88
	1978	3,534	+ 128	3,381	+ 139	98
East Germany (marks)	1960	558		453[c]		79
	1978	985	+ 77	954[c]	+ 111	96
Poland[e] (zloty)	1960	1,560		1,267[f]		74
	1978	4,686	+ 200	4,974	+ 293	101
Rumania (lei)	1960	854		731		82
	1978	2,011	+ 135	1,994	+ 173	99
Czechoslovakia (crowns)	1960	1,365		1,113		77
	1978	2,517	+ 84	2,434	+ 119	94

a Excluding collective farms and private sector.
b Including fishery; all socialist farms. With regard to the percentage increases, Yugoslav inflation must be borne in mind.
c Excluding apprentices and working pensioners.
d Fully employed labor on state farms proper (VEGs).
e Net wages.
f Including employees, but not members, of cooperative and collective farms.
g Percentage increase, 1977 over 1963.

Sources: For Yugoslavia, *Statisticki godisnjak, 1968*, p. 176, and *1978*, p. 133; for GDR, *Statistisches Jahrbuch 1979 der Deutschen Demokratischen Republik*, pp. 106, 163; for all others, *CMEA Statistical Annual, 1978*, pp. 404-406.

1970, this had increased to 86 leva. If fully employed (250–260 days per year), he or she received between 94 and 98 leva in 1971/72,[7] which still was 16–19 percent below the state farm level. In 1971, an existing guaranteed minimum wage of 1.80 leva per day was increased to 2.80 leva for collective farms in mountainous areas.[8] Since 1972, collective farm remuneration has no longer been revealed in Bulgarian statistics. It has become part of the wages in the so-called agro-industrial complexes, and equal to state farm wages in most cases, but differentiated by enterprise sections.

Czechoslovakia: A fully employed collective farmer earned 853 crowns in 1960, 1,685 in 1970, and 2,075 in 1973.[9] Parity with state farm wages was said to have been attained in the record harvest year of 1974. The official figure for net income (probably including labor not fully employed on collective farms) was 1,459 crowns in 1970, and 2,198 in 1977, slightly less than on state farms.

Hungary: Collective farm income for members who are mainly, but not necessarily fully, employed was approximately 1,150 forint in 1960, and 1,450 in 1970. (An estimated deduction has been made for part-time workers.) For all "economically active" collective farmers, the figure was somewhat lower: 1,183 forint in 1970, when the average state farm wage was 2,122 forint. During the same year, a full-time employed collective farmer received 1,885 forint per month, which rose to 3,377 by 1978.

East Germany (GDR): Only Western estimates derived from scanty East German published information are available. According to them, average monthly labor remuneration (excluding private plot income) on collective farms of the "higher" type III amounted to 306 marks in 1961, and to 614 marks in 1973. In type I and II farms, total, including private, incomes were higher than the total incomes in type III farms (increasingly so since 1961), until this type of farm disappeared.[10] Only if private plot incomes are included is it true that collective farm incomes reached the state farm level in 1964 and have been equalling it in subsequent years. During the 1970s, the remuneration system was gradually unified and in the new, 1977, type of collective farms, it has come close to that of the state farms.

Soviet Union: A kolkhoznik, if fully employed (270–280 days per year), earned a monthly average of 50–55 rubles in 1960, almost 90 rubles a decade later (state farm worker wages in 1970: 101 rubles), and 107–111 rubles in 1976. The earlier great disparity between collective and state farm labor remuneration thus diminished up to 1970, but increased again since. By 1978, it was (per day worked) 84 percent of the state farm wage.

Rumania: The disparity between state and collective farm incomes has remained greatest in Rumania. In 1970, a fully employed collective farmer on average still derived only about 500 lei per month from collective work, one-third to two-fifths of a state farm worker's average wage. The actual net average incomes of all collective farm workers, including the underemployed, still were only 544 lei in 1975.[11]

Albania: Almost no meaningful data are known. Collective farm remuneration, presumably for the fully employed, is said to have been about half the average industrial wage in the early 1970s. State farm wages are likely to range somewhere between these two wages levels. Underemployment may be assumed to be widespread on collective farms, perhaps even more than in Rumania. Generally, rural incomes during the 1970s, although lower in absolute terms, seem to have increased faster than urban incomes.[12]

The fully employed collective farmer is not typical for a country like Rumania, where employment on the farm was only about 150 days per year in 1970. Nor is he typical in the USSR where, in this respect, great regional variations exist. From a Soviet source, it emerges that the annual average degree of employment in kolkhozes as compared to state farms was 81 percent in 1965, rising to 89 percent by 1976.[13] While for East Germany, Bulgaria, and the Bohemian and Moravian parts of Czechoslovakia, near to full employment on collective farms may be assumed nowadays, such is not the case for Hungarian collective farms, although underemployment there seems to be less than in Rumania and parts of the USSR. Thus, apart from the disparities in remuneration per unit of labor, to some extent due to differing levels of technical skills, underemployment contributes to keeping collective farm labor incomes below those of state farms.

On the other hand, the five-day week has not been introduced for most collective farmers, as distinct from state employees, so that full employment usually means 270–280 days per year on collective farms and kolkhozes, and 225–235 days in the rest of the economy. Yet state farms, too, have a kind of underemployment, although a less visible one. Many female state farm workers are only seasonally employed, and at peak times, non–farm labor is hired (or sent to the rescue by the state). This enables state farms to have relatively fewer permanently employed workers than collective farms. To these workers, more or less stable wages are paid throughout the year.

Even with rising average incomes, disparities within the agricultural sector are still great, and in fact are often greater than among nonagricultural workers. Thus, 20 percent of Hungarian collective farmers in 1978 earned less than 2,000 forint per month, and 10 percent more than 5,000.[14] In Rumania, livestock personnel in 1975 were guaranteed a minimum labor income of 1500 lei per month, while for field workers it was only 1,000 lei, in both cases under full employment, which is not dominant among field workers[15] (cf. the overall average, above).

In Poland and Yugoslavia, state farm wages played a role for only a minority of those employed in agriculture, and the number of collective farm or cooperative members was almost negligible. The wages of employees (as distinct from members) of the latter two categories of farms may generally be assumed to be of the same order of magnitude as those of state farm labor. However, it is clear that the incomes of individual peasants constituted the bulk of agricultural incomes. Theoretically, one would have to impute part of their income to land rent and capital returns. Yet under the given farm structure and the political restrictions on such incomes, the sums imputed must be small. Moreover, lack of data forces us to neglect them as separate components in total peasant net income.

The disparity between nonagricultural wages and peasant incomes was considerable in Yugoslavia, and only in part resulted from the widespread underemployment of the farm population. In the early 1960s, per head labor income of peasants was only two-fifths of nonagricultural incomes, and with nonagricultural incomes of peasants included, not quite two-thirds.[16] Since then, the disparity has diminished considerably, most of all after the 1965 price reform, but has remained sizable. In 1970, the statistical administration assessed the average net income of a peasant household still at almost 25 percent below that of a full-time worker on a social (i.e., state) farm. Since then, the disparity continued to diminish. By 1976, peasant net income was assessed at 2,496 dinars per month, and another 2,665 dinars from non-agricultural activities. Such incomes, however, refer to the whole household and not to one worker, as in the social sector, where in 1976, the average agricultural labor income was 3,467 dinars.[17]

In Poland, a working peasant's average monthly income available for consumption (i.e., roughly equal to net income) from agricultural activity amounted to 1,781 zloty in 1970. At that time, it was only 5 percent less than that of a farm employee in the public sector, but later on again lagged behind. Only after the new price increases was it almost equal (at 3,539 zloty in 1976) to that of a farm employee. But this equalization refers to full-time workers only. On a per household basis and excluding part-time farming (which accounts for 2 out of 3 million farms with more than 0.5 hectares), peasant net incomes from individual farming in 1976 were still 23 percent lower than those of non-peasant households (excluding those of pensioners).[18] The 1976–80 five-year plan provided for individual peasants' incomes to rise more rapidly than those in the state sector of the economy.

On the whole, one may say that, with the exception of Rumania,

Albania, and the USSR, the level of collective farm labor remuneration under full employment has by now come rather close to that of state farm wages. Moreover part, but less than half, of the remaining disparity may be explained by the generally greater share of technically skilled labor, commanding above-average wages, on state farms. This narrowing, and in part even eliminating, of the gap since the early 1960s indicates great progress toward lifting the discrimination against collective farmers.

Table 9.1 clearly demonstrates that state agricultural wages rose faster than total average wages, or than industrial wages, and that thereby the disparity was in all cases narrowed and sometimes even abolished. Since collective farm remuneration has been coming closer to that in state farms, this has necessarily pulled up the former, although on the whole, it is still below the state farm level, especially when underemployment is taken into account.

The real wage increase was somewhat less than the nominal increase (gross or net). For example, consumer prices in Yugoslavia inflated at an annual average rate of 14 percent during 1963–78. As to CMEA countries, noticeable inflation was officially admitted only in Hungarian and Polish statistics, during 1971–78 at average annual rates of 3.4 and 5.1 percent, respectively. In the other countries, officially admitted inflation was minimal or nonexistent, although a modest or hidden degree of inflation has to be assumed.

When comparing state and collective farm labor remuneration, one must be aware that the state farm wages shown are before income tax deduction (except for Poland), while the collective farm remuneration is after. Collective farms (but not their individual members) pay income tax, which is deducted before the residual is calculated for distribution. (In Rumania this has been so only since the beginning of 1977; and Bulgaria remains an exception in so far as it is the individual collective farmer who pays income tax.) The collective farm formula has, among other things, the consequence that income tax is progressive only in so far as rising profits and/or labor remuneration of a given collective farm as a whole are fined by it, not the disparities among the members' individual incomes. Czechoslovak and Hungarian legislation (in 1974 and 1976, respectively) put special emphasis on this kind of tax progression. On Albanian collective farms, the overall average income tax was 7 percent in 1971, and the percentage increased with increasing income.[19]

On the other hand, transfer payments (social insurance, benefits of the socio-cultural infrastructure in general) favor state farm and other state sector employees. In the early days, the collective farms themselves had to provide such social beenfits for their members, so that in most cases these were minimal, as the farms could not afford large outlays. Nowadays, social insurance benefits for collective farmers are centralized and greatly improved, but are still less than in the state sector. Moreover, collective farmers in Bulgaria, Hungary, the GDR, and Rumania (but not in Poland, Czechoslovakia, and the USSR) have to pay for part of their social insurance out of their individual labor remuneration, which is not the case for state employees.

After the GDR and Bulgaria introduced old-age pensions for collective farmers, Hungary followed suit at the end of 1957. Czechoslovakia was next in 1962 with a more comprehensive system than her earlier rudimentary one of 1952/53, followed by Poland in the same year (1965 also for "lower type" Polish collective farms), the Soviet Union in 1965, Rumania one year later (where the system was restructured in 1977), and Albania in 1972. The Bulgarian system of 1957 was extensively restructured and made more comprehensive by a law of 1975, and all the national systems were repeatedly enlarged in coverage, and their rates raised. Yet generally, collective farmers' pensions are usually smaller than those in the state sector, even where in structure and in age for eligibility the systems have successively been equalized, because the basis for calculating the rates is the earlier lower monetary remuneration of collective farmers.[20] Other kinds of social insurance (health, invalidity, orphanhood, etc.), which in part preceded the introduction of old-age pensions in the various countries, are not examined here for brevity's sake.

Since 1974, Polish private farmers giving their land to the State Land Fund have been entitled to an old-age pension, or land rent. Beginning in 1980, paying social insurance became mandatory for them. Under the 1980 law, they receive a pension also if they hand the farm over to a private heir. The remaining private peasants in Rumania's mountainous areas were included in the pension system in 1978. There is no general and mandatory social insurance for individual peasants in Yugoslavia. They may conclude an insurance contract privately, but the cost is high. However, those who are of eligible age and hand their land over to a socialist farm, receive a rent for the rest of their life, and those who engage in long-term cooperation with the socialist sector are offered the right to a pension.

With collective farm remuneration rising fast, payment in kind became a decreasing part of the total from the mid or late 1950s. However, in absolute terms it decreased not at all, or less than expected; and where it did, the reduction occurred under pressure from above rather than through the collective farmers' own will. This was because the demand for food continued to exceed the supply in the public trade system, particularly in rural areas. The average prices for food on the free market still exceed those of public shops (state and consumers' cooperative), even where supply by the latter is adequate in quantity.

Hungary is a notable exception in that no pressure has been exerted so far to reduce payment in kind. In the early 1960s, under the so-called Nádudvár system, payment in kind was used as an effective labor incentive by making it a fixed percentage of the output produced by small work groups (often families). However, this system was not applied in all collective farms, and not for all branches of their production. Rumania and Bulgaria proceeded along similar lines, with the important difference that in those countries, the percentage was calculated in cash at official prices instead of directly in kind, and was not always paid in kind. Shortages of certain produce, and the preferences of collective farmers, tended to be disregarded, and the incentive effect was less than it was in Hungary. But the other aspect of the Hungarian system, the small (family or kinship) group and its productive performance as the unit to which the remuneration is assigned, was also a salient feature in Rumania.[21] The Hungarian Nádudvár variant diminished in importance in the early 1970s with the ascendance of the Closed Production Systems, but was revitalized in spring 1977, mainly for nonmechanized work and for attracting old-age and juvenile labor. However, an upper limit of such remuneration was set, and the part paid in kind was evaluated in state procurement prices.[22]

According to a Soviet study,[23] the share of payment in kind in total collective farm remuneration, calculated in monetary terms at official prices, diminished in Rumania from 90.2 percent in 1955 to 57.5 in 1961/1962. There is reason to believe that during 1963–1965, this share increased again and then decreased to a level presumably below that of 1961/1962. In Hungary, the decrease was from 72.9 percent in 1954/1955 to 42.3 in 1964; in Bulgaria, from 61.5 (1955) to 11.3 (1963); and in Czechoslovakia, from 37 (1953) to 8.9 (1963). In East Germany, the share is likely to have been less than in Czechoslovakia. The decrease in the Soviet Union was from 60 percent in 1953 to 11 in 1968 and 10 in 1974. But with the considerable rise in overall remuneration since 1968, the absolute value of payment in kind in Soviet kolkhozes must have increased again, if only (for the most part) because of the higher official prices for such produce. Developments since the early 1960s are believed to have been similar for the East European countries.

Although payment in kind on the whole has lost much of its relative weight since the 1950s, when total collective farm remuneration was extremely low, it still has a certain importance, especially in Hungary and Rumania. In addition, sales of farm produce to collective and state farm workers at preferential prices, as often practiced, also constitute a form of payment in kind, especially when the quantities sold are dependent not only on collective farm membership but also on work performance. In Czechoslovakia, payment in kind was said to have dwindled to less than one percent of total collective farm remuneration by 1973. Yet a decision of the Ministry of Agriculture in that year permitted those farms where the private plots either were handed over to the collective or were farmed in common to issue fixed quantities of grain and potatoes to the member households.[24]

An increasing number of collective farm families have one or two members working part time or even close to full time outside agriculture. The resulting income accounts for 10–35 percent of total family income on the average (varying by countries), and in some regions or individual families, may exceed the income from agricultural work.[25] But this is not an aspect of the income from agricultural work, and therefore is mentioned here merely in passing.

In contrast, income from private plot and animal farming is clearly agricultural, being derived within the framework of socialized farms. (Here I am disregarding comparable income of the nonagricultural population and of the few individual farmers left in countries other than Poland and Yugoslavia.) Such private income compensates in part for underemployment in the public sector. A special case is those collective farms in Hungary and the GDR which pay their members cash beyond the labor remuneration in compensation for no longer providing private plots.[26]

Private plot income also alleviates in part the disparity of collective versus state farm wages. But although private plots (and the income derived from them) are generally smaller on state farms, they are not negligible there (except in East Germany and parts of Czechoslovakia). For comparisons with collective farm members' incomes, this fact has to be borne in mind, although almost no data on private plot incomes of state farm workers are available. For Soviet state farm workers, average private plot income was estimated at 516 rubles in 1966,[27] which compares with an average income of 895 rubles from work on the state farm in 1965.

In Hungary, the income of collective farm households from private plots almost equalled that from collective work in 1970 (although by 1975, it had declined relatively to 14 percent less than the latter). In Bulgaria, by 1973 it was 24.5 percent of total income, against 41 percent for collective farm work, the remainder of income deriving from transfer payments, non-agricultural employment, and the like (Bulgarian data on that account beyond 1973 are not meaningful because they refer to a small residual number of collective farms). In the Soviet Union, the figures were 31.9 percent from private plots against 40.3 percent from kolkhoz work in 1970, and 25.2 percent against 44.5 percent, respectively, in 1978.[28] Rumanian collective farmers earned 544 lei from farm work in 1975 and 430 lei from the private plot and animal production.[29] According to V. Bajaja, the share of private plot income in total farm income of GDR collective farm type III members (including so-called land rent, but excluding transfer payments and incomes from outside the farm) was 30.2 percent in 1961 and steadily decreased to 18.1 percent by 1970 (more recent data are not available).[30] The percentages are likely to be similar in Czechoslovakia, but even there, in 1973, the private plot and animal holdings accounted on average for 25 percent of a collective farm household's food consumption. For Rumania, higher shares may be assumed.

Because total family incomes on collective farms rose rapidly everywhere, the shrinking percentages in most cases still implied increases of private plot incomes in absolute terms, especially since the value of such produce increased through rising prices. In some countries, such as Bulgaria and Hungary, even the percentage share grew for a number of recent years.[31]

Including such income, total family income may often equal or exceed that of state employees with comparable work and skills, especially in Hungary, Bulgaria, and Czechoslovakia. Yet it must be added that private agricultural income is in part earned by family members who would not otherwise be working (the old, the young, the invalid). Permanently employed collective farmers often work overtime on the private plot. Moreover, such production is taxed and requires not only labor but also material inputs, although to a modest degree, and these have to be deducted from the income derived. Per hour worked, even the total agricultural income of collective farmers is likely to be considerably below that of state farm and other state-employed labor.[32]

It has been repeatedly stated by Communist leaders that incomes of the agricultural worker should not exceed those of the industrial worker. Two Soviet authors put it in the following way:

> It is obvious that the lagging of peasants' incomes behind those of other social groups, above all the working class, or the surpassing of peasants' incomes over those of non-agricultural workers . . . may become a factor impeding the process of the political consolidation of society, the strengthening of the alliance of the working class and the peasantry.[33]

Where state farm wages approach or equal the level of industrial wages (see Table 9.1), future increases may be expected to obey this tenet. The same is likely to apply where collective farm remuneration has attained the same level (East Germany). It is even claimed that collective farmers' incomes in Czechoslovakia and Hungary already exceed those of industrial workers. But such statements refer to total incomes, including the return from private plots, and therefore to greater labor inputs. This distortion alone reveals the latent misgivings of political leaders about a possible income disparity in favor of agriculture. Equally important is the fact that agricultural labor remuneration in the recent past rose faster than labor productivity, and the policy makers do not want to let this continue.

Notes

This chapter is largely based on a revised and updated version of the writer's article "Labor Remuneration in the Socialized Agriculture of Eastern Europe and the Soviet Union," *Studies in Comparative Communism* 11:1 and 2 (Spring/Summer 1978), pp. 96–120. By permission.

1. An overview of recent forms and developments of collective and state farm labor remuneration in CMEA countries, disregarding individual peasant incomes and payments in kind, has been given by M. I. Palladina, *Oplata truda v sel'skom khozyaistve sotsialisticheskikh stran* (Moscow, 1978).

2. *Incomes in Postwar Europe*, UN/ECE, ed. (Geneva, 1965), chapter 8, p. 29.

3. I. F. Suslov, *op. cit.*, p. 232.

4. G. I. Shmelev and V. N. Starodubrovskaya, *op. cit.*, p. 158.

5. *Kursom martovskogo Plenuma* (Moscow, 1975), p. 316.

6. Apart from the sources specifically quoted and from the national statistical yearbooks, the collective farm data are taken or derived from the following: *The Future of Agriculture in the Soviet Union and Eastern Europe*, R. D. Laird, J. Hajda, B. A. Laird, eds., (Boulder, Colorado, 1977), p. 164; V. Khizhnyakov, in *Khozraschet i stimulirovanie v sel'skom khozyaistve*, A. M. Emel'yanov, ed. (Moscow, 1968), p. 88; Deutsches Institut für Wirtschaftsforschung, *Wochenbericht*, 42/73, p. 381; C. Sporea and K.-E. Wädekin, *Osteuropa* 27:4 (April 1977), pp. 333–336.

7. I. Kostov, *Ikonomika na selskoto stopanstvo*, no. 6 (1974), pp. 25–36 (according to *ABSEES* 7:2 [April 1976], p. 135).

8. *RFER, Bulgarian Situation Report*, 26 November 1971, p. 7.

9. V. Bajaja, *Theoretische Grundlagen und praktische Entwicklung landwirtschaftlicher Betriebsgrössen in der Tschechoslowakei* (West Berlin, 1975), Table 60.

10. From the study by V. Bajaja, "The Organization of East German Agriculture," in *The Organization of Agriculture in the Soviet Union and Eastern Europe*, E. M. Jacobs, ed., vol. II of the present series (forthcoming).

11. *Scînteia*, 7 May 1977 (according to *RFER Romanian Situation Report*, 12 May 1977, p. 16).

12. For the period 1971–75, the increase was 20.5 percent, as against 8.7 percent in urban areas; *Probleme Ekonomike*, no. 1 (1977), pp. 3–23 (according to *ABSEES*, no. 53 [September 1977], abstract 00142).

13. Suslov, *op. cit.*, p. 230 (indirectly derived from the table).

14. *RFER Hungarian Situation Report*, 5 December 1979, p. 25.

15. Ceausescu's speech of 18 April 1977, according to *RFER Romanian Situation Report*, 27 April 1977, p. 8.

16. *Incomes in Postwar Europe, op. cit.*, chapter 12, p. 9.

17. *Statistički godišnjak, 1973*, p. 161, and *1978*, p. 242.

18. *Rocznik Statystyczny 1977*, pp. 41, 68; *Budżety gospodarstw domowych w 1976 r.*, Statystyka Polski, no. 92 (Warsaw, 1977), pp. 11, 112.

19. *Ekonomia popullore*, no. 3 (1972), pp. 9–29 (according to *ABSEES* 2:3 [January 1972], p. 130).

20. For a recent and detailed treatment of Soviet social insurance, including that of collective farmers, see Pavel Stiller, *Die sowjetische Rentenversicherung 1917–1977*, Berichte des Bundesinstituts für ostwissenschaftliche und internationale Studien, 42-1979, Cologne (September 1979), 91 pp. (hectographed).

21. See the relevant Party CC resolution in *Scînteia*, 3 March 1973, which was preceded by widespread practical implementation.

22. *RFER Hungarian Situation Report*, 9 December 1977, pp. 18–19, and 2 August 1977, p. 5.

23. E. V. Rudakov, S. A. Mellin, and V. I. Storozhev, *Ekonomicheskaya reforma v sel'skom khozyaistve sotsialisticheskikh stran* (Moscow, 1968), p. 145.

24. By personal communication; the decision bears the no. 01-183/13/1973.

25. G. I. Shmelev and V. N. Starodubrovskaya, *op. cit.*, p. 178; M. Cernea, *Sociologia Ruralis* 18:2/3 (1978), pp. 116, 118.

26. G. I. Shmelev and V. N. Starodubrovskaya, *op. cit.*, p. 172.

27. E. S. Rusanov, *Raspredelenie i ispol'zovanie trudovykh resursov SSSR* (Moscow, 1971), p. 81.

28. *Statisztitkai Evkönyv, 1976,* p. 491; *Statisticheski godishnik na NRB, 1977,* p. 91; *Narodnoe khozyaistvo SSSR, 1978,* p. 392; G. I. Shmelev and V. N. Starodubrovskaya, *op. cit.*, p. 175.

29. *Scînteia,* 7 May 1977 (according to *RFER Romanian Situation Report,* 12 May 1977, p. 16).

30. V. Bajaja, "The Organization of East German Agriculture", *op. cit.*

31. G. I. Shmelev and V. N. Starodubrovskaya, *op. cit.*, pp. 165–168.

32. See the explicit statements by the Hungarian authors K. Benda and I. Somogyi, *Munkaügyi Szemele,* no. 11 (1973), and J. Moharos, *Tudomány és Mezögazdaság,* no. 1 (1975) (as quoted in *Hungarian Agricultural Review,* no. 3 [1974], p. 29, and no. 1 [1976], p. 36, respectively).

33. G. I. Shmelev and V. N. Starodubrovskaya, *op. cit.*, p. 183. Cf. the explicit official statement in the GDR decree of 28 August 1975, *Deutsche Bauernzeitung,* 19 September 1975, Beilage, p. 7.

10. *Agricultural Producer Prices*

IN DIRIGISTIC and centrally planned economies, if prices serve a directing and allocating function, it is toward the immediate producers and the consumers alone. Put another way, prices do not serve as guidelines to the planners and directors of the economy at large. Rather, the planners' and political leaders' intentions determine the levels at which prices are fixed. Such prices may be based on a more or less adequate insight into a simulated market, but they are not autonomous indicators of shortage or abundance of individual goods.

Nevertheless, market forces are at work even under such conditions, not through monetary flows, but through the flow of goods. Shortage or oversupply of certain goods replaces the signal function of prices and makes the planners and leaders realize that under the given conditions, the producers tend to increase or restrict the output of certain goods (of agricultural products in our case), in the entire country or in some regions. The symptoms are the lines of consumers at retail stores and the rising prices on legal or illegal free markets on the one hand, and unsaleable goods on the other. However, such reactions of consumers or producers do not directly influence the set prices. Only when the leaders and planners become aware of the situation and the necessity of action are the price lists changed or other measures taken (subsidizing, in other ways expanding certain production branches, restricting consumption, etc.). The price mechanism may react, but with less flexibility and with a greater time lag than one in which prices directly play an autonomous role.

The fixed prices and the price policy in such economies merely reveal the extent to which the political authorities pay attention to goods shortages and to consumers' reactions and wishes. For example, the generally high producer prices for pork in most of Eastern Europe and in the Soviet Union (high in relation to the price for feed grain) are mainly a result of the demand for meat, combined with high feed consumption ratios and labor inputs in pig farming. Yet producers' prices have not always been so high. Only when rising consumer incomes resulted in excess demand, which seriously imbalanced the food economy, and the political leaders did not want to change the consumers' prices or else raised them only moderately, were producer prices raised to the present high level, and also subsidized in order to stimulate the growth of pork output. The situation is similar for other animal products, whereas the price rise for most crop products has been less.

The food situation in Eastern Europe and the Soviet Union is characterized by high income elasticity and low price elasticity of the demand for animal and other high quality food. The price elasticity of supply is also low, because for a number of main products, agricultural producers are under obligation to deliver the amounts imposed by the plan, whether or not they find the prices attractive. The procurement system hampers price policy in reacting rapidly and flexibly to arising shortages.

Closely interrelated with the price setting system is the organization and the method of planning and directing economic activities. Neither side of the relationship can be changed fundamentally without the other also changing. Thus, the economic reforms of the 1960s (see chapter 11) were "closely connected with the improvement of the price formation also for agricultural products."[1] One of their major aspects was to minimize the disadvantages of the state planning and directing system by a new price policy. However, the reforms were not intended to entirely offset the system, but rather to make it less inefficient and costly. Anything beyond that was excluded. It is under these two aspects—the organizationally conservative and the economically liberal—that the more recent agricultural price policy in Eastern Europe and the Soviet Union must be viewed. Prices became a complementary instrument for offsetting some effects of the structural inelasticities of dirigistic planned economies, but not for changing the system as such.

Price policy in Poland and Yugoslavia is of greater importance than in the countries with fully socialized agriculture. In the former, the state has to deal with millions of small and medium-size agricultural producers, who are more difficult to control. They constitute elements of a market mechanism in an otherwise socialized economy. In such a setting, prices, combined with the marketing monopoly of state and quasi-state (cooperative) organizations, are the most adequate means for directing the agrarian sector.

In principle, Communist governments aim at making producer prices cover the "average production cost" or the "socially necessary expenditure." The latter does not exclude marginal cost as a price-determining consideration if the planners assume that the price shall stimulate output of a certain product in the needed (planned) quantity. Yet marginal cost considerations are acceptable in a dirigistic economy only if it is not spontaneous demand, but the output goal as set by the planners, by which the quantities and the labor cost (and thereby the Marxian "labor value" of the product) are determined. The practical application, however, has been characterized by a contradiction between the principles of average and marginal cost. A Soviet specialist in this field has pointed out that "the value of the agricultural output emerges from the expenditures under the least favorable conditions of production [i.e., from marginal cost—K.–E.W.], but the practice of price formation is oriented towards the average production costs."[2] Nevertheless, the statement does not wholly reflect reality either, because there have been spontaneous developments of demand and supply—of products as well as of labor—that have forced the leaders and planners to adapt their planning of quantities as well as their setting of prices.

Production costs as a price-determining factor can exert their full impact only where and when they are being calculated. As long as labor costs remained a residual claimant on collective farm income, the impact of total production cost, and therefore of profitability, was relatively weak. Cost accounting in monetary terms was permitted and was gradually introduced in all Soviet kolkhozes only between 1956 and 1963. In most of the other countries, less than half the collective farms were applying it by the early 1970s, although in Czechoslovakia, the figure at that time was 65 percent.[3]

Although all the Communist governments claim to adhere to Marxist economic tenets, the theoretical reasoning regarding agricultural price policy and its application are not uniform among and within the countries (not to speak of the deviating case of Yugoslavia). According to a Polish economist, prices have to contribute to a state of balance in the economy; they must take account of shortages of goods; they also have a function in income distribution; and agricultural prices must not be set without regard to the prices in the other sectors of the economy. However, he also acknowledged that "the actual level of prices is an expression of the will and the jurisdiction of the central government."[4]

After the stagnation or even setback of agricultural production during collectivization and the years immediately following, the political leaders and the planners in Eastern Europe made efforts towards improving the food supply of the growing population, especially its urban segment. Therefore, agriculture rapidly had to raise its output and also the share of animal production in it. In order to stimulate output growth of the socialized farms and also, where such still existed, of the individual peasants, the farms had to be given the opportunity to develop initiative and to invest on a large scale.

One possible way for the farms to provide the means for investment was to reduce their labor cost, at least per product unit, if not in absolute terms. Saving on agricultural labor cost by means of extremely low remuneration had proved counterproductive in the past, and was no longer possible to an extent which would have kept labor cost low. The other way, that of reducing the number of workers while raising their previously low remuneration, required prior heavy investment, at least in a protracted initial stage. It has been said that during 1965–1974, Soviet agriculture's capital expenditure per output unit rose by 57 percent, while labor cost per unit of output rose by only 1.2 percent.[5] (In absolute terms, both increased greatly, as output also increased; see below.) Only for the GDR was it contended that compound expenditures per product unit remained stable,[6] whereas in the other CMEA countries, the "tendency of increasing cost per output unit in agriculture" also prevailed.[7]

In theory, it is possible for a dirigistic planned economy to finance increased agricultural investment not through farm incomes, but through a centralized fund and in accordance with directions from above, as was the prevailing practice in the state farm sector. Yet the financing of state farm investments as practiced prior to the period of economic reforms was not a viable model for the whole agricultural sector. On the one hand, the higher (compared to collective farms) labor remuneration in state farms, although providing more incentives from the very beginning, was costly. On the other hand, the centralized allocation of the means of production in this sector proved more and more inefficient. Both of these shortcomings placed a growing burden on the state budget. This situation was one of the reasons for reforming the whole agricultural sector (see chapter 11) instead of adapting the collective farm and kolkhoz system to that of the state farms.

The experiences of the recent past convinced the planners and political leaders in Eastern Europe and the Soviet Union that output increase and cost reduction through greater and more efficient investment, and under the conditions of a modernizing and increasingly differentiated overall economy, were easier to attain by decentralizing the factor allocation and paying higher producer prices. In other words, the farms had to be enabled to invest as they saw fit under their individual conditions. In this way, producer prices could again stimulate and also direct farm production, a function they had hardly fulfilled under the previous low price level.

An important step and precondition was the elimination of the system of two (in part, three and four) price scales for agricultural products. The previous low prices for planned compulsory deliveries and higher prices for over-plan and other special sales to the state were replaced by a uniform scale of procurement prices, which also incorporated the contract prices (mainly applied to cash crops for industrial processing). Special (and mostly lower) prices paid to state farms continued to exist, but were gradually raised so that they either came close to, or equalled the kolkhoz and collective farm prices. When state farms were put on self-accountability *(khozraschet)* during the reforms, that is, had to live with their profits or losses as more or less independent economic units, they were in most cases paid the same procurement prices as the kolkhozes and collective farms.

Compared to individual peasant farms, which for most products also are obliged to sell to the state's trade organization and consumers' cooperatives, the collective and state farms in the past received preferential prices. During the 1960s and early 1970s, the disparity of prices paid to socialist and private producers (including plot mini-farms) was on the whole eliminated, so that the average price increase for the private sector was greater. In Poland, state farms and private peasants were paid the same prices after 1971, but in Yugoslavia, the prices paid to socialist and to private farms still differ (see below). Bonus prices for over-plan production are usually not paid to private producers. However, during the 1970s, Hungary, Bulgaria, and the GDR introduced premium prices also for private sales to state or cooperative procurement agencies (e.g., in Bulgaria, for fruit, vegetables, and animal products).

Besides the public marketing channels, there are the legal free markets. On these, only sales directly to the consumer are permitted. To a minor extent, kolkhozes and collective farms also sell on these markets. Theoretically, the prices emerge from the demand and supply situation, which the state and cooperative trade organizations may only indirectly influence by also selling on the free markets. In practice, upper price limits are often prescribed locally by the authorities. Such markets (called kolkhoz markets in the Soviet Union) exist in all East European countries. In Czechoslovakia, they had almost disappeared during the 1960s without a formal prohibition, but revived to some degree during the 1970s. In the GDR, the quantities marketed are very minor, and private sales (mainly of fruit and vegetables) for the most part go to state and cooperative trade organizations. Yet as an economic "barometer" (as termed by the Pole, W. Herer),[8] the legal free markets are still of a certain importance, as are genuine black markets, which exist but on which only scattered information is available.

The Hungarians and Rumanians were first to eliminate compulsory deliveries and to introduce uniform procurement prices for collective farms, effective in 1957. Only one year later, Hungary made these prices apply to state farms as well. The Soviet Union established such procurement prices for kolkhozes as of 1 January 1958, Bulgaria followed in 1959, Czechoslovakia in 1960, and the GDR in 1964 for crop products and by 1969 for animal products. Poland took this step for her private peasants as late as 1 January 1972. However, most countries reintroduced a second, higher price level for over-plan sales, in one form or another, soon afterwards.

Most governments were intent on keeping the new prices constant. They emphasized that thereby the agricultural producers would be enabled to plan and calculate for a longer term, but the idea also was not to disturb macro-economic planning by price changes. Nevertheless, price changes (in most cases, increases) have been the rule rather than the exception. An attempt to let agricultural producer prices, especially those for nonbasic foods, fluctuate freely, also in the socialized sector, was made in Hungary and Czechoslovakia under the reforms of the 1960s. This policy was not pursued energetically, and the other countries of the CMEA refrained from copying it. On average, 90–95 percent of agricultural producer prices are still fixed centrally. The exception is Hungary, where centrally fixed prices account for only 50 percent of agricultural producer prices, although they cover the most important products. For the other farm prices, the 1968 reform provided for either free fluctuation (about ten percent of the prices), or more or less flexible regulation.[9] The Yugoslav economic reforms of 1965 envisaged an increase by roughly one-third for the overall average of prices actually paid to agriculture, and the state fixed them in so far as it guaranteed the individual peasants a bottom price for the major products, which in 1977 was lowered or made more flexible. This price was offered by state or cooperative procurement organizations in case the market price threatened to drop below the fixed bottom. For the other peasant farm products, price formation was wholly left to market forces.

The desire not to let producer and consumer food prices become a factor of inflation put a limitation on stimulating output growth in agriculture by means of higher prices. By Soviet and East European as well as Western standards, the prices were already too high in view of the fact that socialized farming is large or giant scale and theoretically should therefore be more efficient than small peasant farming. An immediate raising of producer prices paid to agriculture, and consequently also raising consumer retail prices, to a level granting profitability to socialist farms and higher incomes simultaneously to the agricultural population, would therefore in most cases have led to strong inflationary tendencies. When later on it became necessary all the same to raise producer prices for the sake of farm profitability, production stimulation, and income policy, the means to protect the nonagricultural consumer was heavy subsidization. The subsidy program affected not only retail prices (the main item), especially for animal products, but also investment, current inputs, farm organization measures, preferential credits, etc.

In view of this situation, most CMEA governments beginning in 1965 felt compelled to reintroduce two price tiers for agricultural products in order to stimulate output growth and at the same time not to let total procurement costs rise too much. Again, bonus prices were offered not only for better quality, but also for over-plan procurements. Taken together, these two kinds of bonuses raised the overall procurement cost of a given product by an average 5-10 percent above the basic price level.

In this way, the old problem reemerged that farms were interested to receive "slack plans", that is, low procurement quotas, in order to be able to sell a greater share of their produce at the over-plan bonus prices. Two main approaches were applied to avoid this. One consisted of calculating the plan and then deriving figures for over-plan output and/or sales on the basis of the "attained level" of the preceding years. The second, and this was the main approach in the GDR, was to offer the bonus to farms that agreed to "strained plans," often in the form of additional delivery contracts. In the second case, the additional sales beyond the originally imposed plan, or beyond the quantities calculated on the basis of the "attained level," had to be announced by the farms in advance and coordinated with the plans of the procurement agencies. Either no bonuses or small bonuses were paid for sales beyond the imposed or announced and contracted quantities. Sometimes both approaches toward overfulfillment of plans were combined to prevent "slack plans" and consequent excessive and unforeseen over-plan sales.

In the Soviet Union, neither kind of approach was generally applied, although the problem was much discussed. The Rumanian system remained similar to the Soviet. A major fault of this type of unmodified two-tier system is that overall average prices actually paid rise in good to excellent harvest years, and decline in unfavorable years. Thus, they tend to fluctuate inversely to what a market mechanism would bring. Instead of rising in a year of shortage of a given product, they fall and worsen the farm's losses, while in a year of abundance, they tend to lead to a sur-profit for a farm. In order to avoid this, in a current year with very good (or bad) harvest prospects, the basic plans may be changed upwards (or downwards) for individual regions or farms. Such action not rarely occurs in the Soviet Union, even with the so-called firm delivery plans, and of course makes nonsense of local or on-farm planning for those affected.

A special feature of the Soviet agrarian price system is the regional differentiation by products. This approach began to be systematically applied in 1958 for grain, and since then has been extended to most major products. As one Soviet author put it: "The zonal price mechanism predetermines the regulation of land rent relations between the state and the collective farms [and also the state farms—K.-E. W.]; it helps to redistribute the differential rent between them in definite proportions and also to some extent levels out the economic conditions of farming on a regional scale."[10] Because of the great regional differences in climate, soil, and labor and capital supply on the vast Soviet territory, the zonal price differences have to be very great, otherwise they would not really exert the compensating effect that the government refrains from achieving by taxes or levies. For example, grain prices in the late 1960s were about three times higher for the Baltic Republics and Northwest Russia than those paid to state farms in the best grain areas of Southeast Russia. An unwanted consequence was that Baltic farms strove to increase now profitable grain production, sometimes at the expense of livestock rearing, thus hampering regional specialization and defeating the purpose of the zonal price policy. Less pronounced regional price differentiation is applied in the GDR, Poland, and Bulgaria. It is recognized, however, that "mechanical evening out of the profitability of all products in all regions should be avoided since such an approach may cause damage to the interest of rational territorial specialization of farms."[11]

The Czechoslovak and Hungarian dotations (negative taxes, see p. 218) for producing on less fertile land also have a price aspect, as their payment is connected with product quantities sold to the state. In Hungary, preferential prices for the marginal producers are paid in addition.

The Yugoslav agrarian system has no comprehensive and mandatory delivery or procurement plans. A considerable part of the output, especially of animal products, vegetables, and fruit, is being marketed by the private peasants either through direct sales to the consumer or (mostly on the basis of delivery contracts) through the state and procurement organizations, which hold an oligopoly. Marketed crop products, especially grain, come from socialist farms for the most part, whereas the private peasants either consume, or feed, or process most of their crop products on the farm. Exceptions are typical cash crops such as oilseed, tobacco, etc. The procurement prices paid to farms in the social sector are set by the state, and private peasants are guaranteed a price which is lower, but by no more than 15 percent, than those paid to state and cooperative producers. On the traditional peasant markets, where no intermediate trading is permitted, the prices fluctuate freely and, as a rule, are higher than the procurement prices. On 3 August 1979, all consumer prices and also those paid to social farms were temporarily frozen, which resulted in shortages mainly of meat and milk supplies.

Since the early 1960s, and especially during the second half of that decade, the officially fixed agricultural producer prices in the CMEA countries have been raised in consecutive, sometimes rather great, steps. Thereby, the price increases noticeable in all of Eastern Europe and the Soviet Union in the mid-1950s were resumed. This earlier wave had started in the Soviet Union only half a year after Stalin's death. The overall price level for Soviet agricultural produce (excluding the prices paid to state farms) attained an index of 354 (1952 = 100) by 1964 (for the earlier period, see Table 4.1, above). Between 1964 and 1979, Soviet agricultural producer prices rose by a further 71 percent (see Table 10.1).[12] During 1952–64, this rise was possible mainly by reducing the state's sur-profit from retail sales. However, that source has been exhausted since then.

Table 10.1   Indices (1970 = 100) of the Overall Level of Nominal Agricultural Producer Prices in Poland, Hungary, Czechoslovakia, Yugoslavia, the GDR, and the USSR

	1960	1964	1965	1967	1970	1973	1975	1977
GDR[a]	72.4	84.3	86.3	89.4	100	105.8	106.0	106.7
Hungary	69.0	...	...	78.8	100	112.6	115.3	128.1
Poland	81.2 (1959)	...	...	95.9	100	119.7	136.5	183.0
Czechoslovakia	...	...	...	94.3	100	105.1	103.7	105.7 (1976)
Yugoslavia	...	40 (1963)	...	79	100	205	224	287
USSR	...	66	79	...	100	102[b]	...	113 (1979)
kolkhozes	...	72	82	...	100	103	105 (1974)	...
state farms	...	61	75	...	100	104	106 (1974)	...
private sector (purchases by public sector)	...	70	71	...	100	103	...	...

[a] Average prices actually paid, but excluding the special premia for output increase.
[b] The inconsistency that the overall index rises less than the indices for the three sectors is contained in the source, which bases its indices on a Soviet publication.
Sources: For the USSR, *Economic Survey of Europe in 1976, Part II* (New York, 1977), p. 64 (but indices for 1974 according to Gumerov, *op.cit.*, p. 125, and index for 1979, author's own estimate); for the GDR, *Statistisches Jahrbuch 1979 der DDR*, p. 262; for the other countries, *Economic Survey of Europe in 1978, Part I* (New York, 1979), p. 65 (Table), combined with FAO *Production Yearbook, 1976*, Table 116, and (for Yugoslavia in 1977) *1978*, Table 114.

The overall level of agricultural producers' prices in Czechoslovakia has risen only slowly since 1967 (see Table 10.1). In autumn 1979, it was announced that state purchase prices for cattle and milk would be increased by 16 percent as of 1 January 1980, while the prices for fodder mixes would rise by 13 percent, and those for fertilizer by 15 percent (both still subsidized).[13] Since 1960, Hungary has raised her overall price level for agricultural products almost every year, especially in 1968, when the agrarian system was reformed and the price level increased by 9 percent, according to the FAO index (see Table 10.1). The most recent increase, effective as of 1 January 1980, brought another procurement price increase of 11 percent, mainly for animal products.[14]

In Poland, the agrarian system was not changed or reformed during the 1960s, and the price increases were moderate throughout the decade (according to the FAO index). Yet when in 1971 (effective as of 1 January 1972), the low prices for compulsory deliveries were abolished, the index rose by almost 12 percent. The average Polish grain price remained constant during 1972–75, but was raised by another 40 percent in 1976. Other prices paid to peasant producers also rose in that year. State purchase prices increased only moderately, or even decreased (for grain) in 1977, while free-market prices for potatoes rose by 25.5 percent.[15] Price hikes for pork, beef, and milk followed in June and September 1978.

The rise in agricultural producer prices in the GDR was relatively steady after the completion of collectivization (1960). An overall acceleration of price rises was noticeable in 1963 and 1964. Potatoes especially had a big increase in 1962, with continued rises at a slower rate until 1969. Prices for slaughter animals also marked a strong increase after 1962, and became extreme in 1969.[16] After that, there was no change in the basic prices for the main products (except for another rise in potato and milk prices by 1971). However, new bonus prices for certain qualities of meat or, paradoxically, for farms which introduced modern large-scale animal production were inaugurated in 1973. Therefore, the average prices actually paid increased, although not those for milk, eggs, and crop products. By 1973, the "socially necessary expenditure" in East German collective farms was said to be covered at 92 percent by the prices, as against 83.5 percent in 1970.[17] Beginning in 1976, a number of producer prices were raised and others lowered, to ensure that the overall level remained unchanged while the price ratios among products were improved. These basic prices remained stable in 1977.

The Bulgarian price system was reformed after the Soviet model during 1964–1966, and animal products were especially favored. The UN Economic Commission for Europe estimated that the overall index of Bulgarian agricultural producer prices went up by 7 percent during 1965–70, and by another 15 percent during the subsequent quinquennium. The preceding rise, during 1960–65, must have been very steep, by roughly one-third.[18] Comprehensive price lists, published only after 1967, show a roughly stable price level for most main products since then, except for a fall for oats and poultry meat, a moderate rise for potatoes and slaughter cattle, a significant increase for mutton, and a sudden upwards move for pork in 1973 and 1974. More increases, except for eggs and sheep, followed during 1975–77, and were most sizable for maize, potatoes, pigs, and goats.[19]

Rumania seems to have kept the basic prices for the main products stable during 1971–76. A rise in prices paid for beef, pork, and lamb in 1973 was compensated by lowered prices for poultry, eggs, and fat. (The writer has no aggregated information on Rumanian prices of the preceding and the subsequent periods.) The very steep rise of the nominal prices in Yugoslavia is essentially a consequence of the 1965–67 reforms and of inflation. Prices for industrially produced inputs in agriculture increased even more for a few years after 1967, mainly because the state subsidies for such inputs were discontinued, except those for fertilizer. After 1970, the price scissors closed somewhat, that is, the rise in producer prices paid to agriculture since then became faster than those for industry supplying agriculture. This development was again reversed during 1973–76, and resumed in 1977.[20] During 1979, in spite of the price freeze of 3 August, agricultural producer prices rose by 24 percent, and a planned further increase by 12 percent was announced for 1980.

For the CMEA countries, a small or moderate inflation rate may be assumed. The overall index of prices for means of production and services supplied to agriculture, as officially provided by Czechoslovakia to the FAO, implies an increase of only 3.1 percent by 1970 as compared to 1967, and another rise of 5.7 percent up to 1976. Warsaw and Budapest indicated more pronounced increases of 2–4 percent per year, which were much surpassed in 1976 and after. It follows that in these three countries, the prices for industrial inputs rose at roughly the same rate as those for agricultural output. However, the consumer price index for the agricultural population in Poland and Hungary (no such information is available for Czechoslovakia) increased at a slower rate. Generally, agricultural wages rose sizably (see chapter 9), and total labor costs of the farms increased in spite of diminishing labor inputs. Consequently, the rise in agricultural producer prices, as far as it was not wiped out in real terms by higher prices for industrially produced material inputs, mainly benefited personal incomes in agriculture, at least since 1967, and to a lesser extent the financial situation and the investment capacities of the farms.

Soviet kolkhozes and state farms during 1965–1974 suffered from higher prices for construction materials and labor, fuel, transport, etc., which caused them additional expenses totalling 17 billion rubles, or 3.5 percent of their total receipts from sales to the state. On the other hand, they saved an equal amount through changes in their income tax assessment, state financing of land improvement works, lower prices for electrical energy supply, and annulment of debts. The various increases of the producer prices paid to them raised their gross receipts (apart from those for increased quantities) by 111 billion rubles for the whole 1964–1974 period. Of this sum, 52 billion went into wage increases and improved social transfers, while 59 billion was used for "expanding production," not only for investment, but also for generally higher current expenditures for the increased output. Higher prices of fertilizer and machinery purchases were compensated for by subsidies on these goods, but this did not apply, or applied only in part, to new types of machines and to spare parts.[21]

Gumerov's data on prices and profits of kolkhozes during 1964–1974 show that for a number of products, the increases of prices exceeded those of costs by a total of 3.7 billion rubles, yet for grain, sugar beet, sunflower seed, potatoes, vegetables, and wool, combined, the rise in production expenditures exceeded the rise of prices by a total of 2.2 billion rubles. In view of an overall rise of the price level by more than half, this is hardly a satisfactory result. In a breakdown by regions or predominant output mixes, results may appear better in some cases, and worse in others.

To compare the agricultural producer prices as such among the various Communist countries, or to compare them with Western prices, is neither possible nor would it yield meaningful results. Even if the currencies in Eastern Europe and the Soviet Union were freely convertible, the differences of price structures, for agriculture as well as for its supplying industries, including construction, would prevent sound comparisons. Instead, the structures (ratios) of the main agricultural producer prices may be compared (see Table 10.2), especially where such products are utilized on the farms, and are part of the cost factors of other agricultural products, such as—most important—feed (grain and potatoes). The structures also give an idea of the comparative preferability of one product to another, such as milk versus beef, or beef versus pork. For most of the European Communist countries, the relevant price data since the mid-1960s are available more or less complete, with the one total exception of Albania.

As distinct from prices and costs in monetary terms, these ratios are comparable within and across the socio-economic systems, and therefore two Western countries were included: Denmark (before and after her joining the European Common Market), and the United States for 1977/78. The world market prices were deliberately left out, as they reflect a very specific "residual" market and have little to do with agrarian policies or the natural and economic conditions within a given country. It should be pointed out, though, that over a longer period of time, the average world market price of feed barley was 10–15 percent lower than that of wheat (and one-third lower than first quality hard wheat), that up to the mid-1970s the price for beef used to be higher than for pork, and the ratio of feed barley to pork underwent heavy fluctuations.

In many cases the major Soviet and East European price ratios continue to deviate considerably from those prevailing in the West. For the reasons pointed out at the beginning of this chapter, they directly reflect the intentions of the agrarian policies, and only indirectly those of the supply and demand situation or the cost structure. To a great extent, this also applies to Yugoslavia, where of the products in Table 10.2, only milk does not have a bottom price guaranteed by the state. In Poland, the prices on the free market (added in Table 10.2 in parentheses) are heavily influenced by the state procurement prices, yet deviate from them. Of course, the prices in Denmark and the USA are also highly influenced by policy measures (least so in Denmark before she joined the Common Market).

Among the East European countries, including the Soviet Union, the differences in the price structures have been quite significant. (This also applies to the retail food prices, which cannot be dealt with here.) The wide ratio of the feed to the pork price is striking in both the Soviet[22] and the East German cases, and in Czechoslovakia before 1968. In comparison, the feed:pork ratio in Bulgaria (before 1974), Hungary, and Poland (after the price reforms of 1968 and 1971, respectively), and in Rumania and Yugoslavia looks more "normal." In Denmark, the pork price is for slaughter weight, so a comparable ratio in live weight would be less than Table 10.2 seems to indicate, and, before joining the Common Market, would have been similar to the Polish ratios.

A low comparative value was attributed, until the early 1970s, to potatoes in Poland, and a rather high one in the GDR and Czechoslovakia, although the three countries are similar in so far as great quantities of potatoes are directly consumed by both the human and pig population. In Hungary and the Balkan countries, however, potatoes are considered a vegetable rather than a basic food, and are highly priced. The high potato price in the Soviet Union seems in part due to the fact that potatoes are grown mainly in regions which are generally favored by the system of zonal prices.

As long as farms are not only large but highly diversified, they are less affected by irrationalities of the price structure. At least in part they can compensate for losses in one production branch by profitability in another. The case applies in peasant as well as in socialized farms up to the 1970s. Yet the more that farm specialization is aimed at and becomes

Table 10.2 Price Ratios of Some Main Agricultural Products in 1965, 1972, and 1976 (of average prices actually received per 100 kilograms, in the national currency)

Nation	Year	wheat[a]: feed barley[a]	feed barley[a]: pork[c]	potatoes[b]: pork[c]	beef[d]: milk[e]	beef[d]: pork[c]
USSR[f]	1965	...	1f:14.5	1:19.8	1:0.14	1:1.2
	1974	...	1f:15.8	1:20.4	1:0.12	1:0.9
Yugoslavia	1967	1:0.76g	1g:10.8c	(1.14.8)g	1:0.19	1:1.5c
	1972	1:0.93g	1g: 7.15	(1: 8.4)g	1:0.17	1:0.8c
	1977	1:0.94g	1g: 8.06	(1: 9.4)g	1:0.21	1:1.2c
Rumania	1973	1:0.80g	1g: 9.65	(1:11.4)g	1:0.20	1:1.1
	1975	1:0.80g	1g: 9.85	(1:11.8)g	1:0.17	1:1.0
Bulgaria[h]	1972	1:0.86g	1g: 9.67	(1: 7.5)g	1:0.22	1:1.0
	1977	1:0.88g	1g:13.24	(1: 6.5)g	1:0.25	1:1.2
Hungary	1965	1:1.30g	1g: 4.93	(1: 7.6)g	1i:0.24	1i:1.3
	1972	1:0.93g	1g: 8.48	(1: 8.3)g	1:0.16	1:0.9
	1977	1:1.02g	1g: 9.01	(1: 6.9)g	1:0.17	1:0.8
Poland[j]	1965/66	1:0.91 (0.82)	1: 5.87 (6.09)	1:20.0 (17.6)	1:0.24 (0.32)k	1:1.6 (2.3)k
	1971/72	1:0.98 (0.84)	1: 7.41 (7.48)	1:27.3 (19.7)	1:0.20 (0.25)k	1:1.6 (2.0)k
	1977/78	1:0.93 (0.89)	1: 8.53 (7.78)	1:17.6 (15.6)	1:0.19 (0.21)k	1:1.5 (1.9)k
Czechoslovakia	1965/66	1:0.64	1:13.0	1:16.9	1:0.20	1:1.1
	1972	1:0.86	1: 9.77	1:17.1	1:0.18	1:1.0
	1977	1:1.1	1: 9.41	1:14.4	1:0.20	1:1.0
GDR[l]	1965	1:0.99m	1:12.2	1:28.3	1:0.19	1:1.5
	1972	1:0.93m	1:14.5	1:21.2	1:0.17	1:1.1
	1977	1:0.91m	1:15.3	1:20.0	1:0.16	1:1.0
USA	1977/78	1:0.89	1: 9.3	1:11.1	1:0.28	1:1.15
Denmark	1971/72	1:0.88	1:10.9c	1:10.7c	1:0.18	1:1.3c
	1976/77	1:1.10	1: 8.64c	1:10.4c	1:0.18	1:1.28c

a Wheat of all kinds (in Rumania, contract price); for the USA, Poland, and Bulgaria, the average price of all kinds of barley was substituted for the feed barley price; for the GDR, it is the average basic price for all kinds of barley, including brewing barley; for Czechoslovakia, it is the basic, not the average, price of feed barley; for Rumania, the standard contract price for feed barley.

b Potatoes: average price for all kinds of potatoes in the USSR, GDR, Hungary, Czechoslovakia, USA, and Denmark, excluding early potatoes in Yugoslavia, Rumania, and Czechoslovakia; late varieties and including the bonus prices paid in Bulgaria; excluding early, seed, and industrial use potatoes in Poland.

c Live weight, with the exception of Yugoslavia (in 1967), and Denmark. Average prices for all kinds of pork (but in Poland, for fat pigs; in Bulgaria, excluding piglets).

d Live weight, average prices for all kinds of beef and veal; in the GDR, including all other kinds of meat, but excluding pork; in Yugoslavia and Hungary, beef only from beef cattle; in Poland, Hungary, Czechoslovakia, USA, and Denmark, excluding calves.

e Per 100 kilograms in USSR and Ukraine (milk equivalent), Bulgaria (3.6 per cent fat content in 1972; 3.4 in 1977), GDR (3.5 per cent fat content), USA and Denmark (milk sold to dairies in the last two countries); per 100 liters in Czechoslovakia (3.6 per cent fat content, basic price), Yugoslavia, Hungary, and Poland.

f Only kolkhozes, average prices actually paid, for all grains.

g In countries where domestic production of maize plays an important role, the following price ratios for maize of all kinds are calculated from the same sources:

		wheat[a]: maize	maize:pork[c]
Yugoslavia	1967	1:0.68	1:10.8[c]
	1972	1:0.98	1: 6.86[c]
	1977	1:0.77	1: 9.76[c]
Rumania	1973	1:0.78	1: 9.92
Bulgaria	1972	1:0.95	1: 8.75
	1977	1:1.16	1:10.10
Hungary	1965	1:1.60	1: 4.41
	1972	1:0.94	1: 7.41
	1977	1:1.14	1: 7.93
USA	1977/78	1:0.76	1:10.87

For these countries, the potato:pork ratio was put into parentheses, because potatoes there are practically not used for feed.

h No price information earlier than for 1969; since then, and until 1972, the producer prices in Bulgaria remained almost unchanged; statistically, the feed barley and feed maize prices actually paid declined in 1970 because beginning in that year, grain which was exchanged for concentrate feed and for which no bonuses are paid, was included in the calculation of the average.

i Beef price in 1966.

j Average prices for all purchases of the state from private peasants; in parentheses, free market prices.

k Cattle on the free market are only "old cows for slaughter".

l Average prices actually paid for all deliveries and sales, excluding those from state farms (VEG), including bonus prices, but not the bonus on output growth.

m The ratio of the fixed basic price, which remained unchanged over the years, was 1:0.94 (feed barley only).

Sources: For USSR: 1965, A. Emelʹyanov, Ekonomicheskie nauki, no. 11 (1974), p. 101; 1974, R.M. Gumerov, op. cit., p. 47. For all other countries: Prices of Agricultural Products and Fertilizers in Europe 1968/69, Annual ECE/FAO Price Review, no. 19 (New York, 1970), Prices of Agricultural Products and Selected Inputs in Europe and North America 1975/76, Annual ECE/FAO Price Review, no. 26 (New York, 1977), and Prices 1977/78, no. 28 (New York, 1979).

a reality in Eastern Europe and the Soviet Union, and the more that the large socialist farms begin to concentrate on only a few main products, the more the individual farm wins or loses from an irrational or a changing price structure. The official price formation has to take this into account, if, corresponding to the more recent policy priorities, the greatest possible number of farms is to become profitable, and if it is to be avoided that either the state must make up for their losses, or else certain segments of the agricultural population have to bear them through low wages and incomes. At the same time, the specialized farms are to be given a greater possibility of decision making in adapting to the price structure and thereby in aiming at enterprise profitability under the local circumstances. Within such parameters, the price structure becomes as important as the price increases as such. Formerly, animal production was least profitable, and now is to be expanded, so it was only logical that in most cases its prices were raised more than those for crops. Even so, only crop production is said to be profitable on the average in the CMEA countries.[23]

Agricultural producer prices, compared to incomes and other prices, have become quite high in recent times, especially for meat. Therefore, one might have expected an increase in consumer prices, too. As late as the early 1960s, agricultural subsidies of the magnitude outlined below were considered unacceptable by the CMEA countries. Therefore, during 1962–64, retail meat prices were raised in most countries.[24] Except for Hungary and, of course, Yugoslavia, and the abortive attempts in noncollectivized Poland, this was not repeated up to 1979, obviously for socio-political and more general reasons of policy.

The strong reactions of the population against retail price increases for meat and milk in the Soviet Union in 1962, and in Poland in 1970, 1976, and 1980 are well remembered by the policy makers. In Hungary, however, similar consumer price increases were enacted with more consideration for public opinion and caused no such disturbances. In the event, the overall food price index rose by 6.9 percent in Poland from 1974 to 1977, whereas the percentage was 5.7 in Hungary for the same years.[25] As a substitute for general price increases, the Polish government, beginning in summer 1976, established "commercial shops" where quality foods were more easily to be had, but at higher prices. The Rumanian government announced price increases for vegetables, potatoes, and fruit in state and cooperative stores in autumn 1976.[26]

As late as 1979, it became obvious that consumer food price rises were being discussed intensively in East European Party and government circles, and implementation followed not only in Hungary but in Rumania and Bulgaria. Bucharest raised the prices for fruit and vegetable preserves, for some kinds of fish and fish products, as well as for sugar products, in May 1979. The Bulgarian move in November was more drastic (prices for a number of important foods rose by about two-fifths), although mitigated by sizable wage and transfer payments increases. In July 1979, Hungarian consumer food prices went up by an average of 20 percent (meat by 30), after the price of rice had jumped by one-half in January of the same year.

With official consumer prices for most basic foods and animal products relatively stable in most countries, the otherwise unavoidable results for the processing industries and the trade network were compensated for by severing the link between these two price categories. For a while, the state was able to do so by diminishing the prescribed profit margins to be achieved in processing and trading and the formerly high turnover tax, or by varying the turnover tax rates among individual products. Increasingly, however, price subsidies became a necessity.

In the Soviet Union, subsidies amounted to as much as 11.2 billion rubles for meat purchased by the processing industries, and 2.8 billion in the case of milk, by 1972 (the two most heavily subsidized products). For animal products alone, the subsidy increased to 19 billion by 1975. The total subsidy, including that on industrial products sold to agriculture, was 22.9 billion in 1975, or 6.3 percent of the net material product (national income in Soviet definition).[27] It has continued to increase since then. In Poland, the total state subsidy bill on food cost 120 billion zloty in 1977 (which was equal to 7 percent of the national income), 180 billion in 1978, and an expected 250 billion zloty (a record) in 1979. Already in 1977, the state paid 0.73 zloty of subsidies for every zloty's worth of food bought retail.[28]

For the GDR, it was said that without state subsidies, food prices would have been 28 percent higher than they actually were. Between 1966 and 1972, an effort was made to shift the emphasis from subsidizing agriculture's inputs to subsidizing food retail prices, but the former remained sizable all the same. The latter averaged 4.85 billion marks per year during 1960–63, 4.4 billion during 1966–70, 6.5 during 1971–75, and reached 7.4 billion marks by 1976, which was near to 5 percent of the national income.[29] The Hungarian food subsidy bill in recent years consisted mainly of direct and indirect subsidies for the agricultural producers. The total was 25.3 billion forint in 1978, as against 12.2 in 1968.[30] Subsidies additionally paid to the processing industry and trade network may have been of similar or somewhat smaller magnitude than the state subsidies on food in Czechoslovakia, which in 1976 made up 20 percent in the overall retail food prices.[31] By the mid-1970s in Czechoslovakia, subsidies accounted for one-third of the consumer price of milk, 59 percent of butter, and 53 percent of beef.[32] Albanian agricultural producers' subsidies were said to have amounted to only 27 million leks for the three years 1967–69, but to "hundreds of millions" during the six years 1970–76.[33] No food or agricultural production subsidy figures are available for Bulgaria and Rumania.

The value of subsidies to agriculture in Yugoslavia doubled (in real terms) during 1959–1964, and then decreased as a consequence of the 1965–67 reforms. The growth in subsidies in nominal terms after 1970 is in part explained by inflation, but their increased share of national income reveals a rise in real terms as well. A portion of the subsidies is fixed and paid by the federate republics. The Yugoslav federal government paid 0.6 billion dinars in 1965 and as much in 1970, and this made up 3.4 percent of the national income in the first and only 2.2 percent in the second case. However, by 1976, the percentage had risen to 7.6 (=4.4 billion dinars).[34]

Subsidies on agricultural producer prices, including those paid to the processing industry and trade organization for keeping consumer prices low, probably form the major part of all subsidies directly and indirectly extended to the food sector. But other forms of subsidizing agriculture are also applied, such as preferential prices for inputs bought by farms (fertilizer, machinery and equipment, processed feed, electrical energy, fuel, etc.). Other forms are levies or "negative taxes" for equalizing differing natural conditions of farming, preferential credit terms, etc. They are impossible to aggregate and quantify exactly. Indeed, it is doubtful whether in some cases they do not rather benefit the industries producing inputs for agriculture.[35] In any case, it may be stated confidently that the share of direct food price subsidies has been increasing, except in Hungary. Some of the above data are likely to comprise part of the other subsidies. One must add that agricultural price subsidies are well-known in non-Communist countries as well. Yet the fact that socialized agrarian policy cannot do without subsidies either, and even increased them to heights much above those for Western countries, deserves special mention.

Summing up, it may be stated that the high level of agricultural producer prices (high in comparison to labor incomes in general, and those of agricultural labor in particular) is caused by the desire of the Communist governments to stimulate output growth mainly for animal products and (not dealt with above) certain cash crops. Yet the development during the past fifteen to twenty years shows that the agricultural price policy in Eastern Europe and the Soviet Union is geared toward establishing more balanced price ratios, as distinct from the overall price levels. These ratios are economically more viable for producers and, among other things, by the 1970s began to resemble Western price structures. This is a clear improvement over the first postwar and the precollectivization years, when price policy also served to discriminate against those products, mainly livestock, which were predominantly produced in the private sector of agriculture. More or less (most evident in Yugoslavia), prices are nowadays also fixed with a view toward the private producer. At any rate, the prices and price ratios and their more recent changes demonstrate one thing: the notion that socialized farms can produce without regard to problems of cost and profits has been given up, if it was ever fully adhered to. So far, however, this had been achieved at the expense of staggering subsidies.

Notes

1. J. Borodin and G. Schaschajew, *Internationale Zeitschrift der Landwirtschaft* (Sofia/East Berlin), no. 1 (1967), p. 30.

2. R. M. Gumerov, *op. cit.,* p. 14.

3. M. E. Bukh, *Problemy, op. cit.,* p. 153.

4. A. Woś, in *Integratsia sel'skogo khozyaistva s drugimi otraslyami narodnogo khozyaistva v stranakh-chlenakh SEV* (Budapest, 1976) (hectographed), p. 117.

5. *Pravda,* 6 April 1976.

6. *Kooperation,* no. 3 (1976), p. 140.

7. M. E. Bukh, *Problemy, op. cit.,* p. 87.

8. *Contributed Papers read at the 16th International Conference of Agricultural Economists,* Papers on Current Agricultural Economic Issues, published by the University of Oxford Institute of Agricultural Economics (Oxford, 1977), p. 213.

9. M. E. Bukh, *op. cit.,* p. 154, and B. Csikós-Nagy in *Reform of the Economic Mechanism in Hungary,* István Friss, ed. (Budapest, 1969), p. 138.

10. I. Lukinov (USSR), "The methodology of forming prices . . .," *Policies, planning and management for agricultural development,* Papers and reports, Fourteenth International Conference of Agricultural Economists (Oxford, 1971), p. 243.

11. V. Boyev (USSR), "Pricing as a tool . . .," *The future of agriculture,* Papers and reports, Fifteenth International Conference of Agricultural Economists (Oxford, 1974), p. 203.

12. For prices of individual major products in 1965 and 1971 through 1977, see *USSR Agricultural Situation: Review of 1978 and Outlook for 1979,* USDA, supplement 1 to WAS-18, Washington, D.C. (April 1979), p. 29 (Table 15). For the USSR in 1979, see A. Stolbov, *Sel'skaya zhizn',* 23 June 1979, p. 2.

13. *Život strany,* 22 October 1979 (as quoted by *RFER Czechoslovak Situation Report,* 14 November 1979).

14. *Népszabadság,* 6 and 10 October, and *Magyar Mezögazdaság,* 10 October 1979 (as quoted by *RFER Hungarian Situation Report,* 26 October 1979).

15. *Prices of agricultural products and selected inputs in Europe and North America 1977/78,* Annual ECE/FAO price review, no. 28 (New York, 1979), p. 93.

16. For a detailed study on GDR agricultural prices, see Theodor Berthold, *Die Agrarpreispolitik der DDR* (West Berlin, 1972), especially the graph on p. 127.

17. M. E. Bukh, *Problemy, op. cit.,* p. 152.

18. G. I. Shmelev and V. N. Starodubrovskaya, *op. cit.,* p. 147, indicate 46.7 percent for the whole period 1960–71.

19. *Prices of Agricultural Products and Fertilizers in Europe,* Annual ECE/FAO price review, no. 21 (New York, 1972), p. 63, *Prices of Agricultural Products and Selected Inputs in Europe and North America 1974/1975,* Annual ECE/FAO Price Review, no. 25 (New York, 1976), p. 75, and *1977/78, op. cit.,* p. 73.

20. *Statistički godišnjak, 1978,* pp. 204, 206.

21. For the above, see Gumerov, *op. cit.,* pp. 48–49, 127, 169–196, 231.

22. It would be still greater in the USSR if Soviet feed barley prices were available instead of the average price for all grains. Thus, in the Ukraine where little hard and quality wheat and little brewing barley is being grown, the wheat:feed barley ratio was 1:0.73 in 1973, and the feed barley:pork ratio reached an extreme of 1:20.9 (calculated from price data in *Prices of Agricultural Products and Selected Inputs in Europe and North America 1973/74,* Annual ECE/FAO price review, no. 24 (New York, 1975), p. 96). In part, the feed:pork price disparity is explained by the low feed:meat conversion ratio in Soviet animal farming. Thus, in 1971, for all kolkhozes and state farms, on average more than 9 feed units (equivalent of 9 kilograms of oats) were needed for one kilogram of gain in weight of a pig (*Ekonomika sel'skogo khozyaistva,* no. 5 [1972], pp. 36–42).

23. M. E. Bukh, *Problemy, op. cit.,* p. 152.

24. *Osnovnye cherty i zakonomernosti mirovoi sotsialisticheskoi sistemy sotsializma* (vol. 2 of *Mirovaya sotsialisticheskaya sistema*) (Moscow, 1967), p. 271.

25. *Economic Survey of Europe in 1977, Part 1, op. cit.*, p. 154.

26. *Neuer Weg* (Bucharest), 17 October 1976. For the previous prices, see *Buletinul Oficial*, no. 37, 19 March 1969.

27. V. G. Treml, *Agricultural subsidies in the Soviet Union*, USDA, FER no. 15, Washington, D.C. (December 1978), pp. 9, 12; I. F. Suslov, *op. cit.*, p. 193.

28. *Economic Survey of Europe in 1977, Part 1, op. cit.*, p. 153; DIW (Deutsches Institut für Wirtschaftsforschung): *Wochenberichte*, no. 22/79, p. 234.

29. *Economic Survey, loc. cit.*; DIW:*Wochenberichte*, no. 42/73, p. 380, and no. 5/78, p. 56; Chr. Krebs, *Die wirtschaftlichen und sozialen Zielsetzungen für die Landwirtschaft der DDR und deren Realisierung*, FS-Analysen, 6-1976 (West Berlin, 1976), pp. 35–39, 45–48.

30. *Mezögazdasági adatok I. 1979* (Statistztikai idöszaki közlemények. 452. kötet), p. 10 (for this item, I am indebted to Dr. E. Antal, Giessen); *Figyelö*, no. 73 (1973) (according to *ABSEES* 5:1 [January 1974], p. 3).

31. *Rudé právo*, 23 July 1977.

32. *RFER Czechoslovak Situation Report*, 8 September 1976, p. 12.

33. M. Kaser and A. Schnytzer, in *East European Economies Post-Helsinki*, Joint Economic Committee, U.S. Congress (Washington, D.C., 1977), p. 604.

34. Drugi Deo, in *Poljoprivreda u privrednom sistemu*, Z. Bejtulahu, ed. (Belgrade, 1976), p. 113; Ivo Kuštrak, in *Ekonomika proizvodnje hrane* 24:7–8 (July–August 1977), p. 27. I owe thanks to Dr. I. Lončarević, Giessen, for these two references.

35. For the imputation and data problems involved, in the GDR case, see Chr. Krebs, *op. cit.*, pp. 45–48.

11. The 1960s: A Decade of Reforms

The General Setting

PLANNING AND DIRECTING agricultural production in the Soviet Union and its European CMEA partner countries has basic characteristics inherited from Stalin's reign which must be outlined to ensure an understanding of the nature of the reforms of the 1960s. Jerzy F. Karcz, in his pioneering essay, aptly termed the earlier system "command agriculture."[1] Even so, its economics contained elements of indirect guidance such as prices, wages, and investment policy. Stalin did not object to economic efficiency and individual striving towards higher labor income. Rather, he sought to utilize these forces, providing they did not conflict with his vision of economic development nor with state control. However, this policy applied mainly to industry, which relative to agriculture received high priority in capital goods supply and labor income, those two incentive prerequisites which could effect better performance.

Agents of the discriminatory policy toward agriculture were mainly the rural state administration and the procurement organization. They acted according to national and territorial plans, which were called "laws," binding everybody, and their demands on agriculture were formulated in the procurement plans. The deriving mandatory control figures, which detailed the plans from the center downwards to the district administrative levels and farms, concerned areas sown to individual crops, numbers and composition of the livestock herds, land productivity, capital inputs, ways and timing of the execution of certain work, and many other specific requirements. Not only was the delivery of products to the state prescribed in detail and held under control, but also, in order to ensure fulfillment of tasks, so was the production process itself. As an example of the minute detail involved in this sort of planning, as late as in 1955, Bulgarian state farms had six hundred mandatory plan figures to fulfill.[2]

The Party organization took part in conveying such plans and controlling their execution. In agriculture more than in other branches of the economy, the rigid and centralized planning was unable to take into account the diversity of climate, soil, and social and economic conditions at the local and regional level.

Many of the plans were very ambitious and impossible to fulfill with the available means and incentives. Therefore, the farms and their workers often tried to fulfill only part of the plan obligations, neglecting those on which less official emphasis was put or which made them incur losses. To prevent such behavior, stricter control was applied from above, and the apparatus and jurisdiction of the state and Party organs expanded correspondingly. Concomitant was an equally logical extreme centralization of this apparatus, which, in order to remain manageable, had to be specialized by branches under a number of ministries concerned with agriculture (such as the ministries for state farms, procurements, soil improvement and water management, in addition to that for agriculture in general), each with its own organization down to the province or district level. Even for the amounts of produce in kind and cash income distributed among their workers and members, the farms had to obtain the consent of the state and Party administration. Making optimal use of the production factors and potentials of agriculture proved increasingly difficult for the command economy of the Stalinist type.

The system worked more or less, so long as soil and unskilled labor continued to be available in abundance. Yet the supply of these resources dwindled and the demand placed on agriculture in a developing economy grew more diversified and sophisticated. Consequently, greater numbers of highly skilled farm managers and workers became necessary to produce an expanding and improving output, and the Stalinist organizational models ceased to yield the desired results. Under such changing circumstances, greater and more efficient capital inputs, coupled with a more productive utilization of land and labor, became indispensable. Farm production processes required modernization, the spatial factor allocation had to be improved, and economic planning needed more flexibility.

Soon after Stalin's death, steps toward reforms were taken in Moscow and in other Communist capitals, signalling that such insights were widespread by that time. Not only the 1960s (which will mainly be dealt with in the present chapter), but the late 1950s as well (see chapters 4 and 5), can be characterized by a number of reforms. However, the earlier reforms were overshadowed by political influences and events of a different character: the liberalization under the early New Course (1953–55) in general, the continuation of collectivization in Eastern Europe, conflicts within the national leaderships, and upheavals in Poland and Hungary. They were not part of a consistent and coordinated reform concept for agriculture, and actions often proved erratic and contradictory. Yet the basic tendency of these early steps may be considered part of the first reforms which prepared the ground for the following decade.

It was generally recognized that more attention had to be paid to agriculture and to the needs of the rural population. The "only" problem was how and to what degree this belief was to be enacted. There was no question of endangering the existing political structure with daring reforms combined with a far-reaching economic liberalization. The economic mechanism was an integral part of the state economic and administrative system, and the people directing the mechanism were state employees and part of this system, down to the enterprise level. They managed the economy and profited from it in their careers and incomes, and therefore had a vested interest in improving its performance, sometimes even in removing some members higher up in the hierarchy, but surely not in destroying and wholly rebuilding the established organization. Although kolkhoz or collective farm managers were not state employees, they were in most cases Party members, and if not, certainly owed their position to state and Party.

The shortage, in proportion to demand, of the supply of almost all agricultural products made reforms not only urgent, but also difficult to carry out. Food reserves were lacking, so waiting a certain period for the productive effects of a comprehensive, consistent, and gradual reform was a luxury that could not be afforded. Instead, quick results from part measures were required. The problem was similar in the provision of capital goods to agriculture. As such supply was short, and the leaders preferred not to leave the capital inputs to an unplanned mechanism, they believed that a centralized allocation system would be the most rational. The shortage resulted from the inadequate capacity of industry in general (in itself partly caused by the planning system), and, as far as industries producing capital goods for agriculture were concerned, from failure to give sufficient priority to agriculture's needs. The shortage became more acute as the needs of a modernizing agriculture became more sophisticated.

Thus, a vicious circle ensued: centralized planning caused shortages of resources because it prevented optimal use, so that production was not equal to demands; the inadequate productive performance made it difficult to reform the centrally planned economy and to utilize resources more efficiently. In practice, the circle could only be broken if one approached the problem from all sides: in planning methods, in administrative and enterprise structures, in capital supply, in developing initiative by appealing to the enterprises' own interests, and last but not least, by providing incentives for labor through higher and more performance-oriented remuneration.

The central problems of immediate interest to the state were production planning and controlled marketing, which had been the basis, indeed the determining factors, of "command agriculture." Greater elasticity and more regard for local interests and conditions in the planning stage were preconditions for more efficient use of inputs and for positive effects from changes in enterprise and administration organization. Decentralization was an evident way to try to achieve this. Measures in this direction were taken under Khrushchev, beginning with the 1957 Soviet reform of industry and construction.

A series of reforms of all branches of the economy was enacted in the Soviet Union and Eastern Europe during the early and middle 1960s. Their inception was dominated by Khrushchev's ideas and measures. The reforms varied by individual countries, especially after Khrushchev's ouster, but they had the common intents of making the centralized planning system less rigid and administration more flexible, and of changing the terms of trade in favor of agriculture. As the macro-as well as the micro-economic management organs with their specialized bureaucracies formed part of the state administration, any reform of the planning and controlling system (including the control figures of plans) inevitably involved the state apparatus and its organization.

Following an initial stage, characterized by the hectic 1962–1963 Khrushchevian reorganization in the Soviet Union, the reform wave gained momentum and more steadiness by the middle of the decade. Somewhat changing their trend, the reforms began to acquire a more comprehensive aspect in Soviet and East European economic policy. These second stage reforms, which did not yet concern the collective farm sector, were officially decreed in the GDR in 1963 (June), in Bulgaria in 1964 (January), in Czechoslovakia, Poland, and the USSR in 1965 (January, July, and September, respectively), in Hungary in 1966 (May), and in Rumania in 1967 (October). Important changes toward greater liberalization in general ensued in Czechoslovakia during 1967 (not wholly repealed for economic management after the 1968 Soviet intervention) and in Hungarian economics during 1966–67, where the "New Economic Mechanism" was finally enacted in January 1968.

A comprehensive reform of the whole system of the Yugoslav economy was enacted during 1965–67 on the basis of the resolutions of the VIII (1964) Party Congress. Instruments of a market economy were applied to a greater degree than in Poland, not to speak of the other countries. For Yugoslav agriculture, the main intent of the reforms was to equalize the conditions of production to those prevailing in the rest of the economy and to eliminate the income and price disparities which existed to the disadvantage of agriculture. (As to the concomitant price policy, see chapter 10.)

The basic idea of the reforms was the same among the CMEA countries and differed in nature from the goals in Yugoslavia. The impulses originated not just from Moscow. The overall reform process in the CMEA was influenced also by steps first taken by the other countries. The movement was then more or less coordinated through consultations and conferences on governmental and Communist Party levels, also within the CMEA organization. During each country's practical implementation of agreed principles, the diversity of the national agrarian sectors has rather increased, and some differences have come to the fore.

As two Soviet authors put it: "The essence of the economic reforms in the CMEA member countries consisted of introducing a new mechanism of planning and of stimulating production. . . . It was intended to achieve the economic stimulation of production on the basis of an enhanced role of economic [i.e., indirect, not directly administrative] methods of directing, to increase the role of levers such as price, profit, premium, and others in the production process."[3] The reforms were often revised, changed and restricted, or resumed in subsequent years. In particular, the "forms of evaluation of the results of economic activities then underwent essential changes," i.e., by again de-emphasizing "profit" as a success indicator and "complementing" it with others, for example, net output[4] or "normative" net output. In actual practice, gross income remained the most important single indicator. A relativization of "profit" was logical under the given circumstances, because the price formation was not left to market forces, or in Yugoslavia and Hungary, only in part. As the decreed price reforms did not fully reflect the prevailing scarcity relations of goods, enterprise profit could not effect macro-economically optimal structures of production.

With a lag of between one and three years, the reform spread out to the collective farm sector. More than elsewhere in the economy, the reforms in agriculture were measures of economic policy concerning the place and weight of this sector. They did not fundamentally change the basic forms of socialized agricultural production. (The concept of the so-called horizontal and vertical cooperation and integration, which at a later stage superseded the attempts at systemic reform, will be dealt with in the subsequent chapter, with some inevitable overlapping in period and subject. Equally, there is also some overlapping with chapter 6, which deals with the actual developments of the sectors of farming during the period, but not with the reform concepts as such.)

The impetus of the reforms of the 1960s in the agriculture and food sectors of the CMEA countries diminished in the beginning of the following decade. Officially, however, the reform period was not declared to have come to its end. Specialized publications continued to discuss the problems which the reforms were destined to solve. In the practical implementation, details were revised and ways were sought to increase efficiency.

The underlying ideas also were discernible in Yugoslavia, in spite of significant differences of methods and instruments, and in Poland. The changes in the Yugoslav and Polish state farm sectors were in a number of aspects similar to those in the other countries, because state farms in all of them were organized and directed in a similar way. For individual peasant agriculture, however, comparable underlying motives and trends necessarily had to be put into effect in ways which differed from those applied to collectivized farming. Although the economic activities of the peasants remained more narrowly state-determined than in Western agriculture, a certain liberalization with regard to this sector was unmistakable in both Yugoslavia and Poland. Output growth as a priority goal encouraged higher prices and better supplies for peasants in order to give them both the incentive and the possibility to produce and market more. In Poland, the new leadership under Gierek (after December 1970) realized that the precarious supply and demand situation for food necessitated a more benevolent policy towards the individual peasants. Pragmatic measures, which may not be called a systemic reform of the agrarian system, were therefore enacted. Thus, in both Yugoslavia and Poland, the measures adopted temporarily favored the peasants, as producers as well as income-earners, in spite of the still valid socio-political goal of a "socialist transformation of agriculture" in both countries. By contrast, comparable measures to improve agriculture in the other Communist countries were almost exclusively directed at collective and state farms and their labor force.

Albania reorganized the governmental and administrative apparatus, but abstained from changes in the economic system, labelling as "revisionist" any reforms in such a meaning of the word. The reorganization resembled a mobilization campaign, and was first announced in an "Open Letter" of the Party's Central Committee on 4 March 1966. Shortly afterwards, the collectivization of the remaining parts of the peasant sector followed. The second stage began in 1969 and was accompanied by raising the agricultural producer prices. The resulting increases in collective farm members' incomes apparently compensated in part for the decline of their private incomes, which was due mainly to simultaneous restrictions on private plot activities. The continued planning of unrealistically high growth rates revealed that the leaders stuck to a Stalinist growth model, or perhaps to one influenced by Chinese economic thinking of the time. Another factor which favored setting goals for output growth at almost any price was Tirana's aiming at autarky in feeding the population. In view of the high rate of population growth in Albania, such autarky required a rapid food production growth.

More than any other CMEA country, Hungary has successfully established decentralized decision-making structures, direct economic relations between producers, processors, and consumers of products, and indirect guidance by the state, with direct interference "only in exceptional cases."[5] In this, Budapest was surpassed only by Prague during the tragic interlude of 1967–69. Once "orthodoxy" was reinstated, the

Czechoslovak reformers were accused, as far as collective farming goes, that by too great an emphasis on farm autonomy and its truly cooperative elements, they had adopted a rightist, destructive, and anti-socialist stand. Their critique of the pre-1967 agrarian policy now was labelled "revisionist." Worst of all, they were accused of wanting to weaken the role of the state in controlling agriculture and propagating a market mechanism instead,[6] although these were not in fact their aims.

The most elaborate system of combining strict central control with elastic forms of plan execution at the local and farm level was introduced by the GDR in 1963. It underwent some revisions, and by 1968 attained a form which at that time was considered definitive. However, a new general concept was promulgated beginning in 1969. Since 1970, more and more dirigistic elements have again permeated the system.

Rumania only hesitantly relinquished dirigistic management of the economy. In their practice, though not always in their theory, the Rumanian leaders were closest to Soviet methods. The Bulgarian attitude was similar, but beginning in 1968, Sofia's concept emerged as the most radical. The earlier, timid Bulgarian reform yielded to an ambitious plan to reorganize agriculture completely on the basis of "agro-industrial" cooperation and integration (attempted on such a scale for the first time among Communist countries).

As a consequence of the measures of price policy (see chapter 10), money as a success indicator and a stimulus for output growth gained in importance. In this way, one of the fundamentals of the earlier command economy in agriculture lost part of its impact. Not wholly eliminated was the compulsory procurement of agricultural produce at prices below actual production cost plus incentive profit. Residuals of such command production without adequate rewards for the producers nowadays are a result not so much of an overall price level that is too low, but rather of cost/benefit ratios among individual agricultural products or among regions, which may make certain production "advantageous" or "disadvantageous" for the farm (through prices or levies, taxes and subsidies).

The increase of agricultural incomes is only indirectly connected with reforms and has been dealt with in chapter 9. Yet new systems of incentive payments for better work and management performance have a direct bearing on reforms because they are aimed at improving economic performance. All CMEA member countries introduced premium funds of enterprises and farms, no longer as part of the wage funds, but financed out of profits, especially out of those above the planned level. For agriculture, this new venture originated in the state farms, but soon spread to collective farms in various forms along with the expansion of the general reforms. The additional premium and bonus payments out of farm profits were aimed at increasing the interest of both managers and workers in achieving such profits. At the same time, such financing of premium payments avoided a reduction of the funds available for basic remuneration. Thus, two kinds of premium payments emerged, one financed out of the wage fund, the other through a special premium fund. Since the premium funds, especially in state farms, in part depended on the numbers of employees, they did not stimulate reducing labor inputs and cost.

In a penetrating survey and analysis of the reform period from the early 1960s up to the mid-1970s, W. Brus commented: "In all Soviet-type East European countries, except Czechoslovakia in 1968, economic reforms . . . were conceived by the party leaderships as changes in the economic mechanism only; any impact on wider issues, especially the political system, was to be prevented."[7]

Changes at the Farm Level

The main issue was how the state and Party organs and their central leaders should guide the farms and other enterprises in new, more flexible, ways. Indirectly, this also touched upon the question of enterprise and farm autonomy. Whether spelled out or not, all reform measures had to weigh both aspects. It cannot be doubted that the reforms everywhere diminished the tutelage over the agricultural producers. If nothing else, fewer and more flexibly formulated plan obligations led to greater autonomy from the superordinated administration. The practical application of the reforms, however, was often countered by other provisions to tie the farms into the planning system, for example, by the obligation to conclude advance contracts with the processing industry and the procurement agencies. An actual, though incomplete, farm autonomy emerged only in Yugoslavia and Hungary.

Up to the mid-1960s, farm organization was among the areas touched by the reform measures. However, later in the decade, macro-economic aspects received more emphasis, and changes in farm organization were relatively neglected in comparison. In the initial period, the reorganization of central agencies was paralleled by the organization of production unions and associations of cooperating farms, often including industrial processing enterprises. Specialization in the largest possible units was thereby to be achieved without amalgamating the farms into still bigger farms, although in practice the amalgamations continued. The concept of association has been implemented in varying ways and degrees and has grown into a new development of integration (see chapter 12). The East German Cooperative Association (Kooperationsgemeinschaft [KOG]) of farms was an early example of horizontal integration, sometimes including also a state farm, with the participating farms retaining their legal identity. KOGs were first created in 1965, and by 1971/72 developed into forms of closer integration.

In 1965, Yugoslavia enacted a new Basic Law for the General Agricultural Cooperatives (opšte zemljoradnička—or, in Croatian, poljoprivredna—zadruga=OZZ) and other cooperatives. They were given more management autonomy, within the limits of the overall socio-economic system and the indirect influence of the Communist Party. The equivalent of a state farm in Yugoslavia increasingly was called a self-administering social farm, an "economy," or "combine." Since the 1965 reforms, these farms differ from the OZZ mainly in their size and output mix. The organizational and juridical differences have lost almost all importance under the general system of "Self-Administering Socialism," which in the socialist sector of Yugoslav agriculture was introduced in principle in 1954 and was afterwards gradually put into practice. On the whole, the privileges of socialist, as against private, farms were reduced by the reforms enacted during 1965–67.

The changes in Poland's collective farm sector were not of a fundamental and organizational character since its statutory setup was made more elastic after 1957 and especially since 1961, when the new Model Charter gave a general framework rather than a detailed prescription for the charters of collective farms. Changes in this sector concerned mainly its practical economic activities. Speaking of reforms, the further development of the Agricultural Circles deserves mentioning. Beginning in 1965, they were supplemented by Inter-circle Machinery Centers for leasing and repairing machines, which, since late 1972, in most cases have been merging into larger associations of Agricultural Circles of several villages. By this time, their membership comprised 55 percent of all Polish peasants, and they owned more than one-third of the country's agricultural tractor park.

Changes in the organization of the collective farm or kolkhoz sectors in the other CMEA countries also were minor, if the reform of the wage system is disregarded. However, it may be seen as part of the reform that the theoretical autonomy of the collective of farm members in decision making was taken a bit more seriously than before. Modifications in this sphere were specified, without much conviction, in the new Model Charters for collective farms, as adopted in Rumania (1965 and again in 1972—unpublished—and finally in 1977), in Bulgaria and Hungary (1967), in the USSR (1969), in Czechoslovakia (1975, replacing the Model Charter of 1961), and in the GDR (1977). A comparative survey of those recent charters is not possible here for lack of space.[8] Already since the late 1950s, convoking the general meeting of kolkhoz and collective farm members became difficult in the increasingly large farm units. An assembly of elected representatives from the farm sections therefore replaced the general meeting for most purposes (cf. p. 71). This innovation was incorporated in the new model charters and counteracted the feeble democratization process.

The internal organization of collective and state farms was somewhat changed, more or less experimentally in the beginning, but soon took definitive shape. Above all, the "production principle" of specialized farm brigades or sections for mass-production of certain crops or animal output was intended gradually to replace the older "territorial principle." With the latter, work had been organized according to territories, which often were those of whole collective farms before their amalgamation. In Rumania, the formation of "production farms" within the collective farms started as late as 1971, and many of them were soon criticized for being organized, in actual practice, not along true specialization lines.[9]

Apart from Rumania generally, and apart from some of the huge Bulgarian agro-industrial complexes since 1979 in special cases, the four-tier internal organization as it formerly existed in most state farms and the larger collective farms was replaced by one of three tiers: central management; brigades or sections; and work groups. Thereby, the former tier between central management and brigade was replaced by special central management departments for animal or crop production, technical services, etc. With growing average sizes, the four-tier organization might again have acquired some attraction, but shrinking labor numbers at the middle and lower levels, and stricter specialization of assigned tasks, seem to have made it unnecessary even then. Management of the now giant and slowly specializing farms was thus facilitated, but not so the responsible decision making at the middle (brigade or section) level. In Hungary, it was reported that "the present low level of management decentralization has become an obstacle of development in a great number of cooperative [i.e., collective] farms"; and for the state farms, it was found that "the frequency ratio of the decisions of operational unit managers, made under exclusive authority and competence, generally still is less than 1 per cent", the other 99 accruing to the farm's central managers.[10] (As to Yugoslav internal farm organization, see pp. 86–88.)

Within the sections or brigades, each work group (*zveno* or *otryad* in Russian) was to receive more independent tasks and some very restricted decision-making powers. These lowest units became increasingly smaller in labor numbers as labor in agriculture generally diminished and more capital inputs per worker and group became available. Such developments took various forms and names in the individual countries. Hungary tolerated the functioning of very small kinship or friendship groups of manual labor within the large socialist farms. Rumania went a similar way after 1969, which by 1977 led to a law on such groups (echipă) in collective farms, with labor remunerated by group. In the other CMEA member countries, there have been only some first indications of comparable developments, and it is doubtful how far these will be permitted to go. The "family link" (*semeinoe zveno*), which has recently received much attention in the Soviet press, is very different from the manual *zveno* of the late 1940s and also from a new semiautonomous and abortive form tried in 1962. It usually now consists of tractor and harvester drivers, and fulfills assigned tasks at given times. Its members have little operational autonomy and are remunerated individually.

Up to the mid-1960s, Albanian collective farms had only one level of subdivision, called the brigade. Afterwards, when more collective farms were amalgamated and their size increased, a second unit below the brigade became necessary. The *hallkë*, which parallels the Soviet *zveno*, was created. It seems unlikely that more autonomy for the Albanian brigade or the *hallkë* was intended or discussed.

In the state farm sector, the reforms aimed to relieve the state budget from subsidies and to make the farms produce more efficiently. More emphasis was put on economic (independent) accounting (in Russian, *khozraschet*) in all enterprises, among them the state farms. Theoretically, this was not new, but now it was to be implemented "fully" (as the Russians worded it), which meant that the state farms should live with their losses or profits and rely less on the state budget. State farms were expected to be profitable, and to finance production and investment out of their own resources. In recompense, they would now pay only a fixed part of their profits into the budget. There remained, it is true, state farms for which financial losses were part of the production plan imposed on them. Yet the majority were intended gradually to be separated from public finances.

This did not prevent the state from financing certain investments of the state farms wholly or partly, if this was provided for in the plan. The same applied, although to a far smaller extent, to some collective farm or kolkhoz investment, such as major land improvement works or infrastructural outlays. For most capital assets, the state farms now had to pay a levy, which for all practical purposes was interest paid on the capital which they used and which was owned by the state. The term "interest" was avoided, though. In most countries, the interest rate was very modest: 1–2 percent of the original price or of the book value (according to the most recent general re-evaluation, and not deducting depreciation). In Yugoslavia, a tax on the capital stock of socialist farms had the same function.

In the GDR, the rate of capital levy was fixed within a range of 0.5–3 percent, and in 1973 was incorporated into the more comprehensive "economically justified levy" (ökonomisch begründete Abgabe), which also included the land tax of collective farms and the ephemeral levy on the land (Bodenfondsabgabe) paid by GDR state farms during 1968–1972. Where the "natural conditions of production" were especially unfavorable, this levy could become negative, being turned into a general farm subsidy, the "Bodenfondszuführung."

As a means to foster integration into the economy at large, specialized administrations (trusts, associations, etc.) were formed for state farms on a regional and/or national level during the 1960s, mainly for poultry, sugar beet, cattle feeding, wine, and comparable industry-related farming. Where collective farms or kolkhozes were included in the process, these usually were converted into or joined to the state farms.

In sum, the reformed state farms were not to be wholly independent enterprises, although they were allowed and obliged to make decisions on their own, within parameters defined by legislation, and to bear the consequences. Their superordinated administration system was in part shifted to the provincial level, and thereby decentralized, although not for farms with a very distinct specialization. Such relative autonomy was similar to what had long since been valid, at least in theory, for the collective farms and kolkhozes. In order to enable the state farms to act in this way, the producer prices paid to them were gradually raised to those of collective farms, with a few exceptions such as farms with planned losses. Yet even in the excellent harvest year of 1973, the Soviet state still had to transfer 5.2 billion rubles to the state farms out of the public budget.[11]

The reforms in the Polish state farm sector, which began by the mid-1960s, were also aimed at greater financial and planning autonomy at the farm level, stimulation of productivity, and introduction of truly mass-scale production. In part this was achieved. In addition, beginning in 1962, state farms received more budget subsidies and other help from the state, and their agricultural area was expanded out of the State Land Fund. However, reshaping of state farm management and organization went into effect no earlier than by July 1972. The system of wages and ·salaries in Polish state farms was reformed in two steps in 1963 and 1971, and mainly concerned premium and bonus payments. Beginning in 1971, the limitation of premium payments to a percentage of a state farm's wage fund, as is the case in Soviet state farms, was lifted. These now were calculated on the basis of net farm income per hectare instead of physical yield. Yet such paying out of a farm's profit was rather limited in practice, since most Polish state farms still bear losses or earn only small profits. The sector as a whole is unprofitable, and this situation is not likely to change in the near future. In part, but not wholly, this may be explained by the special tasks assigned to Polish state farms, such as taking over land from peasants, producing on less fertile soils, and paying wages close to the average wages in the economy at large.

With both producer prices and volume of output increasing, farm incomes rose. However, the risks from losses due to harvest failures or other unforeseen setbacks also increased greatly, necessitating agricultural insurance cover. Agricultural insurance has long existed for kolkhozes and collective farms in all the countries (with the possible but unknown exception of Albania), and is compulsory in all of them except Yugoslavia and Hungary. During the 1960s and more recently, it was made more comprehensive and elaborate. The insurance company is, of course, state-owned, and the state fixes the premia and compensation payment rates, including those for voluntary insurance beyond the prescribed volume. For state farms, too, insurance was introduced in most countries, but not until recently (effective as of 1 January 1979) in the Soviet Union, and not as yet in Rumania (and probably Albania). The older system of all profits of state farms going into the state budget and all losses being covered out of it, in practice was a complete insurance in itself. When state farms were made self-accounting units, as long as they were not insured they had to form their own reserve fund and, in addition, transfer above-average or above-norm profits to the administration (or trust, association, etc.) of state farms, enabling these to compensate other farms for losses incurred.

The advantages of an insurance system over an administrative redistribution according to losses—as theoretically would fit a socialist society—lie in the fact that the possible compensation becomes calculable for the farm, and the premium payments more or less correspond to the various risks. Moreover, the assessment of the compensation is less exposed to wilfullness or incompetence, if a specialized agency instead of a general administration takes charge. In Yugoslavia, Rumania, Poland, and Czechoslovakia, also individual peasants and private plot holders may voluntarily conclude contracts of harvest insurance. The general logic is obvious: the more autonomy and self-responsibility accrues to farms, the more society and the economy at large are interested that the farms themselves see to it that risks and damages do not have repercussions beyond the local exigencies.[12]

Income taxes on kolkhozes and collective farms were levied, during Stalin's time and later, on their gross income, including the produce used on the farm (for feed, remuneration in kind, etc.). They weighed on low- and high-profit kolkhozes equally, and were geared to maximum deliveries to the state and to saving on production expenses (including labor remuneration) and investment. In the Soviet Union, in 1957, most production and amortization expenses, but not labor remuneration, were exempted. Beginning in 1966, as part of the reforms, kolkhoz net income, after deducting also the basic labor remuneration, was made the basis of tax calculation. In 1970, the income tax, although still so named, was turned into a profit tax on that part of kolkhoz net income which exceeded the production expenses by more than 15 percent, and on labor remuneration above a fixed minimum. It was the intention of these Soviet reforms to free the low-profit farms from most of the tax burden and to enable them to guarantee a minimum of labor remuneration. High-profit farms benefitted from an upper limit of tax amounting to no

more than 25 percent of their profit. The idea was to stimulate both
output growth and cost reduction per unit of output.

Tax reforms in the other CMEA countries with collectivized agriculture were enacted along the same lines, but took various forms and in
part used other indicators for tax assessment. Thus, the "economically
justified levy" in the GDR replaced all earlier taxes and levies. It was
based mainly on gross farm income per unit of land (differentiated
according to soil quality) and on average remuneration per worker,
thereby combining gross and net income and land value criteria. In
order to deter collective farms from remunerating labor at higher rates
than the authorities considered appropriate, that part of the tax which
was based on labor incomes (to be paid by the farm as a whole) in recent
times was assessed on an increasingly progressive scale, not to speak of
more direct means (remuneration norms and other financial rules issued
from above). In the GDR, it may go up to as much as 40 percent of the
total remuneration above a fixed minimum of 8,000 marks per year per
worker. During the 1960s, Czechoslovakia and Hungary, in addition to
income tax (which in Czechoslovakia up to and including 1974 was based
on gross farm income), introduced a land tax on collective farms in
order to stimulate improved economic planning and land allocation.
The Czechoslovak and Hungarian land tax strongly varied, depending
on the quality of a farm's agricultural land. On very poor soils, it may be
zero or even negative, that is, it may turn into a subsidy. In 1974, both
countries added a profit tax to such levies, and Hungary raised its rate in
1978 in order to restrict "excessive" labor remuneration. The taxes serve
not only to compensate for differences in soil, climate, land relief, etc.,
but also to enable farms to implement production directives where
otherwise these would cause them to incur losses. The character of the
East German "economically justified levy" is similarly complex. Rumania
abolished its land tax in 1968, but reintroduced it by 1977, while the
Soviet Union has levied such a tax on private land use only.

It is inherent in such approaches that the variations of the levies by
farm or region are determined not only by objective factors but also by
prior performance, which in part depends on merits or deficiencies of
planning, management, and work execution. Thereby, these may become part of the basis for tax, levy, or subsidy assessment.

State farms did not have to pay taxes (with the exception of Czechoslovakia), although the levy on productive assets (and in the GDR, the
short-lived levy on state farm land) were a kind of substitute. Labor
income tax had to be paid by their wage earners individually. More
recently, the new concept of agro-industrial integration in itself, where
state farms (enterprises) and collective farms join for integrated ventures, has been acting towards a unified tax system. The East German
levy of 1973 was such a uniform tax. For Bulgarian agro-industrial

complexes, a uniform net income tax for all member farms was introduced in the same year. A land tax was discussed in Bulgaria during the
1970s, but not implemented.

In Yugoslavia and Poland, peasants' incomes are taxed. As far as these
originate in independent agricultural activities, they are not actually
reported and assessed, but estimated on the basis of the farm's land and
its natural quality. Such a tax in essence is more on land than on income,
except that as a percentage of estimated income, it increases along a
progressive scale, more than the size and quality of a given farm's land
would imply. Partly a land tax and partly actually an income tax, its
mode of computation is directed against the supposedly wealthier peasants. Warsaw in 1962 increased the tax, but ten years later reduced it
again in favor of the bigger peasant farms.

Macro-Economic Management and Planning

The basic idea of central direction of agricultural (and other) production
and procurements according to the leadership's plans and priorities has
nowhere been abandoned, but the procedures and institutions have
undergone changes. The manifold and confusing details of the changes
in the organizational structures and links above the farm level can be
dealt with here in only the very broadest outline and meaning.[13]

The Soviet reforms were limited to some slackening of the reins, and
essentially changed only the administrative setup. In agriculture, the
state farms and the beginnings of horizontal and, to a lesser extent,
vertical integration of farms (or local groups of farms) were affected.
Rumania similarly did not really abandon command planning, in spite of
repeated changes in the administrative system and the formation of
various local farm unions and associations. Czechoslovakia retained
some of the liberalization of the years 1967/68 in the collective farm
sector, at least up to 1971, but did not basically abandon the system of
interference from above. Bulgaria, and more recently the GDR, organized gigantic agro-industrial complexes, and thereby an intermediate
planning level where strict orders from the center are transformed into
internal distribution of tasks. Where the complexes lacked narrow
specializations, they merely replaced the former general administrative
units as far as agriculture was concerned. However, where greater
specialization of the complexes and of the farms within them was
brought about, the methods of directing them from the center had to be
adapted suitably.

Hungary, more than any other of the countries with collectivized
agriculture, liberalized the planning system and macro-economic management. One may speak of indicative planning and corresponding
parameters, leaving to the Hungarian farms to act within them as they
see fit. If and when the overall results deviate from the plan, the

parameters, such as prices, taxes, levies, credit allocation, etc., are changed by government fiat.

Yugoslavia, and later on also Poland, wholly relinquished direct and compulsory planning of the peasant farm sector and merely continued to set, and at times to change, the economic and institutional parameters. Such indirect influence on the volume and assortment of private agricultural production was only partly successful. However, the same may be said of the practice in countries with socialized production and more direct and compulsory planning.

The administrative measures of Khrushchev's time (1961–1963) emasculated the powers of the ministries of agriculture and conferred them on other, more specialized agencies throughout the CMEA countries. Later on, the trend was reversed, and reforms were concerned mainly with centralized guidance. With regard to the overall economy, a Soviet author considered this necessary because "in a number of countries," the danger arose that "the effectiveness of centralized guidance declined," so that it became urgent to "make it more perfect."[14] For streamlining, the number of branch ministries was again reduced and a number of branch-oriented directive organs (under various names) was formed. The jurisdiction of some of the new ministries spread over several branches in order to overcome divisions between the subordinated branch administrations. As the agricultural reforms were enacted with a time lag, they were still at a rather early stage when the reorganizations of the central state authorities, among them those for agriculture, took place.

Between 1967 and 1973, the reorganized ministries of agriculture obtained jurisdiction also over the food industries. In most countries, they also took over procurement, supply of technical inputs, and forestry and water management, but not the relevant planning agencies. The unification was to demonstrate that agriculture was no longer to be considered a branch isolated from the rest of the economy, and especially not from its backward and forward linkages. Apart from the USSR, where most publications took a positive stand towards the idea and merely pointed to the immense Soviet territory as an obstacle to its implementation, this reorganization took place in all countries with collectivized agriculture (although Bulgaria included the agricultural machinery industry only as late as 1974). In Albania, the drastic reduction of the number of ministries during 1966, including the incorporation of the ministry of water management into that of agriculture, probably was not motivated in the same way as in the countries of the CMEA.

The organization of agricultural administration was not uniform among the countries, and moreover underwent changes over time within each. A common denominator was the greater emphasis now placed on the intermediate level of administration and planning. In the state farm sector, trusts and production associations often created product-oriented branch administrations, either centralized or regional ones. Local or regional associations or councils of collective farms and kolkhozes, headed by their own National Union or Council, were similar, but had fewer administrative functions. The Bulgarian collective farms, which had been headed by the National Association of Cooperatives since 1967, were again placed under the Ministry of Agriculture and Food Industries and the local general administrative organs in 1970. Similarly, the Bulgarian state farms again came directly under the Ministry, and their district administrations were abolished. The Polish state farms were entirely removed form provincial (województwo) control in 1974, and placed under one central state farm administration. Rumania formed associations and councils as early as 1966, and frequently reorganized them under various names. Most recently, in 1979 and early 1980, a General Directorate for Agriculture and the Food Industry (Directiá generală pentru agricultură si industria alimentară) has been organized in each county. They are supposed to unify and, allegedly at the same time to decentralize the management of the whole sector at the local level, although under the guidance of the Ministry of Agriculture, which probably will continue to exert strict central influence. Soviet kolkhoz councils in a hierarchical order up to the all-Union Council came into being in 1969. Czechoslovakia instituted district associations of collective farms in 1967, which later came under severe criticism, and a state-directed National Council in 1972.

In practice, such councils or associations either were, or soon became, an additional control instrument of the state over the collective farms, least so in Hungary. There, the provincial associations of collective farms, not rarely several in a *komitat* according to their members' specializations, were also accorded the function of spokesmen for farm interests in relation to the state. The government and administration controlled their compliance with the legal rules for their activities, and, of course, also used them for influencing the farms. The National Union of Cooperative Farms, which does not have open directive power over the provincial associations, may even propose central legislation. By contrast, the Kolkhoz Council of the Moldavian Soviet Republic to a large degree, as far as kolkhozes were concerned, replaced the republic's ministry of agriculture and thereby turned into a special kind of administrative instrument.

The Polish administrative reform, which was carried out in several steps during 1973–5, engendered considerable changes in rural areas, but bore no direct relationship to the agricultural reforms of the 1960s. The administrative unit of the small village (*gmina*) was replaced by the *gromada*, which in most cases comprised several villages, and between these and the province (województwo) the district level of administration was eliminated. In parallel, the service organizations for agriculture were reorganized into larger units, especially the Agricultural Circles into associations of such Circles. Other interfarm service organizations were in most cases incorporated into these associations if they were not organized in larger organizations of the state, or of countrywide trade and services cooperatives.

It has been generally recognized, and in recent times openly admitted in Eastern Europe and the Soviet Union that an assessment of the differential land value (Marxian differential rent I) is necessary in socialized agriculture, too, if uneconomic allocation of this factor of production is to be avoided. As land in those systems may not be purchased or sold (with a few minor exceptions), and therefore has no market value, other ways have been sought to assess its economic value, without convincing success so far. Taxes, levies, and subsidies are used as a substitute in order to make up for advantages or disadvantages of different land quality or location. Soviet zonal prices are merely another approach to the same basic problem. On-going experiments, which in the Baltic Republics have already gone beyond the experimental stage, further differentiated the Soviet zonal prices down not only to districts but also to groups of farms, in order better to levy the differential (not the absolute) land rent. As yet, the issue has not been solved in a way applicable for the whole USSR. Similar problems exist in Poland, Hungary, and the GDR, where prices also are differentiated by regions or individual farms, although less markedly than in the USSR.

The bulk of agricultural investment was to be financed out of the rising farm incomes due to higher producer prices, although in practice, most of the latter went into raising agricultural labor incomes. Later on, the progressive scale of taxes on labor remuneration was used as an incentive for investing. In the GDR, while for state farms and the new-style (1977) collective farms, percentages of net income to be assigned to various funds, including investment, were fixed by government or local administration, the collective farms were induced in indirect ways to assign as much as possible for investment. Part of the investments of state and collective farms continued to be financed out of the public budget, directly or in the form of subsidies on inputs bought by farms. Yet as a share of total farm investment, state "dotations" declined.

In compliance with the basic ideas of the reforms, the volume of credits expanded, although not as much as the financing out of the collective farms' own means. Formerly, they received less credits from the state, and most of these were short-term, for covering current expenses such as fertilizer purchases or labor remuneration. In the

Soviet Union, advance payments from the procurement organization served this purpose for the most part. Since then, the role of credits has grown, and interest rates have increasingly been recognized as a means to channel scarce resources into those investments which seem most desirable to the authorities. The interest rate still was too low to be really effective, but has been slowly raised. In the GDR, where it was highest, a level of up to five percent p.a. was attained, while for investment which was specially favored by the central decision makers, credits could be granted at two percent interest.

To the degree that money was able to determine the allocation of goods needed by agriculture, the credits provided for a greater degree of flexibility. Yet the state retained the decisive influence because the state-owned banks were the only source of credits (including the channelling of development loans from the West, as in the case of Rumania), and usually issued them only for specified purposes complying with the overall plan. Moreover, financial resources could in this way be directed to those inputs which were physically available and to the approved production orientation. Under conditions of technological modernization, with its greater need for industrially produced inputs, credit and investment policy turned into an increasingly important tool of guidance.

A logical price and credit policy should have changed from a system of allocating inputs in physical terms (planned quantities and assortment of machinery, construction materials, chemicals, etc.), as inherent in a command economy, to one of allocating in monetary terms, in other words, to replace administrative allocation by trade. An early first step in this direction was the abolition of the Machinery-and-Tractor Stations. However, the socialized economies have been unable to afford a free trade in producer goods because their supply, in most cases, does not meet their demand. Moreover, it seems to be feared that a free exchange of producer goods would deprive the central authorities and planners of their control over the flow of important goods, and thereby centralized planning might become impracticable. The still prevailing situation means that the farms are not really free to buy what they need and cannot always refuse to buy what to them seems expensive or less desirable in comparison with other inputs.

All the same, the command allocation of industrially produced inputs has lost some of its rigidity since 1965. A decisive restriction on free purchasing of inputs for agriculture was still that the overall economic plans, and not the sum of farm orders, determined what and how much the supplying industries were to produce. Where industrial products are not wanted by the farms but continue to be produced, the industry does not have to scrap them and suffer the corresponding losses. A widespread practice is that goods desired by farms are supplied only if they buy other goods also, or that credits are preferentially given for purchases of such less desired goods.

The economic reforms of the 1960s and the early 1970s were intended to move away from detailed delivery plans in physical units of agricultural products and toward more general and flexible sales obligations, some of them formulated only in monetary terms. For this purpose, the number of binding plan figures was reduced, but during the 1970s was again enlarged somewhat. In Hungary, the area sown to grain remained the only compulsory indicator referring directly to the production program of a given farm. The main control figures, or farm action parameters, imposed by the center were for state procurements, prices (with some room for local variations within the national average), availability of nonagricultural inputs, and labor productivity increase. The physical production figures of the national plans were restricted to basic products, given in procurement, not in output, quantities. In the GDR, one further step made "attained level" of each farm the basis from which its new targets were calculated, globally assessed in "grain units," into which the individual physical output items were recalculated and summed. (The grain unit was a more objective yardstick than the monetary value, which suffered from discrepancies of price ratios and from its absence for a number of products, e.g. forage, where no official price was set.) Later, the number of basic products for which physical procurement targets were set was again expanded, and the whole approach was soon discarded. In the other CMEA countries except for Hungary and, since 1980, Bulgaria, the number of products planned in physical units also increased again, often unofficially by local fiat. ·This applied to state farms more than to collective farms, as the former are under the jurisdiction of their own administrative system.

In Albania, the reorganization of the administration, as enacted in 1970, diminished the number of mandatory control figures for state farms, and also introduced some decentralization. However, the remaining control figures were still sufficient to determine all major farm activities. The same held true for the collective farms, at least for those of the 1971 "higher type."

The reform of the planning system in Yugoslavia differed from that in the CMEA countries not only because of the small-scale peasant nature of her agriculture, but also because of the specifics of "Self-administering Socialism," which became part of the Constitutions of 1963 and 1974. The federal center is responsible only for defining the parameters of economic activities, implemented through budgetary and credit policies and through market regulations. The lower social and political entities (republics of the Yugoslav Federation, provinces, and communities) then set the targets and determine which have to be achieved by market forces and which by consultations and agreements among the economic and socio-political entities in a given territory. The latter kind of actions are intended to exert influences comparable to competition because of the pluralism of the decision-making processes. In addition, they are an essential feature of the Yugoslav federative and self-administrative structure. Thus, the plans outline the measures for attaining the production targets, while the economic entities, among them the private farms, act independently within these parameters. In spite of this decentralization, the federal government still exerts a decisive influence on economic activities in the country, especially in foreign trade and in its use of financial instruments. Below the federal level, the lesser economic and socio-political strength of one group (in agriculture usually the producers) in the consultations and agreements sometimes makes its freedom of actions a mere formality.

It has been correctly stated by Soviet authors that if the central organs want to concentrate only on the main proportions of the agriculture and food industry complex and its development, they must increase the farms' autonomy.[15] In general, in the CMEA countries, fewer control figures are imposed on collective farms and kolkhozes than on state farms. For state farms, the control figures coming from the central planning institutions are complemented and augmented by those issued from the administration of the state farm sector. Often the local agricultural administrations proceed similarly with regard to the collective farms. On the whole, the complaint recurred during the 1960s and early 1970s that there was still too much petty tutelage, in spite of the intentions of the reforms.

In a book review, two Soviet jurists pinpointed a number of "important juridical problems" for agriculture. Should the plans of the state procurement and purchasing agencies determine the production specialization of farms or regions, or should not rather "a scientifically founded specialization play the decisive role in fixing the State's plan tasks?" Also, under which guiding idea(s), for example, economic rationality, geographical allocation, manageability, and easiness of planning, should there be a reduction in the number of products for which sales plans were to be imposed on farms? Finally, what should be the regulations and conditions under which farms would be permitted to exchange plan obligations among themselves?[16] The last question is especially interesting. It points to the desirability of allowing each farm to produce and deliver only what best it can, and to do this in excess of its plan obligation, if in exchange another farm takes upon itself those plan obligations for which the first does not have good natural and other preconditions. In this way, the plans imposed from the center would be corrected in the sense of better territorial allocation and greater specialization, without the overall quantity of each product being changed. Such a procedure, also between districts and whole provinces, was at times discussed in specialized Soviet publications, but was not officially approved. Other proposals, to impose on farms only plans for their total production expressed in money terms (or grain units, see above, for the GDR), were not accepted either. As a high-ranking Soviet official, in his book on price policy, later stated unequivocally, "The system of directing agricultural production comprises an interconnected complex of administrative and economic instruments. . . . In this system, macroeconomic planning takes a first and decisive place."[17]

Such an orthodox view continued to prevail in Rumania, Bulgaria, and Albania. However, in Yugoslavia, Poland, Hungary, Czechoslovakia (up to 1970), and the GDR, more or less important deviations from this practice occurred, although the principle itself was not challenged. Where associations of farms grow into much greater production complexes, integrated horizontally or vertically, they must eventually encompass a whole district, and an exchange of plan tasks within this unit is possible. This would allow for greater flexibility, even though the tasks continue to be spelled out in physical units instead of in global value terms. Bulgarian agro-industrial complexes may turn out to be a case in point.

In the Soviet Union, it was decreed early (in 1955) that procurement plans (no longer output plans) should be conceived and elaborated in an information flow between lower and higher levels. The center should convey preliminary global figures, to be broken down by regions. Knowing them, each farm should assess its possibilities, and propose its contribution to fulfillment. The individual proposals (farm "offers" in the GDR) would be collated by the superior administrations and planning organs. At the center, the plans would be worked out into one comprehensive plan, corrected to achieve overall requirements, and returned through the various administrations to farms to form their obligatory plans. As a result, the production potentials of the farms, as conveyed to the center, would be taken into account, and then the total demand, as anticipated or stipulated by the center, would determine the final plan and the assignment of tasks to each farm. The failure of practice to correspond to theory became obvious when nine years later, on 20 March 1964, almost literally the same procedure again had to be declared. In principle, the idea was adhered to also in the other CMEA member countries and has remained valid, at least in theory, since then. It is mainly the district administration which fulfills the task of reconciling the farms' production plans with the regional state procurement plans, and correspondingly with those for investment and specialization. As late as 1978, the Soviet district *(raion)* administrations were still said to be better in transmitting and issuing orders than in planning territorial management.[18]

If the proposals of the producers and the goals envisaged by the central state and Party leaders did not coincide, the total demand, as planned "above," always took precedence over the output offered from "below." Moreover, the political principle of "democratic centralism" provided various and effective ways for manipulating the "information" from "below" at the initial stage, to go along the lines implied by the preliminary figures issued from the center. In view of the prevailing scarcities and the ever-recurring nonfulfillment of plans, it would have been a surprise if the (unpublished) proposals from "below" had not been increased "above."

In addition to the final plans, the producers were expected voluntarily to work out complementary, mobilizing "counter-plans" for above-plan output. They had to announce these at the beginning of a given period, and get them agreed by the authorities as a binding self-obligation, so as not only to produce more, but also to make such additional production part of a "firm" (or "hard") plan and thereby calculable for the planners. Of course, nonperishable produce in especially great demand, such as grain, was readily accepted in any case. The "counter-plans" are not identical with the "above-plan" plans, which most governments imposed from "above." The latter were part of the overall obligatory plan. Higher prices were paid for the quantities sold both "above-plan" and under "counter-plans."

By making premium payments dependent on the fulfillment of plans, less so on unplanned overfulfillment, the governments—most consistently in the GDR—sought to prevent farm managers and workers from trying to receive "slack" plans, which were easy to fulfill, in order to sell produce at over-plan prices.

If procurement targets are not set so high that fulfilling them requires mobilization of practically the whole production potential, and if they are not laid down for each product separately, so that substitution among products is possible, then such plans may leave the producers some room for manoeuvering with their production program. Where the procurement plans were not imposed compulsorily and in physical quantities, substitution for earlier control was generally aimed at by making producers and procurement or purchasing agencies conclude delivery contracts. The idea was that the total of all plans and contracts would add up to the total envisaged by the central planners. Czechoslovakia and the GDR went farthest in this direction. In principle, it was intended that only the procurement and purchasing agencies should have obligatory plans, and that they would fulfill them by concluding the corresponding contracts with the farms as their autonomous partners. This usually applied to the "above-plan" and the "counter-plan" sales mentioned above.

Concluding contracts, however, was bound to be just another term for imposing plan obligations if the procurement agencies were in a much stronger position than the farms. This often was the case, since they influenced, directly or through other state agencies, the supply of scarce inputs, credits, and transport facilities. Also, they could manipulate the quality assessment, and thereby the price paid, for delivered farm products, etc. Moreover, they were supported by the local or regional administrations, which had great interest in reporting plan fulfillment for the territories under their jurisdiction. Only in Hungary did the collective farms seem to have real bargaining power in concluding contracts. This is illustrated, among other things, by the fact that Hungarian farms are compelled only for some basic products (bread grain, cattle and pigs for slaughter, and a few more) to sell exclusively to the monopsonistic procurement agencies of the state. For the rest, they may sell to other buyers (stores, hotels, hospitals, factory canteens, the processing industry, etc.), although not to private persons. Since 1 January 1978, they may even buy certain farm products from other producers for resale on their own account.[19]

Most of the planning and directing of procurement and/or sales formerly also applied to the relations between individual peasants and the corresponding agencies. Yet in Yugoslavia, the monopoly of the General Agricultural Cooperatives (OZZ) on marketing peasant output, except for direct sales to consumers on the traditional peasant markets, was abolished by the 1967 Basic Law on Trade. Since then, sales by peasants to other organizations and enterprises, though not to private traders, have also been permitted. In practice, however, the OZZ have retained their dominant position in trade. Other levers of the OZZ on the peasants were also somewhat weakened when their monopoly on all forms of cooperation with individual peasants was lifted in 1965. Since then, peasants may cooperate, on the basis of joint activities and commercial contracts, with any social enterprise and organization in agriculture, processing, and service.

In Poland, the compulsory sale of basic food products by the peasants was abolished as late as 1 January 1972, together with the lower prices paid for such sales. This diminished the elements of dirigistic planning, but the marketing system remained under strict state control, with monopsonistic state or cooperative industries and organizations, except for direct peasant sales to final consumers. As has been stated elsewhere, "the control figures of the plan are not binding on the peasants, but they have this character towards the institutions and organizations serving agriculture."[20] Apart from prices, among the strong levers of the state on the peasants in Poland are the instruments of budgetary and credit policy, land ownership and utilization policy, and supply of machinery and fertilizers.

Although it is not a systemic reform, the relatively recent practice of conveying the plan figure to each farm for several (in most cases, five) years has been an important development in the practical implementation of plans. This gave the farms a more stable basis for their own medium-term planning. Under the earlier system, farms were told every year by the local authorities what was expected in terms of production and deliveries. In practice, though, frequent current corrections during the plan period still occur, in most cases upward, particularly when it appears that five-year-plan targets otherwise would remain unfulfilled. Rumania has been especially notorious for current upgrading of plan figures.

A new element in planning, not only for agriculture, was the construction of long-term economic plans, so-called General Plans of Development, or perspective plans. Those referring to agriculture were for the whole food sector, including the backward and forward linkages of agriculture. Such long-term plans envisaged developments up to 1980 or beyond, but the planning period was later divided into long-term (twenty years) and medium-term. The GDR in 1963 was first to announce a long-term plan up to 1980. In Czechoslovakia, the next fifteen to twenty years were envisaged, while in Poland, Rumania, and Yugoslavia, the period was 1975–1990. Bulgaria issued general directives for fifteen to twenty years. Whether such long-term planning was based on a correct assessment of all production factors and of needs (demand) remains an open question. In none of the countries has the elaboration of comprehensive long-term plans been successfully concluded. So far they exist only for part-branches of the food economy, as the more ambitious, comprehensive approach proved to be too complicated.

Given the difficulties in constructing long-term perspective plans, attention soon switched to making prognoses for the economies in total and, as a precondition, for their individual branches. During 1966–1970, the effort was intensified, especially with a view to mutual coordination of plans within the CMEA. Each member country organized special institutions for this purpose, and as a time horizon, the quinquennium 1981–1985 was agreed upon (for some branches and developments, 1990–2000). In addition, during 1971–1975, the "perspectives of socioeconomic development" received more attention, in the form of establishing "development criteria to encompass social goals and needs as determining motives for the formulation of the technological-economic tasks of the perspective development."[21] Simultaneously, the objects and time horizons of such prognostication were differentiated according to partial fields of application and to periods (fifteen to thirty years, ten to fifteen years, less than ten years). Prognostication was to become a continuous process, during which the time horizon would be moved further ahead, as the time of elaborating the prognoses advanced. Thus, Soviet institutions by 1978 were elaborating plans for agriculture up to the year 2000. The prognoses of CMEA member countries were said to be distinct from "bourgeois" prognoses in so far as they not only would foresee future developments more exactly, but also would be used as instruments for active influence of central state management on the economy and society.[22]

While such prognostication would continue as a permanent process, its results at each given stage would become part of the long-term and the medium-term development plans. Prognostication would help to coordinate the five-year plans among the CMEA countries. Yet distinct from the five-year plans, the perspective plans were meant to be flexible and adaptable to changing conditions. They would provide parameters for the central planners and leaders, and not fix tasks for the individual partners in the economic process. The actual results and the impact of such long-term prognoses, plans, and (since 1975) "target programs" are hard to assess.

Below the level of long-term prognostication and planning, the five-year plans play "the leading role in the system of economic plans"[23] and are subdivided into annual plans. They continue to be strongly dirigistic, although "economic levers" (indirect guidance) have begun to become more important in them. Hungary has advanced most in this direction, but the GDR and Czechoslovakia have also made some progress, although with methods different from the Hungarian ones. In each country, actual developments have nearly always forced the governments to change a number of plan figures during the plan period.

It was emphasized over and over that "scientifically founded" volume and composition of output and the corresponding plan control figures, as well as the possible supply of capital inputs to agriculture and reduction of labor inputs, should be taken into equal account for overall plans and for the control figures which coordinate the individual farm interest with that of general plan fulfillment. This was called "complex planning," and it has been said that it was best attained in Hungary.[24] Theoretically, this is a basic condition of any planning, but in practice is hard to achieve. Input-output analysis and mathematical techniques, introduced in the CMEA countries during the 1960s, did not fulfill the hopes for more consistent central planning, as economic growth made the central planners' tasks more difficult.

"Complex planning," which ties the development of the entire agriculture and food sector into a consistent whole, was an idea rather than a reality, and the interests of the various part-branches still tended to determine the formulation of the plans. Bukh, the Soviet author who made this statement for all CMEA countries, added that planning, flexibility, and the ability to adjust to changed circumstances also require a "precise differentiation of functions between the central and the local planning organs," that is, some decentralization.[25]

The principles and time horizons of planning as outlined above were valid for all CMEA member countries, and in this regard the Soviet Union and its East European partners proceeded uniformly. It cannot be said, however, that such uniformity was due to impulses originating solely from Moscow. Yet with the exception of Hungary in late 1979, no more measures followed toward a liberalization of the planning and management system, nor toward new forms of farm organization, decentralization of decision making, and stimulation of local or farm initiative. The Soviet new economic reform of July 1979 lacked any such element and essentially was not geared at agriculture. Rather, the reins of central direction and assignment of tasks in terms of physical instead of aggregate indices were again tightened, for example, in the Soviet Union, the GDR, and Bulgaria. Even Hungary, in 1972 and again in 1976, for a while put some restrictions on the decision-making powers of enterprises (among them the farms). However, some of the achieved progress was retained in all countries, and Hungary quite recently, as of 1 July 1979, recommenced the decentralization of agriculture, although not for the state farms, by conferring additional decision-making competence to the county and farm level.

The relationship between planning and directing state organs on the one hand, and the producers on the other, was not essentially changed under the reforms, in spite of the modifications. The state and administration still do not have to justify their actions and orders vis-à-vis the producers. The overall economic interest is defined by the state and the Party, and takes precedence in formulating the plan and other tasks of the producers. The farms are obliged to produce and deliver what they are ordered to, even when this runs counter to their interests or when the overall parameters fixed by the state would allow for other farm behavior. Since the reforms, more attention has been paid to these interests, yet the mechanism of the command economy has remained in force. Except for Hungary, Yugoslavia and Poland, central state influence on agricultural production merely acquired new forms and was not decisively diminished. By 1980, Hungary stood out as the only CMEA country which continued along the path of her economic reforms of the 1960s.

Notes

1. "An organizational model of command farming" in *Comparative Economic Systems, Models and Cases*, Morris Bornstein, ed., 2nd ed., (Nobleton, Ontario, 1969), pp. 278–99.

2. E. V. Rudakov, S. A. Mellin, and V. K. Storozhev, *op. cit.*, p. 89.

3. V. Doroshenko and L. Mel'nik, *Ekonomika Sovetskoi Ukrainy*, no. 4 (1974), pp. 72–73.

4. Conference report by A. Mikhlaev in *Voprosy ekonomiki*, no. 11 (1975), p. 154.

5. A. Czepeli-Knorr, in *Integratsia sel'skogo khozyaistva s drugimi otraslyami narodnogo khozyaistva v stranakh-chlenakh SEV* (Budapest, 1976), pp. 224–25.

6. Jaroslaw Suchánek, *Právní regulace združstevňování československého zemědělství* (Prague, 1977), pp. 263–65, and, by the same author, *K obecné platnosti Leninovy agrární teorie*, Rozpravy ČSAV, no. 2 (Prague, 1974). I owe thanks to Dr. Z. Huňáček, Giessen, for this reference.

7. Włodziemierz Brus, "The East European Reforms: What happened to them?", *Soviet Studies* 31:2 (April 1979), p. 258.

8. Z. Huňáček, Giessen, is preparing a volume of texts and comparative comment.

9. *Probleme economice*, no. 10 (1973), pp. 48–55 (according to *ABSEES* 5:2 [April 1974], p. 230).

10. K. Kiss, *Vezetés a mezögazdaságban, az élelmiszeriparban, az erdészet-fairbarpan*, no. 1 (1973), pp. 25–30, and H. Szederkenyi, *ibid.*, no. 5 (1972), pp. 311–21 (according to *Hungarian agricultural review*, no. 3 [1973], pp. 30, 32).

11. R. M. Gumerov, *op. cit.*, pp. 213–214, 223.

12. For the only comprehensive Western treatment of agricultural insurance in Communist Europe, see E. Schütte, *Die Versicherung landwirtschaftlicher Betriebe in der Sowjetunion mit Berücksichtigung der anderen kommunistischen Staaten Osteuropas* (West Berlin, 1979).

13. For coverage of the whole period, see E. M. Jacobs in a number of consecutive articles (cited in the bibliography at the end of the present book), which are the best available source.

14. P. Kuligin, *Ekonomicheskie nauki*, no. 8 (1975), p. 72.

15. See, for example, M. Bukh, *Planovoe khozyaistvo*, no. 4, (1974), p. 150. Many other authors have made similar statements.

16. *Sovetskoe gosudarstvo i pravo*, no. 1 (1972), p. 157.

17. R. M. Gumerov, *op. cit.*, p. 22.

18. M. E. Bukh, *Problemy, op. cit.*, p. 150.

19. *Népszabadság*, 20 November 1977 (according to *RFER, Hungarian Situation Report*, 12 December 1977, p. 20).

20. A. Woś, in *Integratsia, op. cit.*, p. 257.

21. *Sovershenstvovanie upravlenia ekonomiki stran SEV. Na primere NRB, VNR, GDR, PNR, SRR i ChSSR*, R. N. Estigneev, ed. (Moscow, 1974), pp. 106–107.

22. V. Derzhavin, *Ekonomicheskie nauki*, no. 8 (1975), p. 63.

23. *Sovershenstvovanie upravlenia, op. cit.*, p. 137.

24. M. E. Bukh, *Planovoe khozyaistvo*, no. 4 (1974), p. 151.

25. M. E. Bukh, *Problemy, op. cit.*, pp. 143, 146.

12. *"Industrialization" and Integration of Agriculture*

Basic Concepts

AT THE END of the 1960s and during the early years of the following decade, when the reforms were losing their impetus and were in part being retracted, a new idea to improve agricultural efficiency found expression. After some preliminary discussion, the goal of "agro-industrial" cooperation and integration, also referred to as "industrialization" of agriculture, was officially proclaimed. In essence, this development aimed mainly at a technical and organizational transformation. In this, it differed from the preceding reforms, which were directed mainly at improving the economic mechanisms of the then existing farming structure. However, a clear dividing line cannot be drawn between both processes in either their timing or their substance. Thus, the exclusion from the preceding chapter of a discussion on agro-industrial integration may appear arbitrary to some. Nevertheless, the shift in policy emphasis was so unmistakable that we believe it justifies separate treatment in spite of some overlapping.

A number of fundamental similarities apply to both the 1960s and 1970s, but they concern general economic priorities and corresponding resource allocation rather than the reforms, as distinct from the integration process, in agriculture. The most important common characteristics were the rising agricultural producer prices and incomes and also the great agricultural investments, although as a share of total investment, the latter decreased in all CMEA countries except the USSR in comparison to the late 1960s. These common features were indispensable for the reforms as well as for "industrialization" and integration. They were destined to serve not only output growth under conditions of decreasing labor numbers, but also technological modernization, combined with the integration of agriculture in the economy at large.

At the international level of the Council of Mutual Economic Assistance, a parallel process was inaugurated by the July 1971 "Complex Program for the Further Deepening and Perfecting of the Cooperation and Development of the Socialist Economic Integration of the Member Countries of the CMEA." Although ambitious in their theoretical framing, the measures and perspectives outlined in its 11th section were rather modest. Nevertheless, it is intended that they will ultimately lead to a single comprehensive food sector of all CMEA countries. This would in part include Yugoslavia, who has been participating since 1964 in those activities of the CMEA which she deems to be of interest to her. This international agro-industrial complex, as it is envisaged, would be one of specialized close cooperation, optimal spatial allocation of production, and planned exchange of final or processed products, as well as inputs, in the whole agriculture and food sector.[1]

This project was for the distant future, but efforts were soon made to coordinate among the member countries the national processes of modernization of agriculture in a common theoretical reference framework. In May 1974, a council of experts of the CMEA agreed upon and recommended a terminology. According to this, there would exist "inter-farm organizations" (cooperation) and, if the cooperation is more intensive, "inter-farm associations" and "unions" (integration) of a horizontal type, as distinct from vertical "agro-industrial integration."

In the CMEA and the West, "horizontal" is understood to entail agricultural producers at roughly the same stage in the production process, without nonagricultural linkages, so that greater and more specialized production units within agriculture come into being. By contrast, "vertical" integration (or vertical cooperation) involves farms and forward or downstream (processing, distribution) and/or backward or upstream (supply of nonagricultural inputs and services for agriculture) linking enterprises of an industrial type.

Disregarding the details of names and definitions, which are not quite uniform among the CMEA countries, one gets a good idea of the CMEA's generally accepted difference between cooperation and integration from N. P. Alexandrov's (Moscow) explanation. Cooperation, according to Alexandrov, consists of joint activities up to and including the creation of a commonly owned enterprise where the partners themselves remain juridically and also partially economically independent. On the other hand, integration, again according to Alexandrov, is a closer union or amalgamation, up to and including the partners' loss of juridical independence, when the farms or enterprises turn into sections of a new, larger whole. In CMEA terminology, both phenomena may be either horizontal or vertical. In Western terminology, integration is understood primarily as being vertical (i.e., a joint activity of farm and nonfarm partners in the agriculture and food industries complex, whether in a loose or a very close form), whereas cooperation usually, but not necessarily, implies a horizontal kind of joint activities (i.e., among farm partners).

Beyond these basic aspects, manifold contents and forms in the East European countries and the Soviet Union are discernible. One characteristic is that special attention is paid to whether it is the farm(s) or the industrial processing enterprise(s) (forward linkage) that determine(s) the character of a complex, and whether the association (or amalgamation) is one of farms (or enterprises) of the same juridical ownership category (state farms and enterprises on the one hand; collective farms, also private peasant farms in Poland and Yugoslavia, on the other). Another distinctive characteristic is the preference for organization on the so-called production principle (i.e., according to one or a few main agricultural products) instead of on the still preponderant territorial principle (uniting farms or enterprises of a given administrative unit or geographical region). Such characteristics may apply to cooperation as well as to integration (in Alexandrov's definition), and to the horizontal as well as the vertical kind.

In the Soviet Union and Eastern Europe, the "combination of agriculture with manufacturing industries," as called for in the Communist Manifesto of 1848, was for a long time interpreted merely as the all-round implementation of mechanization and industrial methods in the agricultural production process of large and gigantic farms. Upon entering the 1970s, a concept closer to the original was developed, assigning great importance to "amalgamation" with the forward and backward linked branches of the economy and recognizing that the two aspects have to be combined. As Bukh put it, "closely connected with the industrialization of agriculture is the formation and development of an agro-industrial complex."[2]

Under Western political and socio-economic conditions, such an integration came about spontaneously, starting for the most part soon after the Second World War (first in the USA), and was conceptualized as "agribusiness" or "vertical integration of agriculture." According to a Soviet author, "in the GDR and Czechoslovakia, the course toward industrialization of agriculture [i.e., the great increase of investment and other industrially produced inputs] was taken in the mid-1960s, while in the other European CMEA countries [this occurred] at the end of the 1960s and the beginning of the 1970s."[3]

Specific forms of integration in the West as well as in Eastern Europe and the Soviet Union in fact existed long before they were seen as such and theoretically conceptualized. They were conditioned by the nature of the product and the necessity or desirability of its immediate processing, for example, sugar beet and sugar, milk and dairying, grape growing and wine production. More recent developments, as brought about by new technologies of processing, were fruit and vegetable growing and processing, or large-scale production, processing, and packaging of meat.

Vertical as well as horizontal integration usually are facilitated by the existence of large farms, but are possible also through cooperative and other organizational forms uniting small farms with enterprises with forward as well as backward linkages. In contrast to the very early (1848) theory, the practical implementation in Communist countries was for a while confined to farm amalgamation and mechanization, while vertical integration lagged far behind the reality of American agribusiness. At the start of the 1970s, the more comprehensive idea was introduced under the notion of creating (micro-economic) "agro-industrial complexes" or "associations" (unions) or a macro-economic—national or at least regional—"agro-industrial complex." This development is illustrated by Bukh's data (Table 12.1), where bloc II designates agriculture proper, and blocs I and III are the backward and forward linkages, respectively. Comparing the USSR and USA, another Soviet study found that around 1970, only 25 percent of the American labor force was occupied in bloc II, and the other 75 percent in blocs I and III, while in the Soviet Union, the proportions were reversed. More recent figures are over 60 percent for Soviet agriculture proper, and 7 and 32 percent for blocs I and III, respectively.[4]

Because of the low labor productivity in agriculture and its higher level in the other branches, especially in the industries supplying inputs to agriculture, the shares in employment give a somewhat distorted picture. Calculated in total gross product produced by the macro-economic agro-industrial complex in the mid-1970s, the share of blocs I and III was considerably greater (in current prices), for example, 59 percent in the USSR, 68 percent in Czechoslovakia, and 63 percent in the GDR.[5] By such indicators, according to the numerical definition of the French author, L. Malassis (obviously accepted by M. E. Bukh and two Czechoslovak authors), only Czechoslovakia and the GDR ranked as highly industrialized countries, while the Soviet Union ranked among the relatively industrially developed ones.[6]

Of agriculture's total material production expenditures, the share of the costs of inputs supplied by non-agricultural branches in the early 1970s was 37 percent in the GDR, and by 1973 was 32 percent in Poland and 54 percent in Hungary.[7] The rapid increase of this share during the preceding decade shows the investment growth and the "industrialization" of agriculture in the technical sense of the word, although it was in part caused by rising prices for such inputs, as in Hungary.[8] Correspondingly, the growth rates of agricultural machinery and fertilizer production were above the average of all industries and also higher than those of agricultural output. Of the marketed output of agriculture, 55 percent was industrially processed at that time in the Soviet Union, and 70 percent in the GDR.[9]

American agribusiness and West European vertical integration were increasingly dealt with and analyzed by Soviet and East European specialized publications towards the end of the 1960s. Obviously, they were considered models, although their "capitalist" character was denounced and was to be turned into a "socialist" one. Yet a similar,

Table 12.1 Share of Labor in the Agricultural and Nonagricultural Branches as Percentages of Total Labor in the Macro-Economic "Agro-Industrial Complex," by CMEA Countries

	Bloc II[a]			Blocs I and III[a]		
	Mid-1960s	Early 1970s	Mid-1970s	Mid-1960s	Early 1970s	Mid-1970s
USSR	83[b]	66[b]	...	17[b]	34[b]	...
Rumania	92	89	84	8	11	16
Bulgaria	86	83	79	14	17	21
Poland	88	84	81	12	16	19
Hungary	82	78	70	18	22	30
Czechoslovakia	69	58	58	31	42	42
GDR	58	54	47	42	46	53

[a] For the definitions, see above.
[b] The share of bloc I is overstated and that for bloc II understated in comparison to the other countries, because the Soviet figures include machinery repair and supply workers of "Sel'khoztekhnika" in bloc I.
Source: M.E. Bukh, *Problemy, op. cit.*, p. 116.

spontaneous development was not possible under the conditions of Communist central planning. At an international symposium in Budapest in 1976, a Rumanian participant stated, "Integration [in Eastern Europe] is not a spontaneous process, on the contrary, it is achieved on a planned basis, as a result of the growing role of State superstructure."[10] In such countries, only the central political authority can overcome the inertia of the nonagricultural branch administrations, which on their own have little interest in expending their activities into the long disadvantaged agrarian sector. Therefore, vertical integration in the CMEA countries, except for Poland and the USSR, was initiated by newly formed ministries with jurisdiction over agriculture as well as over the procurement agencies, and the supplying and the processing industries in the whole agriculture and food sector (see chapter 11).

It is especially the backward linkage (supply of industrial inputs and of technical services) where integration of East European and Soviet agriculture lags most, or was initiated only recently. For Hungary, it was bluntly stated that these branches of the economy are not really interested in a closer cooperation with agriculture because on the one hand they are in a seller's market, and on the other, the fixed prices for industrially produced inputs for agriculture fetch them only small profits.[11]

The most important attempt at backward integration has been the organization of "Agro-Chemical Centers." Some were joint ventures of farms, others were state institutions, and still others originated from the cooperation of farms with state organizations. The idea came into being in the GDR in the late 1960s, and soon after in Czechoslovakia. In the USSR, similar organizations appeared in 1973. There they seemed to enter into some competition with the state-owned organization for technical supplies and repairs in agriculture ("Sel'khoztekhnika"), until in 1979, the all-Union association "Sel'khozkhimia" was set up for them.

Disregarding ideology, it might seem that by the very fact that socialized agriculture's large-scale collective and state farms already underwent horizontal integration of a kind (amalgamation of the former small peasant farms), they would more easily proceed towards vertical integration. Yet, as this horizontal integration (i.e., collectivization) was paralleled by rapid and forced overall industrialization under the conditions of postwar capital shortage, agriculture was left with little capital for a genuine modernization of those large-scale farms. Agriculture was assigned the sole task of supplying food and other raw materials to the state-owned industries and, for the most part, was very much separate from the rest of the economy because its means of production were "collective property" rather than "state property." Moreover, the industrial branches most related to agriculture, those producing consumer goods, were assigned a low priority. More or less, in all CMEA countries until the early 1960s, industrial activities of the farms themselves were frowned upon officially or unofficially.

Such facts contributed to increasing the development lag of agriculture and the rural areas, instead of lessening it by integration. According to a Polish author, the total number of "those working in agriculture" in the CMEA countries from 1950 until 1964 decreased by no more than 5.9 million, while those employed "in the branches supplying agriculture and processing agricultural products" increased by 3.5–4.0 million.[12] This is modest progress, to a large part reflecting no more than some diminution of agrarian underemployment and the reconstruction of the prewar agricultural implements and processing industries.

In the early 1960s, when more attention was paid to the capital and income needs of agriculture, the preconditions were created for changing the situation. Even then, under Khrushchev, the envisaged character of a modernized rural economy and society was basically agricultural. The encouragement of interfarm enterprises and of "agro-towns" initially only was to serve agriculture and its population, but not to bring major industrial activities into the countryside. By the mid-1960s, it began to be recognized that "industrialization of agriculture" cannot be a matter only of large-scale farming with sufficient technical equipment, and that exactly by restricting the rural nonagricultural economy, one induced the young and skilled workers and cadres (i.e., those who were needed most for the modernization of agriculture) to migrate to the towns.

The 1961 Program of the Communist Party of the Soviet Union already propagated the formation of "agro-industrial associations" (or unions). However, for a number of years this was interpreted in practice to apply only to the concept of technological "industrialization" of the agricultural production process and of shortening and improving the marketing channels of its products. In contrast, the ideas of Czechoslovakia, the GDR, Hungary, and also in Yugoslavia were more advanced and a few first steps were taken in these countries. The forms envisaged were of a micro-economic character, i.e., associations or unions (increasingly also called "complexes" or "combines") of farms and other enterprises. Toward the end of the 1960s, the terms "agro-industrial complex" and vertical (or horizontal) integration had become widespread, and a number of international conferences on the subject were held beginning in 1971. In retrospect, by 1972, János Márton, an influential Hungarian agricultural economist, stated that in the socialist countries, "vertical integration" received greater attention only in "recent times." It was also a Hungarian, L. Komló, who revealed: "There have been theoretical arguments and also serious political battles everywhere where integration has appeared on the agenda."[13]

The "agro-industrial complex," in a macro-economic meaning of the phrase as a vertically integrated branch of the whole economy, became a concept at the beginning of the 1970s. Since then, it has become a major reality in the GDR, the Soviet Moldavian Republic, and, most conspicuously, in Bulgaria.

Aspects of Implementation in Communist Systems

The practical implementation of measures to overcome the isolation of agriculture from the other, more modernized, sectors of the economy by organizing "agro-industrial complexes" in Eastern Europe and the Soviet Union has several aspects. They are connected with questions of horizontal or of vertical integration, or of both combined.

One such aspect is that of specializing and modernizing the agricultural production process as such on an "industrial basis." For this purpose, the collective and state farms, although already very big, associate for providing the required resources, among them physical capital, needed for extremely large production units. Individual collective or state farms may participate in more than one cooperation project. Cases of such joining of resources may be found most often in certain branches of large-scale animal production (eggs, poultry, meat, cattle and pig fattening), and in specialized crop production (fruit, wine, sugar beet, tobacco, and vegetables, including in greenhouses). Modernization of this kind through capital-intensive concentration (horizontal cooperation and integration) is considered part of a preliminary step towards the comprehensive "industrialization" and vertical integration of agriculture. In crop production, where a high degree of concentration has already been achieved by socialization, it is mainly a question of mechanization and of specialization on still larger land units.

The task is more difficult in animal production, where technological modernization in Western countries, too, made appreciable progress only after the Second World War. Still, progress was marked in the Communist countries as well, with some delay, during the 1970s. By 1975, "animal production sections of an industrial-type character" were said to hold 20 percent of all the CMEA countries' pigs, 60 percent of the meat poultry, to produce 25 percent of all eggs, and less than 20 percent of beef.[14] Depending on the definition in a source, the percentages vary, but generally are higher for the GDR and Czechoslovakia. Large interfarm feeding lots in the Soviet Union still produce only a minor part of all state-procured meat.[15] To this amount, however, must be added quantities from comparable "industrialized" sections of specialized state farms or kolkhozes: the state *Glavzhivprom* system accounted for 14 percent of total Soviet pork and beef output (as distinct from state purchases; probably in 1979), while kolkhoz-owned farms are likely to have contributed much less.[16] In the GDR, it was expected that by 1980, a sizable share (20 percent) of total animal production would be concentrated in big, industrial-type units.[17]

To a greater or lesser extent in all CMEA countries, poultry farming, with its closed production cycle from breeding to slaughter and finishing, is often named as a first-rate example of horizontal integration combined with elements of vertical integration. Units with 100,000 laying hens or 0.5–1.5 million broilers are envisaged, although these have been attained so far in only a few cases. Yet even in the GDR, where already by 1974 or 1975 more than half of all poultry meat was produced by only seventeen farms, one has become aware of the enormous cost of concentrating all animal production in such giant specialized enterprises, and of the fact that it may take decades to achieve this. It may be assumed that those responsible for agrarian policy initially underestimated the cost.

Amalgamating socialist farms into ever larger units, as was done at irregular intervals during the 1950s and 1960s in the whole area, only rarely at the time led to truly specialized large-scale production. In most cases, the result was only an adding of farm sections, which in themselves remained diversified and medium-sized entities under one central management. As a Soviet author admits, "mechanical amalgamation of several independent farms not rarely makes managing the production process more difficult, and the indices of the economic efficiency of the production as a whole go downward."[18] True specialization was impeded by the shortage of capital and the obligation of the farms to deliver a multitude of various products. Therefore, the approach began to be changed by the mid-1960s. Instead of whole farms being amalgamated, now units of the desired size were formed by the cooperation association of special production branches of several farms, without making new farms out of them. They continued to be sections of the original farms, although they were more or less definitively separated from them and formed specialized production units of their own. In the GDR, this concept was adhered to during 1965–1977. In contrast, the Bulgarians and Rumanians, and at times also the Hungarians and the Soviets, for a while still considered the emergence of ever bigger farms a positive phenomenon.

In such cooperation associations, problems are presented by the role and position of the weaker partners. As a rule, the collective farms are in that position vis-à-vis state farms because they own little capital and, therefore, such joint ventures either overstrain their resources or leave them as junior partners with little say in the new enterprise. In addition, there is the danger that in view of the overall still unsatisfactory economic and technological level in agriculture, the scarce resources will be concentrated too much on a number of modern and integrated enterprises, while a great part of agriculture lags ever farther behind. The labor of the lagging sector will migrate, and some of the agricultural land may fall idle. If such discrepancies arise, they will have a negative impact on the food sector as a whole, and in the end may act as a brake on its highly modernized parts, thereby jeopardizing the efficiency of the concentrated investment.

Engaging in nonagricultural activities is another aspect of cooperation among farms. In East European and Soviet practice, it was often combined with horizontal concentration, and gained momentum in the 1960s. Such activities may relate to production and services for the agricultural production process, to processing and marketing of its output, or else to activities which have no direct connection with agriculture. Construction work and the production of building materials is an intermediate case, as tasks may include the erection of farm buildings as well as living houses, schools, medical stations, cultural facilities, etc. Nonagricultural activities of collective farms became widespread and manifold, often beyond the local area, and those which were not related to agricultural production later came under criticism in Hungary, Czechoslovakia, and the USSR. If such activities were engaged in with already existing non-agricultural enterprises, they were part of a vertical integration. But in principle, it was the same phenomenon when, as often happened, only farms participated and thereby entered into non-agricultural spheres. Even within the food sector, such advancing of the agricultural producer along the chain of processing and marketing hit a number of obstacles, either of an administrative (e.g., territorial boundaries in the strictly centralized system) or an economic nature (as a consequence of inadequate infrastructure, the opposing interests of the state-owned industry and trade organizations, etc.).

There is also the aspect of creating jobs for off-season times, when the farms cannot provide employment, and consequently income, for their work force. It was hoped that subsidiary employment would slow the outflow of labor from agriculture and at the same time provide labor reserves for the peak season. In a peasant country like Poland, establishing such nonagricultural but agriculture-related jobs may help to combat rural underemployment in certain regions.[19]

A point can also be made for bringing certain industries and services other than those connected with agriculture to rural areas, so that the villages will be kept alive as economic and social organisms despite their dwindling agricultural population. Under Communist regimes, where large-scale production is considered superior in almost any field, such a rural development strategy is hard to put into reality. All the same, it may find application in countries or regions where the rural population still grows and urban areas can not, or can only with great difficulty, absorb the outflow of people from the villages. Thus, it was considered desirable in Albania to establish industries in the countryside close to the raw material base and abundant labor resources. This was also seen as helping to approach the Marxist ideal of overcoming the differences between urban and rural life.[20] Similar considerations may apply to parts of Yugoslavia[21] and Soviet Central Asia.

The concentration of the agricultural population in large and compact settlements of an "urban type" has some bearing on most aspects of "industrialization" and integration of agriculture and on the development of rural nonagricultural economic activities. It was first and most strongly advocated under the name of "agro-towns" in the Soviet Union during Khrushchev's time, and again in the late 1960s. A relevant decree, "On Regulating Construction in the Countryside," was issued on 2 September 1968, and has remained in force since then. In actual practice, however, the discussions on the subject have ebbed, the notions about preferable settlement size have become more moderate, and also the planning of resettlement and the corresponding actions seem to have become less intensive in recent years. However, the concept as such has remained on the agenda. In other CMEA countries, the concentration and reconstruction of villages and their modernization have also been advocated, mostly in connection with the concentration and specialization of farming. Nevertheless, these have been done less vigorously than in the USSR and with little actual impact so far.

Only in Czechoslovakia and Bulgaria was the scheme pushed more intensively. The numbers of Czechoslovak villages have kept diminishing since 1960, increasingly so since 1975, in part as a consequence of migration but apparently also in connection with purposeful mergers. In Bulgaria, such plans were discussed and in 1972 advocated by Todor Zhivkov himself, when the formation of agro-industrial complexes had come into full swing. Since 1977, the notion of mergers has developed into one of organizing "systems of inhabited places" in rural areas. In parallel, the agro-industrial complexes were reorganized so as to coincide with one such system each, and to serve as its administrative and management center. As a result, the number of complexes, which was only 153 for the whole country in 1974, and 161 in 1977, grew to 338 by 1979. It is doubtful, though, whether real, rather than only administrative, settlement concentration, together with infrastructural and housing modernization, will come about. The cost involved would be immense, and the needs for productive agricultural investment are more urgent.

If nonagricultural activities of an industrial type are taken up by farms or associations of farms, the question of the size of such industrial units arises. The technologically optimal size of such a production unit, for example, a canning factory, often exceeds the quantities of raw materials which the farm or farm association can supply, or the volume of production (e.g., of a building enterprise) that the farms need. Such considerations may have played a role in Bulgaria when the enormous agro-industrial complexes were conceived. Another solution consists in forming ever bigger associations in the course of vertical integration. This was often done in the Soviet Union, where some state farms developed into so-called combines, integrated production units, which for the supply of raw materials had rigid contracts with farms. Or else an already existing industrial enterprise became the integrator and head of the whole, and amalgamated the required number of farms in a more or less stringent form. It is questionable, though, whether in such ways a genuinely integrated new enterprise in its own right, also benefitting the participating farms, comes into being. The question is the more relevant since in the new as well as in the old form of relationships, technological progress has the tendency to demand ever-growing optimal sizes for industry, and thus each farm becomes a relatively smaller partner.

At first, East European and Soviet publications on the subject paid little attention to the question of which partner would be the leader in the micro-economic process of integration (and thereby usually in the newly arising integrated enterprise). But since the early 1970s, this problem has also been tackled. The opinion seems to prevail that with progressing vertical integration, it is the processing partner, the one of the forward linkage, which should act as the "integrator." At the same time, the desire is to avoid what often happens in "capitalist" integration, namely that the producer of the final product moves into a dominating position in relation to the producer of the primary product, or the farm. Some Communist authors recognized, however, that such a danger may be inherent in "socialist" integration as well.[22] In a socialist system, the position of the weaker partner, usually the collective farm, becomes still more inferior because the superordinated agricultural administration in practice has great influence on its management. The administration's influence is less on the state farms and enterprises, which mostly have their own administration hierarchy up to the higher levels.

Nor is unity of interest guaranteed in an enterprise that fully integrates the partners within the food sector (agriculture, industry, and trade) and where the formerly independent farms are sections and/or shareholders of such an enterprise. In fact, each section must have its special tasks and must therefore achieve its own performance indicators within the framework of the "intra-enterprise self-accounting" (in Russian, *vnutrikhozyaistvennii raschet*). This has an immediate bearing on the bonus payments or premia for the workers as well as for the managers of that same section. For the exchanges of intermediate products (e.g., feed) within the enterprise, special accounting prices—not procurement or market prices—are fixed, and these may be less advantageous to one as against the other section.

Moreover, it is questionable whether in the council (board) of a "complex" or "union" the representatives of the collective farms are able to bring their interests better to bear against the economically stronger partner (the leading section) than they formerly could against the superordinated administration or the procurement agency. Possibly the new partner, being the leader within the integrated enterprise, will dominate even more. The question also hinges on whether in each case of integration an economically optimal unit actually comes into being, and whether the degree of efficiency will be higher than before. This cannot be taken for granted, because in practice, other than purely economic considerations emanate from political authorities and influence the formation of a given complex or association. Although it is only rarely admitted in Communist publications, the authorities do seem to exert some pressure.[23] With unsatisfactory economic results, the intraenterprise contradictions are likely to become more pronounced.

If highly technisized, specialized, and forward integrated agricultural production is to operate satisfactorily, it needs a well-developed economic, social and communicational infrastructure (including roads and transport in general). Such exists, at best, in the GDR and Czechoslovakia, in part also in Hungary and Poland and some Soviet regions (metropolitan areas and the Baltic Republics), but not in most of Eastern Europe and the Soviet Union. It is to be assumed that this lack is a major obstacle to modernization and integration of Communist agriculture. It seems that a certain tendency to form associations or complexes within one and the same territorial administration unit, instead of erecting a network of different and territorially parallel units according to the product specialization, was more widespread where the transport system was poor. Good all-weather roads are also required for bringing the workers to the fields and livestock premises, which will be increasingly farther away from the settlements as the farm or integrated enterprise and its sections become bigger and more specialized. The same applies to interfarm and intraassociation or intracomplex transport among the constituent production units.

Specific problems result from the simultaneous existence of a collective (kolkhoz) and a state sector in agriculture. "In spite of their enormous importance, they are the least elaborated upon" theoretically, complained V. Starodubrovskaya, who later in a book tried to fill the gap. But her statements (together with those of G. Shmelev) remained rather hazy. What she sees emerging in Communist agro-industrial integration is not as yet a new category of property, but the co-existence of state and collective property "under one roof", and a factual "rising of the degree of socialization."[24] The basic consensus among CMEA countries was summed up at a symposium in Budapest in 1976 by the Soviet delegate, Alexandrov, in the following way:

> Concerning the social aspect, the course propagated by the Party must be put into effect. It is directed at the gradual coming closer to each other, and subsequent amalgamation [sliyanie] of the forms of collective and state prop-

erty and the gradual formation of property of the whole people on this basis.

In this way, the agro-industrial unions of the mixed type [i.e., those with collective farms or kolkhozes participating with state farms] are one of the concrete forms of transition to a uniform kind of property. In the formation of a mixed union, the principles of kolkhoz democracy and also of the positive organizational experience, of labor remuneration, social insurance and management, as acquired in the unions of the state sector, shall be widely utilized.[25]

However, this formula does not point out what organization is supposed to exert the property rights in the name of the people. Up to the present, the only imaginable bearer of those rights has been the state. For the distant future when, according to the Communist Program, the state will "wither away," the bearer presumably will be the Party or other "mass organizations." If these should continue to be centrally organized, the nature of the property of interfarm enterprises, agro-industrial complexes, and so on will in essence differ little from that of present state property. In this respect, the Yugoslav model of decentralized social organization of the economy is more consistent.

Practically no concrete statements are being made on the final form of "property of all the people." Communist publications do indicate in what respects the organization and activities of collective and state farms are to become similar, but they do not explain or point out why such similarities should eliminate the difference of property forms in terms of differing interests. East European and Soviet practical experience has shown in many ways that even within the state sector, with its uniform category of property, there may exist contradictory interests among the individual branches or their administrations, and equally within the branches (between supplying and receiving enterprises, and between those who are to make payments and those to whom payments are due).

Because of the differences in their remuneration system and in the ways in which plans and taxes are imposed on them, in the kind and volume of credits and dotations they obtain, in the right of disposing of profits earned, etc., collective and state farms do not react identically to state directives and incentives offered. Only if these differences disappear in an integrated enterprise will such problems disappear. That requires uniformity of property rights or, to be more specific, a uniform position vis-à-vis the administration and outside economic partners. As long as the two categories of farms coexist in a less unified enterprise, the state farm or enterprise will be the stronger partner also because of its closer connection with the superordinated state administration, the interests of which do not necessarily coincide with the interest of the individual farm.

Institutionally, the difference of the property category does not present a major obstacle, as long as the cooperation or integration is based on mutual contracts. Yet, full integration is hard to imagine when the partners continue to belong to different categories. That is why in practice (e.g., Soviet practice) state farms more often associate with state farms, and kolkhozes with kolkhozes, or the latter are converted into state farms in the integration process. Another reason why vertical integration tends to advance more within the state farm sector is that this sector's supply of capital is already greater since the state has been more inclined to grant the investment means (in money, as well as in physical supplies) to its own sector.

Attempts have often been made to circumvent these difficulties by inducing collective and state farms to organize joint processing, supply, or service ventures (especially in construction) on a shareholding basis. It has been said that such enterprises, being independent juridical persons and working according to the principle of "self-accounting," represent a "higher degree of socialization" of property.[26] This may be so, but it does not eliminate questions such as: exactly what kind of property is this, who actually dominates in the joint management, and will not the new enterprise soon gain so much weight of its own that the property rights of the shareholders become a mere formality (a question not unknown to cooperatives and companies in the West)? It may then turn out that the shareholders have supplied the capital and labor, and indirectly bear the enterprise risk, but have little influence in the actual decisions concerning the investment quota and contributions, as well as the profit quota.

The degree of economic success of Communist agro-industrial integration is hard to assess. For one thing, there are no quantifiable alternatives, and for another, the development has been too recent to be definitively judged. Moreover, the initial great investments may pay off at a later stage, and it is not clear which results can be attributed to agro-industrial integration alone. In the Bulgarian case of most rapid implementation, gross agricultural growth, not to speak of net output, in fact stagnated during 1968–74. However, in Soviet Moldavia, where integration also has proceeded quickly, gross agricultural output for 1976–78 was 36 percent over that of 1966–70, which compares favorably with the simultaneous all-Union average increase of only 24 percent. Still, Moldavian agriculture obtained above-average investment during that time and, in addition, had more abundant labor than most other parts of the USSR. Therefore, it is doubtful whether Moldavia's better achievement is valid for net output as well, and if so, whether it may be attributed to the integration measures.

Actual Cooperation and Integration in Individual Countries

As with the administrative and organizational changes described in chapter 11, the multiplicity of almost continuously changing forms and institutions of agro-industrial cooperation and integration is confusing, and therefore will be dealt with here only in very broad outlines.[27] It has to be pointed out from the beginning that the term *agro-industrial* in itself implies vertical integration or cooperation. In Soviet and East European agricultural publications, however, the term is not infrequently also used to describe "industrialization" of agricultural production, that is, simply modernization of the agricultural production process in ever-increasing (cooperating or integrating) units. In the subsequent text, at least some elements of actual vertical integration or cooperation are part of what is called an "agro-industrial complex."

SOVIET UNION

Agro-industrial integration in the Soviet Union neither took specially interesting forms, nor did it precede such integration in other countries. The formation of Agro-Chemical Centers also came about later than in the GDR and Czechoslovakia, and made a real start only in the mid-1970s.

From the first, in the late 1950s and during the 1960s, the combination of agricultural and industrial activities consisted mainly of creating either interfarm enterprises or subsidiary small-scale industries on farms. Since the middle of the 1960s, the undertakings were increasingly organized as joint ventures of a number of kolkhozes, soon also on a territorial basis encompassing all farms of an administrative area. Among projects, building and other activities subsidiary to agriculture predominated, while "industrialized" feeding lots gradually became numerous. Significantly, it has been an ever-recurrent complaint that many interfarm building enterprises in the USSR, mostly organized on a province (*oblast'*) level, tend to neglect their farms and to do much construction work on more profitable urban and industrial sites. This casts doubts on the community of interests between the constituent farms and their practically independent joint ventures.

The creation of agro-industrial complexes during the 1970s was concentrated on certain branches and regions, such as fruit, vegetables, and wine in the Moldavian and Azerbaidzhan Union Republics and in Krasnodar province; sugar beet and wine in the Ukraine; and meat and milk production in Belorussia, the Mari ASSR, and Voronezh and Penza provinces. It has been reported that more than half of Soviet fruit and vegetable processing is now being done in agro-industrial complexes, and that 55 percent of wine-processing capacities is owned by state farm factories.[28] Moldavia has become the most conspicuous case of such integration since the early 1970s, and in spring 1973, the republic's Kolkhoz Council was given responsibility for policy making, operational management, and overseeing agro-industrial integration in the republic's kolkhoz sector.

Various forms of interfarm (kolkhoz-kolkhoz, kolkhoz-state farm, and state farm-state farm) enterprises and unions came into existence. Wholly amalgamated enterprises remained rather few, except for the areas indicated above, and were mainly organized within the state farm sector, not rarely by converting kolkhozes destined for inclusion. A special form is the previously mentioned "sovkhoz-zavod" (state farm factory), where one state farm is the integrator and concentrates on the processing of produce to be supplied by integrated and/or contractually linked farms. It is often maintained that the integrated associations, unions, or enterprises are highly profitable compared to simple farms, but it has to be added that, as a rule, they get preferential treatment with financing and physical supplies of industrially produced inputs.

On the whole, it has to be assumed that the weak material and socio-cultural infrastructure (roads, buildings, communications, training facilities, living conditions, retail trade, etc.) in the Soviet countryside in many, and large, regions does not yet favor "industrialization" (which often simply means technological modernization) and integration. The major emphasis is on "intensification" (i.e., increasing crop and animal yields) and on specialization (by farms and groups of farms), and less on vertical integration. So far, the overall growth rates of Soviet agricultural production have not revealed a specially positive impact of agro-industrial integration, even less so if one considers net ("national income") results, with material inputs deducted. It strikes the observer that the most recent pronouncements, such as Brezhnev's at the November 1979 Central Committee plenum, have remained almost silent on agro-industrial integration. A recent article by one of the USSR's deputy ministers of agriculture, A. Ievlev, "On the Way of Cooperation," even seemed to sound rather a defensive note.[29]

BULGARIA

In Bulgaria, the first few experimental Agro-Industrial Complexes were organized after late 1968, following the July session of the Central Committee and a speech by Todor Zhivkov. The process started on a mass scale in spring 1970. In spite of the name, the great majority of complexes have been horizontally integrated. The few vertical types were characterized by the dominance of industrial processing enterprises (e.g., the first one in sugar production, founded at Russé in January 1973), and were called Industrial-Agricultural Complexes. But this form has not expanded very much. The main efforts have been concentrated on forming horizontally integrated agricultural production units within administrative district confines for mass-scale specialization. As a rule, to each Agro-Industrial Complex belonged an Agro-Chemical Center and a machinery repair shop.

In practice, the member collective and state farms were placed completely under the direction of the management of the complex, which in turn received directives from the central state organs (since 1976, the National Agro-Industrial Complex). The levels and systems of labor remuneration and taxes, and the calculation of the profit and decisions on its distribution and on investments, which in the early stage were differentiated within each complex, increasingly became uniform, to resemble those of the former state farms. Although a minority of complexes consisted exclusively of state farms (plus industry), in part through conversion of collective into state farms, as many as 38 percent were "mixed" (composed of collective and state farms) by 1975. At that time, retaining the juridical entity and independence of the constituent farms, which had prevailed for a while, was considered a brake on further development. In February 1975, the Party Central Committee decided that the production principle of organization should replace the territorial principle. In April 1976, the XI Party Congress proclaimed that the merging of collective and state property in the Agro-Industrial Complexes should gain momentum during the current five-year plan (1976–80). Independent collective and state farms were no longer favored as the "main form of agricultural enterprise." By 1977, the member farms of most complexes had lost their independent identities and had turned into territorial subdivisions and/or production branches of the complex.

The formation of a macro-economic National Agro-Industrial Complex, placed under the general guidance of the Ministry of Agriculture and the Food Industries, was decreed on 5 August 1976. A number of specialized production and service enterprises were organized under the National Complex. By the end of 1976, the number of such bodies reached over 300, around twice the number of complexes at the time. It seemed only logical that in 1979, the Ministry and the National Complex were merged into one comprehensive National Agro-Industrial Union. The marketing organization (the Economic Trust) of the National Agro-Industrial Complex (Union) exerted strong influence on the production programs of all local complexes. Apart from mandatory procurement plans for up to eight main products, the complexes had to comply with control figures for their payments into the state budget, for limits on their technical inputs, raw materials, energy consumption, and import needs, and also with the plan of valuta income from the exported part of their output. Beyond that, they were expected to act according to their own economic considerations. By 1977, each complex began to be assigned the task of supplying the population in its district (its "system of inhabited places," see p. 243) with all the necessary food, except where climate and soil made this impossible or economically very disadvantageous.[30] Private plot producers were called upon to contribute to such local self-sufficiency.

In spring 1979, a new reorganization was announced, and Agro-Industrial Unions, locally and with their National Center (which replaced the previous Ministry of Agriculture and the Food Industry), were created. They were designed to promote vertical integration and at the same time to avoid the previous excesses of bureaucratization. Allegedly, joining them was voluntary.[31] It is too early to judge whether they will effect genuine improvement and speeding up of the integration and modernization process.

A special form of integration that has played an increasing role in Bulgaria, and has come into being in other CMEA countries, too, is the "Science and Production Union" (or ". . . Combine", or ". . . Complex"). The first in Bulgaria was organized in 1974. They have concentrated chiefly on large-scale animal production or on vegetables, fruit, and wine. Each is under the guidance of a member research institute and is destined to promote scientific progress in these branches, that is, in production as well as in processing and distribution.

It has been said in Bulgaria, perhaps with more emphasis than in other countries, that integration is also destined to overcome the differences between rural and urban life. To what degree this has been accomplished is hard to judge, if one disregards a few model complexes. At any rate, the outflow of labor was stopped, whatever measures may have caused this effect, but one cannot say how long this halt will last. On the whole, the Bulgarian-type integration was hasty and instigated by direct state interference, so that it is yet too early to discern its lasting, as against the temporary, effects.

THE GDR

The GDR, as a highly industrialized country, had the best economic conditions for agro-industrial integration from the outset. It applied more flexible and sophisticated forms than Bulgaria and the USSR, yet in comparison to Hungary, state control remained much more rigid. From 1965 on, horizontal cooperation associations of farms (Kooperationsgemeinschaften) were propagated and organized under the "New Economic System." They soon comprised the majority of collective farms. A program for organizing agro-industrial complexes was announced as early as 1967. However, the emerging vertical cooperation unions (Kooperationsverbände) developed slowly, and had only every fourth collective or state farm as a member by 1972. As part of a policy favoring "industrialization" in animal production, such farms were not expected, at least not in the beginning, to reduce cost per output unit, and therefore were paid higher prices and received credits at preferential interest rates. Beginning in 1971, support for these associations and unions ceased, and Cooperative Sections for Crop Production (later on, also for animal production) were increasingly formed instead.

Cooperative sections of both types were "self-accounting" enterprises in their own right, but the participating farms retained their juridical and economic identity. Animal Production Sections usually restricted their activities solely to animal farming, for which they received the fodder from the Crop Sections. The average size of constituent farms in Crop Sections was smaller than in all other CMEA countries except Poland, but the Cooperative Sections for Crop Production themselves farmed on average 3,000–5,000 hectares of land contributed by their member farms on a long-term assignment basis. A number of specialized units for animal production, state-owned (such as the Combines for Industrial Fattening—Kombinate für industrielle Mast, KiM) or as joint ventures of collective farms, had no (or almost no) land of their own, and were completely dependent for their fodder supplies.

Beginning in 1977, independent and very large collective farms, specialized in either crop (actually often the former Crop Production Sections) or animal production (without crop land) came into existence. This implied a change of the formal concept, so that only the strict division of animal from crop farming continued to represent an East German specialty. The new collective farms for animal production, in spite of their numerically large herds, by 1979 still had most of their livestock in the existing small to medium premises scattered over the land and villages of the newly organized crop production farms or the old Crop Production Sections (where the latter still existed). The new specialized collective farms were organized in a way very similar to that of the state farms, although the system of labor remuneration retained some collective farm features. Their economic activities seem very much directed by the economic organs of the state.

Most of the Agro-Chemical Centers, which serve more than 80 percent of the East German agricultural area, are cooperative ventures, as distinct from the state-owned County Stations for Agricultural Machinery. The bulk of land improvement and construction work, too, is done by interfarm cooperative organizations.

Agro-Industrial Unions, comparable to those of Bulgaria and the Soviet Union, were first mentioned in 1976. Eleven of them were organized experimentally soon after, and farmed 30,000–40,000 hectares each, i.e., roughly all the farm land of a county (Kreis). They also comprised the local Agro-Chemical Center and County Station for Agricultural Machinery. However, they do not seem to have played a noticeable role in most recent years.

In sum, vertical integration, which made an early start in the GDR during the 1960s, apparently has not progressed much beyond what the general economic level would lead one to expect. A high degree of specialization and mechanization within agriculture seems to have developed instead.

CZECHOSLOVAKIA

In Czechoslovakia, where the economic preconditions were about the same as in East Germany, horizontal and vertical integration beyond the individual farm level began only hesitantly. Instead, practically all farms engaged in some kind of cooperation on a contractual basis, and the increase of average farm sizes through amalgamations between 1973 and 1975 was on a scale second only to Bulgaria (see chapter 6, Table 6.2). In October 1975, the Party Central Committee decided that integration (instead of amalgamations) should make more progress during the 1976–80 plan period, but no major changes were noticeable up to 1979.

HUNGARY

The development of Hungary's average farm sizes was similar to that in Czechoslovakia, although somewhat slower. At the same time, manifold and relatively free and flexible forms of cooperation and integration developed under the 1968 economic reform.

A specific Hungarian venture was started by 1971 in poultry and pig farming at "Babolna" state farm, and was soon extended to crop farming. It consisted in forming so-called Closed Production Systems, where one leading farm, or a profit-oriented special and separate enterprise, applied technologically advanced methods bought from foreign (mainly American) firms, in most cases for maize, on other farms on a contractual basis. This procedure disseminated modern know-how but not necessarily cost-consciousness. By 1975, 60 percent of all maize sowings, and a total of 1 million hectares of agricultural land, were under this scheme, and the goal for 1980 was 1.6 million.[32] By the end of 1978, near to 90 percent of all collective and state farms participated in one or more of the overall 72 CPS, but disappointing results in the course of this over-rapid expansion were also reported.[33]

According to a Hungarian expert, "in building up vertical connections aimed at the processing and marketing of agricultural products, the progress is much slower than desired."[34] Most vertical links up to the mid-1970s consisted in ancillary activities (predominantly in building) of collective farms, and in industrial or trading plants or organizations established by associations of such farms. Participation of state-owned industry and of state farms was small. Heated debates were followed at the XI Party Congress by some authoritative statements. Subsequently, the first large-scale vertically integrated enterprise confined not to only one special crop made its appearance at Nádudvár in spring 1976, and three more followed. The constituent farms remained juridical and economic entities of their own, but through new investment of the integrated enterprise, mixed state-collective property arose. All four such enterprises were still considered experiments even in 1978,[35] and discussions on their viability continued. The cautious approach to vertical integration and the preference for flexible forms of cooperation, combined with emphasis on the continuing importance of collective as against state or some new kind of property in farming, have been characteristics of the Hungarian development since the late 1960s.

In Rumania, vertical integration in the true meaning of the word did not start before late 1976. Until then, various horizontal associations of collective farms (in which state farms could participate since 1970), and a territorially defined system of collective farm councils, were organized. The forms changed several times and in part were connected with reorganizations of the public administration. However, decisive state influence, through the administration of the state farms, was maintained on all levels of such organizations.

By the end of the decade, production specialization of the Inter-Cooperative Associations was receiving increasing attention. Recent measures reorganized the Inter-Cooperative Councils into Unitary State and Cooperative Agro-Industrial Councils. Each county's Council was made responsible for all aspects of agricultural production and investment, including the state farms and the state-owned Stations for the Mechanization of Agriculture. The latter have to organize all work of machines, including those which remain in the ownership of the collective and state farms, which farms retain their legal and economic identity. The economic operation of the Stations returned to that of the old MTS system in so far as they are now paid by the farms no longer on a scale based on work performed, but by a share of the output produced with their help. The new system was embodied in the directives for the 1981–85 five-year plan and for the perspective plan (up to 1990). While this reorganization as such has few aspects of vertical integration, the Stations for Mechanization form a certain backward linkage. The forward linkages were enhanced by dividing the Ministry of Agriculture and Food Industries into two departments, one of which is for the Industrial Processing of Agricultural Products and will have direct contacts with, and influence on, the Unitary Agro-Industrial State and Cooperative Councils (created in early 1979).

The two countries with still predominating peasant farming necessarily went special ways. Belgrade and Warsaw also consider agro-industrial integration as the future of the sector, and at the same time they attribute to it a function of furthering the gradual socialization of agriculture. Cooperation and integration in various forms are to lead the peasants into growing interdependence with the socialist sector and its upstream and downstream linkages, and finally to giving up their farm autonomy (see Chapter 6).

To this end, Yugoslavia has been advocating many kinds of cooperation, which often are rather loose, but in some cases may come close to incorporating the small peasant farms into the social (self-administrating) large, and in part integrated, farms. Within this scheme, it is also possible for peasants to rent their land to the socialist farms or hand it over to them forever, against monetary indemnity or a pension. The initiative for integration can originate from various persons or organizations, from the peasants themselves, from the agricultural or agriculture-related organizations of socialized labor, from local communities, commercial socialist enterprises, even from banks, etc. Cooperation or integration of peasant farms among themselves is also possible and may take all kinds of forms, from cooperatives to group farming.

The very big socialist farms engage in vertical integration, mainly in the fertile plains of Vojvodina and Slavonia, where such "combines" expand their activities into the non-agricultural branches. Some form still larger unions of "combines." There are General Agricultural Cooperatives (OZZs), which neither own nor rent farm land, but are engaged exclusively in agricultural trade and services. As there remains only one category of property in the socialist sector, that of the self-administrating enterprises and farms, the question of state versus collective property does not play a role in Yugoslavia. Whether this has more than formal importance is another matter, as the—not rarely contradicting—group and enterprise interests admittedly play a role in the Yugoslav economy and society.

POLAND

In Poland, the policy of strengthening the ties between peasants and cooperatives (or the socialist economy at large), as inaugurated in 1957 after the renunciation of collectivization, has continued through the Agricultural Circles as a form of peasant cooperation. A parallel policy was that of reviving peasant cooperatives of a traditional type. Since 1972, it has been hoped that voluntary group farming will in the long run develop toward closer production cooperation with socialist units, among them processing and trading enterprises or cooperatives. Forms and ways were encouraged, with some official emphasis, to make peasants engage in such cooperation on a contract basis, mainly in order to tie them into the overall system of planning and managing the economy. In Polish Communist thinking, these were correlates to the reluctantly favored strengthening of the peasant economy.

The Polish agricultural production cooperatives, although still bearing more resemblance than their Yugoslav counterparts to the Soviet-type collective farm, strongly engage in nonagricultural activities, above all in supplying farm inputs for the peasant farms and processing and marketing their produce. By 1975, only two socialist enterprises were called vertically integrated agro-industrial complexes and in fact deserved this name. It has also been reported that the processing industries organized by farms were not important, except for the specific case of distilling alcohol.[36] One gigantic agro-industrial union, encompassing also local food industries and service organizations, is that of Ketrzýn in Olsztyn województwo, which was organized in 1973/74 on the basis of ten state farms and has close economic ties with the private peasants of the area. A. Wós and Z. Grochowski envisaged that more such enterprises, each with 8,000–10,000 hectares of agricultural land, would come into being "in the foreseeable future."[37] Another recent development is the formation of interfarm organizations or unions of state farms, called "agro-combines," which are to put their production "on an industrial basis." The relevant decree was issued in 1976, but progress has been slow since then.

Notes

1. Cf. K.-E. Wädekin, "The Place of Agriculture in Comecon Organization and Cooperation", *Radio Liberty Research,* RL 129/78, 6 June 1978.
2. M. E. Bukh, *Problemy, op. cit.,* p. 12.
3. *Ibid.,* p. 18.
4. M. Lemeshev, *Ekonomicheskie nauki,* no. 2 (1972), p. 68; *Sel'skaya zhizn',* 9 December 1978, p. 2.
5. M. E. Bukh, *Problemy, op. cit.,* p. 114.
6. *Ibid.*
7. Yu. Shintyapin and B. Frumkin, *Voprosy ekonomiki,* no. 2 (1975), p. 83; for Poland, see *Economic Survey of Europe in 1976, part II* (New York, 1977), p. 47; for Hungary, *Integratsia sel'skogo khozyaistva s drugimi otraslyami narodnogo khozyaistva v stranakh-chlenakh SEV* (Budapest, 1976) (hectographed), p. 87.
8. Z. Kardos, *Közgazdasági Szemle,* no. 12 (1978), pp. 1456–65 (according to *Hungarian Agricultural Review,* no. 3 [1979], p. 30).
9. M. E. Bukh, *Problemy, op. cit.,* pp. 113–14.
10. D. Dumitru, in *Symposium on Forms of Horizontal and Vertical Integration in Agriculture,* vol. 2, FAO, ECE/Agri/29 (New York, 1977) (hectographed), p. 375.
11. *Integratsia, op. cit.,* pp. 90–91, 94, 99.
12. B. Struzek, *Internationale Zeitschrift der Landwirtschaft* (Moscow/East Berlin), no. 5, (1966), p. 464.
13. J. Márton, *ibid.,* no. 2 (1972), p. 141; L. Komló, *Acta Oeconomica* 10:1 (1973), p. 75.
14. D. Perun, *Internationale Zeitschrift der Landwirtschaft,* no. 2 (1977), p. 108.
15. *Narodnoe Khozyaistvo SSSR, 1978,* pp. 252, 269.
16. A. Zholobov, *Sel'skaya zhizn',* 8 February 1980, p. 2.
17. H. Kuhrig (Minister of Agriculture and Food Industries), *Neues Deutschland,* 10 January 1977, p. 3.

18. M. E. Bukh, *Problemy, op. cit.*, p. 120.

19. It was recommended for this purpose by J. Tepicht, *Marxisme et agriculture: Le paysan polonais* (Paris, 1973), p. 160.

20. *Zëri i popullit,* 15 June 1975, p. 2; *Rruga e partisë,* no. 5 (1975), pp. 36–45 (according to *ABSEES* 6:4 [October 1975], pp. 130, 138).

21. M. Špiljak (President of the Yugoslav Trade Unions' Federation), *NIN,* 2 December 1979 (according to *RFER, RAD Background Report,* 13 December 1979, p. 3).

22. See, for example, M. Bukh, *Ekonomicheskie nauki,* no. 4 (1972), p. 70; B. Umetskii, *Voprosy ekonomiki,* no. 8 (1973), p. 111.

23. For examples, see E. Jacobs, in *East European Economies Post-Helsinki,* Joint Economic Committee, U. S. Congress (Washington, D. C., 1977), passim; V. Bajaja, "The Organization of Czechoslovak Agriculture", in *The Organization of Agriculture in the Soviet Union and Eastern Europe,* E. M. Jacobs, ed., vol. 2 of the present series (forthcoming).

24. G. I. Shmelev and V. N. Starodubrovskaya, *op. cit.,* pp. 97, 106; for the earlier complaint, see *Voprosy ekonomiki,* no. 12 (1975), p. 108.

25. *Symposium on Forms of Horizontal and Vertical Integration in Agriculture, op. cit.,* p. 124.

26. G. I. Shmelev and V. N. Starodubrovskaya, *op. cit.,* p. 97.

27. For more detailed reports of the numerous changes, see the works by E. M. Jacobs cited in the bibliography.

28. M. E. Bukh, *Problemy, op. cit.,* pp. 127–130.

29. *Sel'skaya zhizn',* 25 January 1980, pp. 1–2.

30. T. A. Vankai, *Progress and Outlook for East European Agriculture, 1976–80,* USDA, FAER, no. 153 (Washington D. C., 1978), p. 8.

31. For further details, see P. Wiedemann, "The Organization of Bulgarian Agriculture", in *The Organization of Agriculture in the Soviet Union and Eastern Europe,* E. M. Jacobs, ed., vol. 2 of the present series (forthcoming).

32. T. A. Vankai, *op. cit.,* p. 24.

33. *RFER, Hungarian Situation Report,* 3 September 1979, pp. 15–16.

34. A. Klenczner, *Gazdálkodás,* no. 7 (1975), pp. 11–18 (as quoted in *Hungarian Agricultural Review,* no. 2 [1976], p. 33). See also L. A. Fischer and P. E. Uren, *The New Hungarian Agriculture* (Montreal and London, 1973), pp. 42–43, and *Integratsia, op. cit.,* pp. 30–31.

35. V. Marillai, *Tudomány és Mezögazdaság,* no. 2 (1978), pp. 45–52, and S. Belák, *Vezetés tudomany,* no. 11 (1978), pp. 5–7 (according to *Hungarian Agricultural Review,* no. 1 (1979), p. 36, and no. 3 (1979), p. 32, respectively).

36. *Integratsia, op. cit.,* pp. 67–68, 127.

37. A. Woś and Z. Grochowski, *L'agriculture polonaise - les dernières mutations* (Warsaw, 1979), p. 108.

Modernization under Communism: A Conclusion

THE COLLECTIVIZATION OF agriculture in most countries of Communist Europe closely followed the model in the Soviet Union as established by Stalin and essentially, though not wholly, maintained under Khrushchev. With the exception of Yugoslavia and Poland, and with some delay in Albania, collectivization was completed by the early 1960s. The main socio-political goal thus being attained, the emphasis of Communist agrarian policies shifted to increasing food and other agricultural output of the socialized farms. In addition, it was soon recognized that an urgent improvement in the productivity not only of land but also of labor and capital, that is, cost efficiency, was required in order better to supply the growing and urbanizing population with more and better food as well as to prove the economic viability of the socialist mode of agricultural production. To be sure, hunger among major population segments had ceased to be a problem before 1960, but the rising urban incomes in those industrializing societies made demand for quality food exceed supply.

In the process of enabling agriculture to meet this demand, three points acquired paramount importance in most countries with predominantly socialized agriculture:

The large-size socialist farms had to be supplied with more capital inputs than had been received formerly by the small peasant farms—capital could no longer be withheld from agriculture.

Labor remuneration for the diminishing number of agricultural labor on the socialist farms had to become enough of an incentive to stimulate efforts comparable to those previously made on the private peasant farms—the low remuneration on collective and, to a lesser degree, also on state farms had to be raised.

The rigid Stalinist system of "command agriculture" had to be made more flexible by relinquishing strictly centralized planning, procuring of produce, and price-setting—some reform of the system was necessary, although at the same time, the leaders were eager to uphold central control and to prevent the re-emergence of "petty bourgeois capitalism" on the farms.

258

As a consequence, the 1960s became a decade when capital transfer out of agriculture was reversed, when higher producers' prices were paid to the farms, enabling them to raise labor remuneration and to invest more, and when the organization of socialized agriculture underwent reforms of a "decompression"[1] character. Poland and Yugoslavia, although hesitantly, made similar concessions to their predominant peasant farm sectors. Only Albania stuck to the old command system.

These changes had some positive effects on the productive performance of agriculture and on the living standards and working conditions of its labor. Even the remaining private sector (i.e., the mini-plots and a small number of peasant farms) in otherwise socialized agricultural sectors profitted from the change. The private sector did not "wither away," as was envisaged in Khrushchev's time, although as a percentage of the growing overall gross output, its contribution continued to decline.

It soon became evident, however, that either the reform had to continue beyond what effective control over agriculture would allow, or else the food supply would lag behind increasing demand and moreover, the inefficiencies of the more or less only cosmetically reformed system would make labor and capital costs of agriculture rise to unbearable heights. Only Yugoslavia and, to some degree, Hungary, continued along the path of "decompression." The majority of the countries, guided by Moscow's ideas, although sometimes anticipating them, embarked on "agro-industrial cooperation and integration" instead, as outlined in chapter 12. Brezhnev repeatedly called this the "main-line track" *(magistral'nyi put')* for the development of agriculture, and during the 1970s it superseded the by then more or less abortive reforms. Thus, technological and organizational modernization, without relinquishing the basic principles of a Soviet-type socialist economy, was resorted to as a means for improving agricultural performance while at the same time avoiding systemic reforms. Some elements of Western agribusiness were adopted, but were meant to work in accordance with Marxist-Leninist methods of exerting economic power and of directive planning. J. F. Karcz, as early as 1970, and with a view to the "present emphasis on integrating agriculture and industry," pointed out that "the discussion appears to focus on narrow, technological efficiency," and that "the fact that merely ordinary competition (not even the textbook variety) is a fundamental component of Western agro-business is probably not sufficiently appreciated at the highest levels in East European countries."[2]

Even among CMEA countries, the notions of agro-industrial integration and of its purpose, and the terms to be applied, differed in fairly important ways almost from the beginning. At an international symposium on horizontal and vertical integration in agriculture held in Budapest in 1976, the differences of conceptualization among symposium participants from the West and from Communist countries were especially illuminating. Moreover, those from Communist countries, including the Yugoslavs, put more emphasis on the interdependence of horizontal and vertical integration, thereby arguing, among other things, in favor of large enterprise sizes. In contrast, Western scholars did not give priority to amalgamation into large enterprises, and pointed out that cooperation of small farms may also play an important role in processing and marketing, in supplying and trading inputs, and in advisory services for agricultural producers. Equally, contractual relations with big trading firms in their eyes may contribute much to integration without eliminating farm autonomy. In addition, they emphasized that economies of scale have their limits, which differ by product and are dependent on quality and veterinary risks. Safeguarding the position of the farmer in the market and stabilizing the markets receive more attention in Western than Communist thoughts on integration, while in Communist countries, the main interest centers on increasing the productive potential and actual output.[3]

It was found at the symposium that Eastern Europe, including the Soviet Union and Yugoslavia, had a broad range of forms of integration, from "rather loose and limited arrangements among independent units of varying sizes up to involved networks of horizontal and vertical links among specialized units and sub-units within one large-scale enterprise." Even in Hungary, with her manifold forms of limited integration (e.g., joint ventures with or without juridical independence and with separate organizations participating, or "simple economic integration"), the gradually evolving large units of the type of close organizational coherence ("agro-combines," "agro-industrial combines," "combines of the food industry") were considered the most progressive form and were aimed at for the future. In general terms, it was stated that in all of Eastern Europe and the Soviet Union, horizontal integration clearly dominated, and vertical integration was still in its initial stage. In the eyes of the Communist delegates, integrated units which have a juridical personality of their own and are of a closed type with a high degree of concentration (very large production units) and of specialization, represent the most progressive form. It was added that besides these, part-integration of various kinds among independent farms and enterprises, and new forms of such, retain their importance. This was said to apply especially to certain activities where the organizing initiative comes from independent large-scale enterprises.

Relations and links among farms and among farms and processing industries based merely on contracts were considered an inferior kind of integration in the East European and Soviet publications of the early 1970s. This view has changed since, and the importance of contracts was generally emphasized in subsequent years, but essentially because they may form an initial stage of a cooperation and integration which later is to lead to stricter organizational arrangements. In fact, recently it was stated that closer and long-term contractual relations led to the "formation of unions" (or associations) as an "objective necessity" already during the early 1960s in the GDR, Czechoslovakia, and Yugoslavia, with some time lag in Hungary, and finally in Bulgaria and Rumania.[4] One has to bear in mind, though, that in the practice of Communist countries (Yugoslavia and Hungary being excepted to some degree), the contents and forms of contracts are strongly influenced by state organs, and terminating a contract of integration is made extremely difficult.

At the Budapest symposium, delegates from the CMEA member countries emphasized the important role played by the state (including its regional and local administration) in advancing horizontal as well as vertical integration, and in this they saw a resulting advantage for economic planning. What they had in mind was not only improved planning of the manifold farming and correlated activities in themselves, but also the greater simplicity and efficiency of the planning procedure with regard to the economy as a whole. Obviously, the large agro-industrial unit is meant to take over some of the planning, and also some of the control of execution and the regulation of the relations of agriculture with its input suppliers and with the final consumers in a territorial or a branch-defined area. On its own responsibility, the unit should remove inconsistencies and defend its interests against those of partners in the various economic processes. But at the same time, according to the views prevailing in CMEA countries, it should strictly follow the aggregated plan directives received from the central state authorities.

Another aspect of integration pointed out at the symposium was the balancing of interests of those participating in integration to the advantage of all. J. Márton, of the Hungarian Research Institute of Agricultural Economics, even admitted possible "conflicting interests" to be eliminated in the integration process. In other respects, too, his points of emphasis slightly differed from those of his Eastern colleagues. Thus, among the main purposes of integration, he named in the first place "improvement of the economic position of integrated activities" (not agents), in second place, the "elimination of conflicting interests," and only in the last resort, "stability in production, marketing and services."[5]

What is expected from integration in countries under Communist rule clearly reflects their politically defined economic and social parameters. The focus is on output growth, cost reduction and, in view of resulting smaller numbers of enterprises, better possibilities for planning. By contrast, in Western industrialized countries, the attention is directed at maximization of profit or of other advantages for the individual participant in integration under an essentially "private economy" structure with small to medium-sized farms and with indirect guidance through a more or less autonomous market. The advantages for socialist planning, however, can come true only if the bigger, more specialized units, with their many links toward similar partners, actually are accorded more competence for planning and directing within their own sphere than the earlier, unintegrated, farms possessed.

The socio-political ideas, as they prevail in all CMEA member countries, were pertinently stated by the Rumanian delegate at the Budapest symposium:

> People belonging to differing social classes, or groups, are employed in the agro-industrial economy, having different training and income levels as well as various forms of existing labour organization etc. In the process of building up and developing the agro-industrial sector, these differences are reduced gradually, the social relations are being transformed. Within the framework of Romania's economy this process represents one of the fundamental principles capable of ensuring the gradual reduction of the existing differences both within the agricultural sector, as well as between this sector and the other material producing sectors. Thus [*sic*] bringing closer to the average incomes of society the incomes of various categories of working people employed in the agricultural and the food industry sectors.[6]

Such a statement spells out the long-term socio-political goals connected with agro-industrial integration in Eastern Europe and the Soviet Union. Yet for the short, and probably also, for the medium term, the economic effect is given priority. Obviously, the idea is to increase agriculture's output rapidly and to slow the outflow of its technically skilled labor without having to question the established socio-political system. However, the Soviet author, M. E. Bukh, estimated that no earlier than by 1990—and then only "basically"—would agriculture's "complex mechanization" be achieved in the CMEA countries, and would its capital stock, per worker or per output unit, attain the industrial level.[7]

Upon entering the 1980s, the modernization process, as well as simply the productive performance, of agriculture in the Soviet Union and most of Eastern Europe lagged critically behind official plans and actual demand. Some speakers and publications of those countries, as well as some observers in the West, used to point to an unusual sequence of meteorologically unfavorable years as a major reason for the frequent harvest shortfalls during 1975–80. Yet one finds it hard to believe that the weather was responsible in three or four out of six years under the diverse climates of the immense territory from the Baltic to the Caspian Sea and Hindukush, and from Berlin to the Pacific Coast. Rather, one suspects that Marxist-Leninist agrarian policies—except for their Hungarian variant—have ingredients which are unpropitious not only for peasant small-scale farming but also for socialist large-scale farming.

As to modernization, it is not by thoroughly reforming the agro-economic system and thereby making it more efficient that the necessary capital is to be generated, but by the public sector supplying the needed inputs for "industrialization" through subsidized prices and investment.

Notes

1. This very fitting term was used by J. F. Karcz in his contribution to *Agrarian policies and problems in Communist and non-Communist countries*, W. A. D. Jackson, ed., (Seattle and London, 1971), p. 179 and passim.

2. *Plan and market*, M. Bornstein, ed. (New Haven and London, 1973), p. 243.

3. For the above and the following, see *Symposium on Forms of Horizontal and Vertical Integration in Agriculture*, FAO, ECE/AGRI/29 (New York, 1977), vol. 1, pp. 4–12, and passim in both volumes.

4. *Agrarno-promyshlennaya integratsia stran SEV* (Moscow, 1976), p. 130.

5. *Symposium, op. cit.,* p. 39.

6. *Ibid.,* p. 366.

7. M. E. Bukh, *Problemy, op. cit.,* pp. 162–63.

Selected Bibliography of
English-Language Publications

This selected bibliography has been compiled for the general reader who lacks knowledge of Slavic and other East European languages, in order to help him find more detailed information on aspects which could not be dealt with, or could be covered only very briefly, in the present book. It is hoped that specialists also will find the bibliography of use.

The numerous sources cited here also serve to refute the often heard complaint that not enough relevant information is available in the West. In fact, three times as much space would have been needed if the author, as he originally intended, had included works in the better known Western languages besides English, e.g., French, German, Italian, and Spanish. Even so, the bibliography is still selective in the sense that newspaper or other small items have been omitted, although this regrettably leads to the fact that only a few articles from Radio Liberty and Radio Free Europe Research Bulletins (abbreviated RFER in the present book) are listed. The reader wanting to find current, up-to-date information should turn to these periodicals, although they deal with agriculture only among other subjects. The same applies to the ABSEES (Soviet and East European Abstracts Series) microfiches published on behalf of the British National Association of Soviet and East European Studies in collaboration with the Centre for Russian and East European Studies of the University of Birmingham, England (up to 1976, in book form in collaboration with the Institute of Soviet and East European Studies, University of Glasgow, Scotland).

For the period before the 1970s, especially in relation to precollectivization times, space limitations forced the author to refer to no more than a few of the many comprehensive and authoritative works available, and to omit a wealth of articles still retaining their value for the specialist.

The arrangement and subdivision of the bibliography is dictated by practical considerations and does not pretend to represent an irrefutable, generally applicable system. Where the number of items is great, as in the cases of the Soviet Union and Hungary, more subdivisions have been devised, in other cases less. Moreover, the subheadings are not strictly uniform but depend on what could be applied to the selected items.

It is important to note that the English-language publications are not all of Western origin. To include many from the countries concerned was meant to enable the reader also to gain an insight into how the developments and problems are viewed there. In particular, Poland, Hungary, Yugoslavia, and Rumania have issued a respectable number of monographs, serials, and journals in English, French, and German. Many of these, especially the more recent, reveal a high scholarly standard without wholly relinquishing a certain political bias. For lack of space, from the Eastern serials and journals not every item could be included. Of Eastern journals, the *Hungarian Agricultural Review* (with abstracts from Hungarian publications) and the *Co-Operative Scientific Quarterly* (a supplement to the Polish Spódzielczy Kwartalnik Naukowy) deserve special mention. A considerable number of articles by Soviet authors have been published in English in American translation journals. Not surprisingly, the number of English-language publications on East German agriculture is small. This is so in part because most Western research in this field has been conducted and published in Germany, and in part because the GDR's own publications are in a Western language (German) anyway.

For the author, the marvellous "in-cataloguized" (i.e., cardexing also the individual articles in periodicals) catalog of the specialized economic library at Kiel (Bibliothek des Instituts für Weltwirtschaft) with its unique inventories was of invaluable help. The present bibliography, however, is not for the librarian or area specialist, who has better ways to find what he needs. Yet it seems that, for a more general English-reading public with an interest in the agricultural affairs of Communist Europe, a comprehensive, cross-national bibliography has not been available so far. Therefore, if the reader feels that it gives a useful overview, the author will readily accept criticism which might rightly be raised by more competent bibliographers.

Index to Bibliography

I. Eastern Europe (including the USSR) as a whole, or more than one country

I.1. Monographs, books, brochures

Adams, Arthur E., and Adams, Jan S. *Men versus systems. Agriculture in the USSR, Poland, and Czechoslovakia.* New York: The Free Press; London: Collier-Macmillan, 1971. V+327 pp.

Agrarian policies and problems in Communist and non-Communist countries. W. A. Douglas Jackson, ed. Seattle, London: Univ. of Washington Press, 1971. VIII+488 pp.

Agricultural policies in the USSR and Eastern Europe. R. A. Francisco, B. A. Laird, R. D. Laird, eds. Boulder, Colorado: The Westview Press, 1980. 332 pp.

The agricultural situation in Eastern Europe. Production and trade statistics, 1970–1975 (Economic Research Service, Foreign Agricultural Economic Report no. 117). Washington, D.C.: USDA, 1976. 15 pp.

Agricultural statistics of Eastern Europe and the Soviet Union 1950–66 (ERS—Foreign 252). Washington, D.C.: USDA, 1969. 110 pp.

Agricultural statistics of Eastern Europe and the Soviet Union, 1950–70 (ERS—Foreign 349). Washington, D.C.: USDA, 1973. VI+106 pp.

Agriculture and national economic planning (Research Institute for Agricultural Economics, Bulletin 27). Budapest: 1970. 115 pp.

Balogh, A. *International division of labour in the agriculture of COMECON countries.* Budapest: Research Institute for Agricultural Economics, 1974. 70 pp.

Berend, Iván T. *Agriculture in Eastern Europe, 1919–1939* (Papers in East European economics. Centre for Soviet and East European Studies, St. Antony's College, Oxford, 35). Oxford: 1973. II+84 pp.

Bergmann, Theodor. *The development models of India, the Soviet Union and China. A comparative analysis.* Assen, Amsterdam: van Gorcum, 1977. XI+255 pp.

Brandt, Karl (in collaboration with O. Schiller and Franz Ahlgrimm). *Management of agriculture and food in the German-occupied and other areas of Fortress Europe.* Stanford, Calif.: Stanford University Press, 1953. 707 pp.

Collectivization of agriculture in Eastern Europe. I. T. Sanders, ed. Lexington, Ky.: Univ. of Kentucky Press, 1958. 214 pp.

Comintern and peasant in East Europe, 1919–1930. G. D. Jackson, Jr., ed. New York, London: Columbia Univ. Press, 1966. 339 pp.

Comparative economic systems: Models and cases. Morris Bornstein, ed. Homewood, Ill.: Irwin, 1965 (revised edition: 1969). X+464 pp.

Conference on agricultural economics 1969 (Research Institute for Agricultural Economics, Bulletin, 24). Budapest: 1969. 109 pp.

Cooperative and Commune. Group farming in the economic development of agriculture. Peter Dorner, ed. Madison, Wisconsin; London: The Univ. of Wisconsin Press, 1977. XIII+392 pp.

Decision-making and agriculture (Papers and reports, Sixteenth International Conference of Agricultural Economists). Th. Dams, K. E. Hunt, eds. Oxford: Agricultural Economics Institute, 1977. XIII+603 pp.

Dovring, Folke. *Land and labor in Europe in the twentieth century.* With a chapter on land reform as a propaganda theme, by Karin Dovring. (Third, revised edition of Land and labor in Europe, 1900–1950) The Hague: Nijhoff, 1965. 511 pp.

Dumman, Jack. *Agriculture, capitalist and socialist.* Studies in the development of agriculture and its contribution to economic development as a whole. London: Lawrence & Wishart, 1975. 256 pp.

East European agricultural serials in EC libraries. A selective catalogue of holdings in libraries in Belgium, France, Netherlands, FRG, UK. I. Clemens-Zeltsman, ed. Brussels, Luxembourg: Publ. by CBT-CNDST, 1977. IX+300 pp.

East European economics post-Helsinki. A compendium of papers submitted to the Joint Economic Committee, Congress of the United States. Washington, D.C.: U.S. Government Printing Office, 1977. XXIV+1427 pp.

Eastern Europe's agricultural development and trade (Economic Research Service). Washington, D.C.: USDA, 1970. VI+62 pp.

Economic development in the Soviet Union and Eastern Europe. Vol. 2: Sectoral analysis.

Zbigniew M. Fallenbuchl, ed. New York, Washington, London: Praeger Publishers, 1976. XVII+409 pp.

Economic developments in countries of Eastern Europe. A compendium of papers submitted to the Joint Economic Committee, Congress of the United States. Washington, D.C.: U.S. Government Printing Office, 1970. VIII+634 pp.

Economic models and quantitative methods for decisions and planning in agriculture. Earl O. Heady, ed. (Proceedings of an East-West seminar). Ames, Iowa: Iowa Univ. Press, 1971. XIII+518 pp.

Economic performance and the military burden in the Soviet Union. A compendium of papers submitted to the Joint Economic Committee, Congress of the United States. Washington, D.C.: U.S. Government Printing Office, 1970. 295 pp.

Economic problems of agriculture in industrial societies. Ugo Pappi, Ch. Nunn, eds. London, Melbourne, Toronto: Macmillan; New York: St. Martin's Press, 1969. XXIX+671 pp.

Economic questions of main agricultural branches. A. Bisztray, M. Kóvári, L. Rácz, eds. (Research Institute for Agricultural Economics, Bulletin, 37). Budapest: 1975. 95 pp.

Ellman, Michael. *Socialist planning.* Cambridge, London, New York, Melbourne: Cambridge Univ. Press, 1979. 300 pp. (pp. 81–111: "Planning agriculture").

The feed-livestock economy of Eastern Europe: Prospects to 1980 (Economics Research Service, Foreign Agricultural Report, 90). Washington, D.C.: USDA, 1973. XVI+140 pp.

The future of agriculture in the Soviet Union and Eastern Europe. The 1976–80 five year plans. R. D. Laird, J. Hajda, B. A. Laird, eds. Boulder, Colorado: Westview Press, 1977. IX+242 pp.

Government, law and courts in the Soviet Union and Eastern Europe. V. Gsovski, K. Grzybowski, eds. London: Stevens & Sons; The Hague: Mouton, 1959. 2 vols. ("Part Seven: Land and Peasant," pp. 1605–1901)

Incomes in postwar Europe: A study of policies, growth and distribution. (Economic Survey of Europe in 1965, Part 2). Geneva: UN Publication, 66, 1967. Separately paginated parts.

Indices of agricultural production in Eastern Europe and the Soviet Union, 1950–68 (Economic Research Service—Foreign 273). Washington, D.C.: USDA, 1969. 22 pp.

Indices of agricultural and food production for Europe and the USSR. Average 1961–65 and annual 1970 through 1979 (Economics, Statistics, and Cooperatives Service, Statistical Bulletin, 635). Washington, D.C.: USDA, 1980. 42 pp. (Issues with corresponding titles were published for earlier years.)

Integration of the agriculture with the national economy in Poland and Hungary. Warsaw: Instytut Ekonomiki Rolnej, 1977. 243 pp.

Karcz, Jerzy F. *The economics of Communist agriculture.* Edited and with an introductory essay by A. W. Wright. Bloomington, Indiana: International Development Institute, 1979. IX+494 pp. (A posthumous selection of major articles.)

Kozlowski, Zygmunt. *The first steps of collectivization of agriculture in Eastern Europe: 1949–1954* (Papers in East European economics, 47). Oxford: St. Antony's College, Centre for Soviet and East European Studies, 1975. 71 pp.

Kozlowski, Zygmunt. *The integration of peasant agriculture into the socialist economies of Eastern Europe since 1948* (Papers in East European economics, 54). Oxford: St. Antony's College, 1978. 136 pp.

Land utilization in Eastern Europe. Béla Sárfalvi, ed. Budapest: Akadémiai Kiadó, 1967. 88 pp.

May, Jacques M. *The ecology of malnutrition in Central and Southeastern Europe.*

Austria, Hungary, Rumania, Bulgaria, Czechoslovakia. New York, London: Hafner 1966. XV+290 pp.

Mellor, R. E. H. *Eastern Europe. A geography of Comecon countries.* London, New York: Macmillan, X+358 pp.

Mieczkowski, Bogdan. *Personal and social consumption in Eastern Europe. Poland, Czechoslovakia, and East Germany.* New York, Washington, London: Praeger, 1975. XXIV+342 pp.

Miller, R. F. *Socialism and agricultural cooperation. The Soviet and Yugoslav cases* (Department of Political Science, Research School of Social Sciences, Occasional Paper, 9). Canberra: Australian National Univ., 1974. III+94 pp.

Mitrany, David. *Marx against the peasent.* Chapel Hill: Univ. of North Carolina, 1951. XVI+301 pp.

Modernization of agriculture: East and West. (XI International Seminar, Urbino, July 3–5, 1975). Milano: CESES (*l'est, quaderni,* 8), 1976 (?). XV+273 pp.

New directions in the Soviet economy. Studies prepared for the Joint Economic Committee, Congress of the United States. Washington, D.C.: U.S. Government Printing Office, 1966. XI+871 pp. (in four parts)

The new economic systems of Eastern Europe. H. -H. Höhmann, M. Kaser, K. C. Thalheim, eds. London: C. Hurst, 1975. XXI+585 pp.

Ofer, Gur. *Effects of intra-socialist trade on industrial goods and efficiency. A case for specialization in agricultural goods* (The Eliezer Kaplan School of Economics and Social Sciences, Department of Economics, Research Report, 95). Jerusalem: The Hebrew Univ. of Jerusalem, 1976. 65 pp.

Osborne, R. H. *East Central Europe. A geographical introduction to seven socialist states.* London: Chatto & Windus, 1967. 384 pp.

The peasantry of Eastern Europe. Ivan Volgyes, ed. New York, Oxford, Toronto, Sidney, Frankfurt, Paris: Pergamon. Vol. I, 1978 (IX+192 pp.), Vol. II, 1979 (IX+232 pp.).

Plan and market. Economic reform in Eastern Europe. Morris Bornstein, ed. New Haven, London: Yale Univ. Press, 1973. 416 pp.

Policies, planning and management for agricultural development (Papers and reports. Fou⁀ ⁀enth International Conference of Agricultural Economists). Oxford: Institute of Agrarian Affairs, 1971: XVI+616 pp.

The political economy of collectivized agriculture. R. A. Francisco, B. A. Laird, R. D. Laird, eds. New York, Oxford, Toronto, Sydney, Frankfurt, Paris: Pergamon 1979. XII+250 pp.

The process of rural transformation. Eastern Europe, Latin America and Australia. I. Volgyes, R. E. Lonsdale, W. P. Avery, eds. New York, Oxford, Toronto, Sydney, Frankfurt, Paris: Pergamon Press, 1980. IX+347 pp.

Reorientation and commercial relation of the economies of Eastern Europe. A compendium of papers submitted to the Joint Economic Committee, Congress of the United States. Washington, D. C.: U.S. Government Printing Office, 1974. V+771 pp.

The role of private plots in socialist agriculture. Annotated bibliography. (Commonwealth Bureau of Agricultural Economics, CAB, series A, no. 7). Oxford: 1971. III+9 pp.

Rural change and public policy. Eastern Europe, Latin America and Australia. W. P. Avery, R. E. Lonsdale, I. Volgyes, eds. New York, Oxford, Toronto, Sydney, Frankfurt, Paris: Pergamon Press, 1980. XIV+327 pp.

Second National Conference on Agricultural Economics, Debṛecen, 8–11 December 1971. A Bisztray, M. Kővári, L. Pósvai, eds. (Research Institute for Agricultural Economics, Bulletin, 33). Budapest: 1973. 114 pp.

The social structure of Eastern Europe. Transition and process in Czechoslovakia, Hungary,

Poland, Romania, and Yugoslavia. Bernard Lewis Faber, ed. New York, Washington, London: Praeger, 1976. XV+423 pp.

The socialist price mechanism. Alan Abouchar, ed. Durham, N. C.: Duke Univ. Press, 1977. 298 pp.

Soviet and East European agriculture. Jerzy F. Karcz, ed. Berkeley, Los Angeles: Univ. of California Press, 1967. XXV+445 pp.

Stanis, Vladimir. *The socialist transformation of agriculture. Theory and practice.* Transl. from the Russian by S. Vechor-Shcherbovich. Moscow: Progress, 1976. 238 pp.

Symposium on forms of horizontal and vertical integration in agriculture (ECE/Agri/29). New York: Food and Agriculture Organization of the United Nations, 1977. 2 vols.

Trend, Harry. *Agriculture in Eastern Europe: A comparative study.* (Radio Free Europe Research, East Europe, May 15, 1974): Munich 1974. (changing pagination of parts)

Vankai, Thomas A. *Progress and outlook for East European agriculture, 1976–80* (Economics, Statistics, and Cooperatives Service, Foreign Agricultural Report, 153). Washington, D. C.: USDA, 1978. II+51 pp.

Walston, Lord. *Agriculture under Communism.* London: Bodley Head, 1962. 108 pp.

Warriner, Doreen. *Revolution in Eastern Europe.* London: Turnstile, 1950. XV+188 pp.

Zagoroff, S. D.; Végh, Jenö and Bilimovich, A. D. *The agricultural economy of the Danubian countries, 1935–1945.* Stanford, Calif.: Stanford University Press, 1955. XIV+478 pp.

I.2. Articles in periodicals and collective volumes

Adams, Arthur E. "The Soviet agricultural model in Eastern Europe." *East European Quarterly* 8 (1974) 4: 461–77.

"The Agrarian Question." *Survey* 43 (1962): 31–43.

Bango, J. F. "Central European peasantry." *Documentation sur l'Europe Centrale.* (Institut de Recherches de l'Europe Centrale, Louvain) 9 (1971) 3: 189–201.

Bornstein, Morris. "Economic reform in Eastern Europe." *East European Economics* (see I.1): 102–34.

Brus, Wlodzimierz. "The East European reforms: What happened to them?" *Soviet Studies* XXXI (1979) 2: 257–67.

Chumichev, D. A.; Berger, Ya. M.; and Avdeichev, L. A. "The natural-historic basis for the development and specialization of agriculture in the socialist countries (outside the USSR)." *Soviet Geography: Review and translation* 6 (1965) 1: 43–47.

Csepely-Knorr, A., and Szakonyi, L. "Development and subvention of agriculture in the European member-states of COMECON." *Studies on the foreign economic factors* (see VI.1): 21–37.

Cseres, N. "Role of Hungarian vegetable and fruit sector in the CMEA." *Acta Oeconomica* 22 (1979) 1–2: 155–70.

Elias, Andrew. "Magnitude and distribution of the labor force in Eastern Europe." *Economic developments* (see I.1): 149–239.

Enyedi, György. "The changing face of agriculture in Eastern Europe." *The Geographical Review* LVII (July 1967) 3: 358–72.

Feiwel, George R. "The standard of living in centrally planned economies of Eastern Europe." *Osteuropa Wirtschaft* 25 (1980), 2: 73–96.

Fuchs, Roland J., Demko, George J. "Commuting in the USSR and Eastern Europe." *East European Quarterly* 11 (1977) 4: 463–75.

Galeski, Boguslaw. "The models of collective farming." *Cooperative and Commune* (see I.1): 17–42.

Gamarnikow, Michael. "Balance sheet of economic reforms." *Reorientation and commercial relation* (see I.1): 164–213.

Hajda, Joseph. "The impact of current policies on modernizing agriculture in Eastern Europe." *Agricultural Policies* (see I.1): 296–310.

Hoffman, George W. "Rural transformation in Eastern Europe since World War II." *The process of rural transformation* (see I.1.): 21–41.

Jacobs, Everett M. "The impact of agro-industrial programs on East European agriculture." *Agricultural policies* (see I.1): 237–62.

Jacobs, Everett M. "Organization and management of agriculture in Eastern Europe, 1967–1974." *Economic development*. Fallenbuchl, ed. (see I.1): 284–305.

Jacobs, Everett M. "Ownership and planning in Soviet and East European agriculture." *The prediction of Communist economic performance*. Peter J. Wiles, ed. London: Cambridge Univ. Press, 1971: 39–86.

Jacobs Everett M. "Recent developments in organization and management of agriculture in Eastern Europe." *East European economies* (see I.1): 333–55.

Jaehne, Günter. "Problems of agricultural integration within the CMEA." *Agricultural policies* (see I.1): 221–36.

Jędruszczak, Hanna. "Land reform and economic development in the people's democracies of Europe." *Studia Historiae Oeconomicae* (Universytet w Poznaniu) 7 (1962): 199–211.

Karcz, Jerzy F. "Agricultural reform in Eastern Europe." *Plan and market* (see I.1.): 207–43.

Karcz, Jerzy F. "An organizational model of command farming." *Comparative economic systems*. Morris Bornstein, ed. Homewood, Ill.: Irwin, 1969: 278–99 (2nd, rev. ed.; the article is not contained in the first edition of 1965).

Karcz, Jerzy F. "Certain aspects of new economic systems in Bulgaria and Czechoslovakia." *Agrarian policies* (see I.1): 178–204.

Karcz, Jerzy F. "Comparative study of transformation of agriculture in centrally planned economies." *The role of agriculture in economic development* (International Journal of Economic Affairs. 3 [1961], 2). London: Oxford Univ. Press, 1961: 237–66 (see also the comment by J. M. Montias und Karcz's reply, *ibid.:* 266–76).

Kleer, Jerzy. "Economic integration within the framework of CMEA in relation to agriculture." *Decision-making* (see I.1): 344–55.

Kolaja, Jiri. "Post World War II and recent regional and rural-urban migrations in Poland and Czechoslovakia." *Polish Western Affairs* 5 (1964) 1: 120–44.

Korbonski, Andrzej. "Political management of rural change in Eastern Europe." *Rural change* (see I.1.): 21–35.

Kosinski, Leszek A. "Statistical yearbooks in East Central Europe." *Zeitschrift für Ostforschung* 23 (1974) 1: 137–47.

Kozlowski, Zygmunt. "Agriculture in the economic growth of East European countries." *Agriculture in development theory*. L. G. Reynolds, ed. New Haven, London: Economic Growth Center, Yale Univ., 1975: 411–50.

Lazarcik, Gregor. "Agricultural output and productivity in Eastern Europe and some comparisons with the U.S.S.R. and U.S.A." *Reorientation and commercial* (see I.1) 328–93.

Lazarcik, Gregor. "Comparative growth and levels of agricultural output and productivity in Eastern Europe, 1965–76". *Eastern European economies* (see I.1): 289–332.

Lazarcik, Gregor. "Growth of output, expenses, and gross and net output in East European agriculture." *Economic developments in countries . . .* (see I.1): 463–527.

Leng, Earl R. "Agronomic problems in Southeastern Europe." *Soviet agricultural and peasant affairs* (see II.1.b.): 190–202.

Loncarević, Ivan, Schinke, Eberhard; and Zilahi-Szabó, Géza. "Farming." *The new economic systems* (see I.1): 411–41.

Moravcik, Ivo. "Canada's wheat export to Eastern Europe and the USSR." *Canadian Slavonic Papers* 20 (1978) 1: 91–114.

Nagy, János. "Changes in the system of economic control and management of agriculture in the European Comecon countries." *Abstracts* (Research Institute for Agricultural Economics, Budapest), 41 (1976–77): 105–15.

Nazarenko, V. "Integration of CMEA countries in the field of agriculture." *Decision-making* (see I.1): 356–64.

Nove, Alec. "Agricultural performance compared: Belorussia and Eastern Poland." *Economic development.* Fallenbuchl, ed. (see I.1):261–83.

Pienkos, Donald E. "Education and emigration as factors in rural societal development: The Russian and Polish peasantries' responses to collectivization." *East European Quarterly* 9 (1975) 1: 75–95.

Rosenblum-Cale, Karen. "The search for economic viability in East European agriculture." *Rural change* (see I.1.): 231–52.

Shmelev, G. "The private household plot in CMEA countries." *Problems of Economics* 22 (1979) 1: 79–100. (Transl. from *Voprosy ekonomiki*, no. 7, 1978.)

"Social security protection for members of farmers' co-operatives in Eastern Europe." *International Labour Review* 81 (1960) 4: 319–34.

Spulber, Nicolas. "Collectivization in Hungary and Rumania." *Collectivization* (see I.1): 140–65.

Spulber, Nicolas. "Eastern Europe: The changes in agriculture from land reforms to collectivization." *The American Slavic and East European Review* 13 (1954) 3: 389–401.

Volgyes, Ivan. "Attitudinal and behavioral changes among the peasantry of Eastern Europe." *Rural change* (see I.1.): 83–110.

Volgyes, Ivan. "Economic aspects of rural transformation in Eastern Europe." *The process of rural transformation* (see I.1.): 89–127.

Wittfogel, Karl. A. "The peasants." *Outline of Communism.* Gerhart Niemeyer, ed. London: Ampersand Books, 1966: 91–159.

Wädekin, Karl-Eugen. "Economic reform and social change in Soviet and East European agriculture." *The Soviet Union and East Europe into the 1980s.* S. McInnes, W. McGrawth, P. J. Potichnyj, eds. Oakville, Ontario: Mosaic Press, 1978: 205–29.

II. Soviet Union

II.1. Agriculture in general

Lydolph, Paul E. *Geography of the U.S.S.R. Topical Analysis.* Elkhart Lake: Misty Valley Publishing, 1979. VIII+522 pp. (agriculture and related subjects on pp. 57–103 and 215–61).

Symons, Leslie. *Russian agriculture. A geographic survey.* London: Bell & Sons, 1972. XI+348 pp.

USSR agriculture atlas. Washington, D.C.: Central Intelligence Agency, 1974. 59 pp. (maps, diagrams, text).

Volin, Lazar, *A century of Russian agriculture. From Alexander II to Khrushchev.* Cambridge, Mass.: Harvard Univ. Press, 1970. VIII+644 pp.

II.1.a. Up to the October Revolution

Atkinson, Dorothy. "The statistics on the Russian land commune, 1905–1917." *Slavic Review* 32 (1973) 4: 773–87.

Blum Jerome. *Lord and peasant in Russia from the ninth to the nineteenth century.* 2nd ed. New York: Atheneum, 1964. 656 pp.

Gill, Graeme J. "The mainsprings of peasant action in 1917." *Soviet Studies* XXX (1978) 1: 63–86. Cf. the critique by Kress, John H. "The political consciousness of the Russian peasantry: a comment on Graeme Gill's 'The mainsprings of peasant action in 1917." *Soviet Studies* XXXI (1979) 4: 574–80, and Gill's reply, XXXII (1980) 2: 291–96.

Gill Graeme J. "The failure of rural policy in Russia, February-October 1917." *Slavic Review* 37 (1978) 2: 241–58.

Kimball, A. "The First International and the Russian Obshchina." *Slavic Review* 32 (1973) 3: 491–514.

The politics of rural Russia 1905–1914. L. H. Haimson, ed. Bloomington and London: Indiana Univ. Press, 1979. X+309 pp.

Robinson, Geroid Tanquary: *Rural Russia under the old régime.* New York: Macmillan, 1967. (1st ed. 1932). VIII+342 pp.

Simms, James Y., Jr. "The crisis in Russian agriculture at the end of the nineteenth century: a different view." *Slavic Review* 36 (1977) 3: 377–98.

II.1.b. Since the October Revolution

Agriculture of the Soviet Union (XIV International Conference of Agricultural Economists). Minsk: Mir Publishers, 1970. 245 pp. (A collective volume written for the conference by Soviet authors.)

Dalrymple, Dana G. "The Soviet Famine of 1932–1934." *Soviet Studies* XV (1964) 3: 250–84, and XVI (1965) 4: 471–74.

Jasny, Naum. *The socialized agriculture of the USSR.* Stanford, Calif.: Stanford Univ. Press, 1949. 837 pp.

Jasny, Naum. "Soviet grain crops and their distribution." *International Affairs* XXVIII (1952) 4: 452–59.

Karcz, Jerzy F. "From Stalin to Brezhnev: Soviet agricultural policy in historical perspective." *The soviet rural community* . . . (see II.7.): 36–70.

Karcz, Jerzy F. "Soviet agriculture: a balance sheet." *Studies on the Soviet Union* VI (April 1967) 4: 108–45.

Karcz, Jerzy F. "Thoughts on the grain problem." *Soviet Studies* XVIII (April 1967) 4: 399–434; cf. XXII (Oct 1970) 2: 262–94.

Laird, Roy D. "Lenin, peasants and agrarian reform." *Lenin and Leninism.* B. W. Eissenstat, ed. Toronto, London: Lexington Books; Lexington, Mass.: D. C. Heath and Co., 1971: 173–82.

Laird, Roy D., and Laird, Betty A. *Soviet communism and agrarian revolution.* Harmondsworth, Middlesex: Penguin, 1970. 157 pp.

Lewin, M. " 'Taking grain': Soviet policies of agricultural procurements before the war." *Essays in Honour of E. H. Carr.* C. Abramsky, ed. London: Macmillan, 1974: 281–323.

Male, D. J. *Russian peasant organization before collectivization.* Cambridge, Mass.: Cambridge Univ. Press. VIII+253 pp.

Matskevich, Vladimir. *Agriculture.* Moscow: Novosti Press Agency Publ. House, 1971. 159 pp.

Miller, Robert F. "Soviet agricultural policy in the twenties: the failure of cooperation." *Soviet Studies* XXII (April 1975) 2: 220–44.

Miller, Robert F. "The Politotdel: A lesson from the past." *Slavic Review* XXV (Sept. 1966) 3: 475–96.

Narkiewicz, O. A. "Stalin, War Communism and collectivization." *Soviet Studies* XVIII (July 1966) 1: 20–37.

Shanin, Teodor. *The awkward class. Political sociology of peasantry in a developing Society: Russia 1910–1925.* London: Clarendon Press, Oxford Univ. Press, 1972. XVIII+253 pp.

Soviet agricultural and peasant affairs. Roy D. Laird, ed. Lawrence, Kansas: Univ. of Kansas Press, 1963. XI+335 pp.

Soviet agriculture. An assessment of its contributions to economic development. Harry G. Shaffer, ed. New York, London: Praeger, 1977. XVI+166 pp.

The Soviet peasantry. An outline history. Transl. from the Russian by S. Vechor-Shcherbovich. Moscow: Progress Publ., 1975. 378 pp.

Volin, Lazar. *A survey of Soviet Russian agriculture* (Agriculture monograph, 5). Washington, D.C.: U.S. Department of Agriculture, 1951. VIII+194 pp.

Wittfogel, Karl August. "Communist and non-Communist agrarian systems with special reference to the U.S.S.R. and Communist China: A comparative approach." *Agrarian policies* . . . (see I.1): 3–60.

II.1.c. Collectivization

Arkhipov, A. *How farmers' cooperatives were organized.* Moscow: Novosti Press Agency Publ. House, 1977. 63 pp.

Biggart, John. "The collectivization of agriculture in Soviet Lithuania." *East European Quarterly* 9 (1975) 1: 53–75.

Ellman, Michael. "Did the agricultural surplus provide the resources for the increase in investment in the USSR during the first five year plan?" *The Economic Journal* 85 (Dec. 1975): 844–64.

Lewin, M. "The immediate background of Soviet collectivization." *Soviet Studies* XVII (Oct. 1966) 2: 162–97.

Lewin, Moshe. *Russian peasants and Soviet power.* London: Allen and Unwin, 1968. 556 pp.

Lewin, Moshe. "Who was the Soviet kulak?" *Soviet Studies* XVIII (Oct. 1966) 2: 189–212. (Cf. the comment by R. Beermann, *Soviet Studies* XVIII (Jan. 1967) 3: 371–75.)

Millar, J. R. "Mass collectivization and the contribution of agriculture to the first five year plan: a review article." *Slavic Review* 33 (Dec. 1971) 4: 750–66.

Millar, James R. "Soviet rapid development and the agricultural surplus hypothesis." *Soviet Studies* XXII (July 1970) 1: 77–93. (Cf. A. Nove's comment and J. R. Millar's reply, *Soviet Studies* XXII (Jan. 1971) 3: 394–401, and XXIII (Oct. 1971) 2: 302–6.)

Millar, James R., and Guntzel, Corinne. "The economics and politics of mass collectivization reconsidered." *Explorations of economic history* 8 (fall 1970) 1: 103–16.

Millar, James R., and Nove, Alec." Was Stalin really necessary? A Debate on collectivization." *Problems of Communism* XXV (1976) 4: 49–66.

Nove, Alec. "The decision to collectivize." *Agrarian policies* . . . (see I.1): 69–97.

Sirc, Ljubo. "Economics of collectivization." *Soviet Studies* XVIII (Jan. 1967) 3: 362–70.

Taagepera, Rein. "Soviet collectivization of Estonian agriculture: The deportation phase." *Soviet Studies* 32 (1980) 3: 379–97.

Taagepera, Rein. "Soviet collectivization of Estonian agriculture: The taxation phase." *Journal of Baltic Studies* 10 (1979): 263–82.

II.1.d. Since Stalin's death

Arutjunjan, J. V. "The State, agrarian policy and the trends of agriculture in the Soviet Union." *Economies et Sociétés* 10 (1976) 7/8: 1631–52.

Bush, Keith. "Agricultural reforms since Khrushchev." *Bulletin* (Institute for the Study of the USSR) 16 (1969) 5: 3–26.

Bush, Keith. "Soviet agriculture in the 1970s." *Studies on the Soviet Union* 11 (1971): 1–45.

Bush, Keith. "Soviet agriculture: ten years under new management." *Economic development* . . . Fallenbuchl, ed. (see I.1): 157–204 (Reprinted in *Modernization of agriculture* . . . (see I.1): 79–122.

Carey, David W. "Soviet agriculture: recent performance and future plans." *Soviet economy in a new perspective* (see below): 575–99.

Carey, David W., and Havelka, Joseph F. "Soviet agriculture: Progress and problems." *Soviet economy in a time of change* (see below): 55–86.

Centrally planned agriculture. China and USSR. A compilation of review articles and annotated bibliography. K. P. Broadbent, ed. (Review publication. Commonwealth Bureau of Agricultural Economics. No. 2.) Oxford 1972. I+42 pp.

Clark, M. Gardner. "Soviet agricultural policy." *Soviet agriculture* . . . H. Shaffer, ed. (see II.1a): 1–55.

Clarke, Roger A. "Soviet agricultural reforms since Khrushchev." *Soviet Studies* XX (Oct. 1968) 2: 159–78.

Clayton, Elizabeth. "Productivity in Soviet agriculture: a sectoral comparison." *Jahrbuch der Wirtschaft Osteuropas—Yearbook of East-European Economics* 2 (1971): 315–28.

Clayton, Elizabeth. "Productivity in Soviet agriculture." *Slavic Review* 39 (1980) 3: 446–58.

Diamond, Douglas B., and Davis, W. Lee. "Comparative growth in output and productivity in U.S. and U.S.S.R. agriculture." *Soviet economy in a time of change* (see below): 19–54.

Diamond, Douglas B., Krueger, Constance B. "Recent developments in output and productivity in Soviet agriculture." *Soviet economic prospects* . . . (see below): 316–39.

Durgin, Frank A. "The inefficiency of Soviet agriculture versus the efficiency of U.S. agriculture: Reality or idol of the mind." *The ACES Bulletin* XX (fall-winter 1978) 3–4: 1–36.

Durgin, Frank A. "The low productivity of Soviet agricultural trade." *Agricultural policies* . . . (see I.1): 27–36.

Emel' ianov, A. "The Party's agrarian policy and structural changes in agriculture." *Problems of Economics* 18 (1975) 7: 83–108. (Transl. from *Voprosy Ekonomiki* no. 3, 1974.)

Feiwel, George R. "The vicissitudes of Soviet agriculture." *Rivista Internazionale di Scienze Economiche e Commerciali* 19 (1972) 8 and 9.

Garaventa, John. "Down on the farm, Soviet style." *National Geographic Magazine* 155 (June 1979): 769–96.

Goldich, Judith G. *USSR agricultural trade, 1955–77: a historical perspective* (FAS-M-289). Washington, D.C.: USDA, 1979. III + 80 pp.

Green, Donald W. "Soviet agriculture: an econometric analysis of technology and behavior." *Soviet economy in a time of change* (see below) 115–32.

Hahn, Werner G. *The politics of Soviet agriculture, 1960–1970.* Baltimore, London: Johns Hopkins Univ. Press, 1972. XIII+311 pp.

Hingley, Ronald. "Home truths on the farm—the literary mirror." *Problems of Communism* XIV (1965) 3: 22–34.

Iglesia, J. C. Sta; Tiongson, F. A.; Paulino, L. A.; Lawas, J. M.; and Mangahas, M. "An over-view of Soviet agriculture." *Journal of Agricultural Economics and Development* (Philippine Agricultural Economics Association) 1(1971) 1: 55–72.

Il'ichev, A. K. "Certain principles in determining the place of USSR agriculture in the international division of labor." *Problems of Economics* 17 (1975) 9: 34–47. (Transl. from *Vestnik Moskovskogo Universiteta. Ekonomika*, no. 2, 1974.)

Johnson, D. Gale. "Agricultural production." *Economic trends in the Soviet Union.* A. Bergson and S. Kuznets, eds. Cambridge, Mass.: Harvard Univ. Press, 1963: 203–34.

Johnson, D. Gale "Soviet agriculture revisited." *American Journal of Agricultural Economics* 53 (1971) 2: 257–68.

Kahan, Arcadius. "Changes in agricultural productivity in the Soviet Union." *Conference on International Trade and Canadian Agriculture.*Ottawa 1966: 345–95.

Karcz, Jerzy F. "Seven years on the farm."*New Directions* . . . (see below): 383–450.

Karliuk, I. "Economic problems in the development of agriculture." *The Soviet Review* 15 (1974) 3: 3–21. (Transl. from *Voprosy Ekonomiki*, no. 12, 1973.)

Kerblay, Basile. "The Peasant" [i. e., The Russian Peasant]*." Survey* 64 (July 1967): 99–107.

Klatt, Werner. "Reflections on the 1975 Soviet harvest." *Soviet Studies* 26 (1976) 4: 485–98.

Klatt, Werner. "Soviet farm output and food supply in 1970." *Soviet Affairs* No. 4 (St. Antony's Papers, 19). London: Oxford Univ. Press, 1966: 104–33.

Laird, Roy D. "Agriculture under Khrushchev." *Survey* 56 (July 1965): 106–17.

Laird, Roy D. "Political and economic trends in Soviet agriculture." *The New Russia.* D. Dirscherl, ed. Dayton, Ohio: Pflaum Press, 1968: 37–54.

Laird, Roy D. "The politics of Soviet agriculture." *Soviet agricultural and peasant affairs.* (see II.1.b): 269–86.

Laird, Roy D. "Prospects for Soviet agriculture." *Problems of Communism* 20 (1971) 5: 31–40.

Laird, Roy D. "Soviet agriculture in 1973 and beyond in light of United States performance." *The Russian Review* 33 (Oct. 1974) 4: 362–85

Laird, Roy D. "Soviet farmer productivity, 1950–70, as measured by a U.S. barometer." *Economic development* . . . Fallenbuchl, ed. (see I.1): 241–60.

Laird, Roy D., and Laird, Betty A. "The widening Soviet grain gap and prospects for 1980 and 1990." *The future* . . . (see I.1): 27–47.

Leversedge, Francis M., and Stuart, Robert C. "Soviet agricultural restructure and urban markets." *Canadian Geographer* XIX (1975) 1: 73–93.

Lukinov, I. "Problems in agricultural forecasting." *Problems of Economics* 14 (1972) 10: 3–18. (Transl. from *Voprosy Ekonomiki*, no. 7, 1971.)

Mezhberg, Iu. "Current problems in the restructuring of the countryside." *Problems of Economics* 21 (1979) 9: 44–61. (Transl. from *Voprosy Ekonomiki*, no. 5, 1978)

Millar, James R. "Post-Stalin agriculture and its future." *The Soviet Union since Stalin.* S.F. Cohen, A. Rabinowitch, R. Sharlet, eds. Bloomington, London: Indiana Univ. Press, 1980: 135–54.

Millar, James R. "The prospects for Soviet agriculture." *Problems of Communism* 26 (1977) 3: 1–16.

Millar, James R. "Transformation of the Soviet rural community." *Scholarly exchanges with the USSR and Eastern Europe: Two decades of American experience.* A. H. Kassof, ed. (forthcoming, 1980).

Morozov, V. *Soviet agriculture.* Transl. from the Russian by Inna Medova. Moscow: Progress Publ., 1977. 220 pp.

Nikiforov, L. "Overcoming socioeconomic differences between town and country." *Problems of Economics* 18 (1975) 7: 61–82. (Transl. from *Voprosy Ekonomiki*, no. 2, 1975.)

Nimitz, Nancy. "The lean years." *Problems of Communism* 14 (1965) 3: 10–22.

Nove, Alec. "Soviet agriculture under Brezhnev." *Slavic Review* 29 (1970)3: 379–410. Cf. comments *ibid.* Jackson, E. A. Douglas. "Wanted, an effective land use policy and improved reclamation," pp. 417–26; Karcz, Jerzy F. "Some major persisting problems in Soviet agriculture," pp. 417–26; Nove, A. "Reply," pp. 427–28.

Osovsky, Stephen. *Soviet agricultural policy. Toward the abolition of collective farms.* New York, Washington, London: Praeger, 1974. XI+300 pp.

Pinvdak, F. *Ideology and economics in Soviet agricultural policies* (Netherlands School of Economics, Centre for Development Planning. Discussion Paper 13, 1971). 16 pp.

Ploss, Sidney I. *Conflict and decision-making in Soviet Russia. A case study of agricultural policy 1953–1963.* Princeton, N.J.: Princeton Univ. Press, 1965. 312 pp.

Prybyla, Jan S. "Soviet economic reforms in agriculture." *Weltwirtschaftliches Archiv* 109 (1973) 4: 644–87.

Sarap, Kailas. "Soviet agricultural development, 1953–58" *Economic Affairs* (Calcutta) 28 (1978) 3: 109–15.

Schinke, Eberhard. "Some peculiarities of the employment of factors in Soviet agriculture." *Agrarian policies . . .* (see I.1): 142–55.

Schinke, E. "Soviet agriculture. An uncertain outlook." *The U.S.S.R. in the 1980s. Economic growth and the role of foreign trade.* (Colloquium Jan. 17–19, 1978, Brussels. Nato-Directorate of Economic Affairs. Series No. 7). Brussels: 1978: 23–30.

Schoonover, David M. "Soviet agricultural policies." *Soviet economy in a time of change . . .* (see below): 87–115.

Schoonover, David M. "Soviet agricultural policies from development to maturity." *Soviet Union* 4 (1977) fasc. 2: 271–96.

Schoonover, David M. "Soviet agriculture in 1976–80 plan." *The future . . .* (see I.1): 79–94.

Severin, Barbara S., and Carey, David W. "The outlook for Soviet agriculture." *The future of the Soviet economy: 1978–1985.* Holland Hunter, ed. Boulder, Colorado: Westview Press, 1978: 100–132.

Shaffer, Harry G. "Soviet agriculture: success or failure?" *Soviet agriculture: an assessment . . .* (see below): 56–105.

Slater, John A. "Soviet grain harvest and imports in perspective." *World Today* (1976) 2: 75–80.

Smith, Enid. "Comparing Soviet agriculture." *Soviet Studies* 16 (July 1964) 1: 82–89.

Soviet agriculture: an assessment of its contributions to economic development. Harry G. Shaffer, ed. New York: Praeger, 1977. XVI+166 pp.

Soviet agriculture: The permanent crisis. R. D. Laird, E. L. Crowley, eds. New York, Washington, London: F. A. Praeger, 1965. 209 pp.

"Soviet conservation law. Land." *Soviet Statutes and Decisions. A journal of translations* 9 (1972) 1: 5–96.

Soviet economic performance: 1966–67. Materials prepared for the Joint Economic Committee, Congress of the United States. Washington, D.C.: U.S. Government Printing Office, 1968. 292 pp.

Soviet economic prospects for the seventies. A compendium of papers submitted to the Joint Economic Committee, Congress of the United States. Washington, D.C.: U.S. Government Printing Office, 1973. XVII+776 pp.

Soviet economy in a new perspective. A Compendium of papers submitted to the Joint Economic Committee, Congress of the United States. Washington, D.C.: U.S. Government Printing Office, 1976. XXXIX+821 pp.

Soviet economy in a time of change. A compendium of papers submitted to the Joint Economic Committee, Congress of the United States. Vol. 2. Washington D.C.: U.S. Government Printing Office, 1979. VI+677 pp.

Strauss, Erich. *Soviet agriculture in perspective.* London: George Allen and Unwin, 1969. 328 pp.

Syrodoev, N. "*Soviet land legislation.* Transl. from the Russian by K. Kostrov. Moscow: Progress Publ., 1975. 147 pp.

Volin, Lazar. "Agricultural policy of the Soviet Union." *Comparative economic systems. Models and cases.* M. Bornstein, ed. Homewood, Illinois: R. D. Irwin, 1965 (2nd ed. 1969): 310–43.

Volin, Lazar. "Khrushchev and the Soviet agricultural scene." *Soviet and East European agriculture* (see I.1): 1–21.

Wädekin, Karl-Eugen. "A survey of Soviet agriculture in 1974." *Modernization of agriculture* . . . (see I.1.): 135–58.

Wagener, Hans-Jürgen. "Sectoral growth: the case of Soviet agriculture." *Jahrbücher fur Nationalökonomie und Statistik* 188 (1975) 6: 512–38.

Whitehouse, F. Douglas, and Havelka, Joseph F. "Comparison of farm output in the US and USSR, 1950/1971." *Soviet economic prospects* . . . (see above):340–74.

Zabijka, Valentine. "The Soviet grain trade 1961–1970: a decade of change." *The ACES Bulletin* 16 (1974) 1: 3–16.

II.2. Soviet statistics, research and education

Adams, Arthur E. "Informal education in Soviet agriculture." *Comparative education review* 11 (1967) 2: 217–30.

Cox, Terence. *Rural sociology in the Soviet Union. Its history and basic concepts.* London: C. Hurst & Co., 1979. VI+106 pp.

Davies, R. W. "A note on grain statistics." *Soviet Studies* 21 (Jan. 1970) 3: 314–29.

Feshbach, Murray. *The Soviet statistical system: Labor force recordkeeping and reporting* (International Population Statistics Reports, Series P-90, no. 12). Washington, D.C.: U.S. Department of Commerce, Bureau of the Census, 1960. 151 pp. (Cf. same author and title with ". . . , since 1957" added, no. 17, 99 pp.)

Gerasimov, I. P., Vendrov, S. L., Zonn, S. V., Kes', A. S., Kuznetsov, N. T., and Neyshtadt, M. I. "Large-scale research and engineering programs for the transformation of nature in the Soviet Union and the role of geographers in their implementation." *Soviet Geography* 17 (1976) 4: 235–45. (Transl. from *Materialy VI s"ezda Geograficheskogo Obshchestva SSSR.* Leningrad: 1975.)

Gillula, James W. *Bibliography of regional statistical handbooks in the U.S.S.R.* (Second edition). Washington, D.C.: Foreign Demographic Analysis Division, Bureau of the Census, September 1980. 99 pp. (mimeograph).

Grossman, Philip. "A note on agricultural employment in the USSR." *Soviet Studies* 19 (Jan. 1968) 3: 398–404.

Grude, E. N. "A model for optimizing the production pattern in the agricultural enterprise." *Matekon* 12 (1976) 4: 88–97. (Transl. from *Ekonomika i Matematicheskie Metody,* no. 5, 1975.)

Jasny, Naum. "More Soviet grain statistics." *International Affairs* 32 (Oct. 1956) 4: 464–66.

Joravsky, David. "Ideology and progress in crop rotation." *Soviet and East European agriculture* . . . (see I.1): 156–172.

Joravsky, David. *The Lysenko affair.* Cambridge, Mass.: Harvard Univ. Press, 1970. XIII+459 pp.

Kahan, Arcadius. "Soviet statistics of agricultural output." *Soviet agricultural and peasant affairs* (see II.1.b.): 134–60 (cf., *ibid.,* pp. 161–68, "Commentary" by L. O. Richter).

Kravchenko, P. G. "Methods of economic analysis and programming in the organisation of large farms in the USSR." *Policies, planning and management* . . . (see I.1.): 166–76.

Medvedev, Zhores. *The rise and fall of T. D. Lysenko.* (Transl. by I. M. Lerner) New York, London: Columbia Univ. Press, 1969. XVII+284 pp.

Petrov, A. "Rural vocational training in the USSR." *International Labour Review* 110 (1974) 4: 319–34.

Raup, Philip M. "Some consequences of data deficiencies in Soviet agriculture." *Soviet economic statistics.* V. G. Treml, J. P. Hardt, eds. Durham, N.C.: Duke Univ. Press, 1972 : 263–78.

Richter, Luba. "Some remarks on Soviet agricultural statistics." *The American Statistician* 15 (1961) 3: 23–27.

Schinke, Eberhard. "Soviet agricultural statistics." *Soviet economic statistics.* V. G. Treml, J. P. Hardt, eds. Durham, N.C.: Duke Univ. Press, 1972: 237–62.

Sergeyev, S. "Methods of economic analysis of large Soviet agricultural enterprises." *Policies, planning and management . . .* (see I.1.): 160–66.

Solomon, Susan G. *The Soviet agrarian debate. A controversy in social science, 1923–29.* Boulder, Colorado: Westview Press, 1977. XVI+309 pp.

USSR grain statistics: National and regional, 1955–75 (Economic Research Service, Statistical Bulletin no. 564). Washington, D.C.: USDA, 1977. III+42 pp. (Tables and short introductory text).

Zaslavska, Tamara [Tatiana] I. "On the main lines of rural sociology in Siberia." *Problems of the development of agriculture and information on the state of rural sociology in various countries.* Warsaw: Institute of Philosophy and Sociology, Polish Academy of Sciences, 1971: 179–93.

Zaslavskaia, Tatiana J. "Toward methodology of systemic study of the countryside." *The integrated development of human and natural resources: the contribution of rural sociology.* Wrocław: Institute of Philosophy and Sociology, PAS, 1979: 9–25.

II.3. Individual crops and animal production

Anderson, Jeremy. "A historical-geographical perspective on Khrushchev's corn program." *Soviet and East European agriculture . . .* (see I.1): 104–28.

Anderson, Jeremy. "Fodder and livestock production in the Ukraine: A case study of Soviet agricultural policy." *The East Lakes Geographer* 3 (Oct. 1967):29–46.

Beaucourt, Chantal. "The crop policy of the Soviet Union: present characteristics and future perspectives." *Agricultural policies . . .* (see I.1): 49–70.

Bryan, Paige. "The Soviet sugar situation 1970–74." *The ACES Bulletin* 16 (winter 1974) 3: 53–62.

Chrisler, Donald. *Livestock feed balances for the USSR* (Economic Research Service, ERS-Foreign 355). Washington, D.C.: USDA, 1973. IV+23 pp.

Evans, Robert B. "Cotton in the USSR." *Foreign agriculture* 23 (June 14, 1976): 2–5.

Goldich, Judith. *Sunflowerseed in the USSR: Production, processing and trade* (FAS M-290). Washington, D.C.: USDA, 1979. 13 pp.

Goldich, Judith G. "U.S.S.R. grain and oilseed trade in the seventies." *Soviet economy in a time . . .* (see II.1.d.): 133–64.

Gray, Kenneth. "Performance and organizational development in Soviet red meat production." *The ACES Bulletin* 21 (1979) 3–4 (fall-winter): 43–65.

Hultquist, Warren F. "Soviet sugar-beet production: Some geographical aspects of agro-industrial coordination." *Soviet and East European agriculture* (see I.1): 135–55.

Hurt, Leslie C.; Barry, Robert D.; and McFarlane, John S. *USSR sugar–today and tomorrow* (FAS-M-284). Washington, D.C.: USDA, 1978. 15 pp.

Jasny, Naum. "The failure of the Soviet animal industry." *Soviet Studies* 15 (Oct. 1963 and Jan. 1964) 2 and 3:187–218 and 285–307.

Jasny, Naum. *Khrushchev's crop policy.* Glasgow: George Outram & Co., [1963]. 243 pp.

Schoonover, David M. "Soviet agricultural trade and the feed-livestock economy." *Soviet economy in a new perspective . . .* (see II.1.d.): 813–21.

Schoonover, David M. "The Soviet feed-livestock economy: projections and policies." *Economic development* . . . Fallenbuchl, ed. (see I.1): 223–40.

U.S. team reports on Soviet cotton production and trade (FAS-M 277). Washington, D.C.: USDA 1977. 30 pp.

Zahn, Michael D. "Soviet livestock feed in perspective." *Soviet economy in a time* . . . (see II.1.d.): 165–85.

II.4. Spatial allocation and regional aspects

Borchenko, N. "The program of integrated development of agriculture in the Nonchernozem zone." *Soviet Geography* 16 (1975) 4: 249–56. (Transl. from *Planovoye Khozyaystvo*, no. 7, 1974.)

Boyev, V. R.: "Problems of agri-alimentary complexes in territories with extreme production conditions." *Decision-making* . . . (see I.1): 263–66.

Bronshtein, M. "Improving the territorial management of agriculture." *Problems of Economics* 17 (1974) 1: 45–60. (Transl. from *Voprosy Ekonomiki*, no. 10, 1973).

Bystrova, N. P. "The problem of developing rural services in Novgorod Oblast." *Soviet Geography* 17 (1976) 2:121–24. (Transl. from *Vestnik Leningradskogo Universiteta*, no. 18, 1974.)

Bystrova, P. "The relationship between the location of services and rural settlement." *Soviet Geography* 21 (1980) 4: 419–27. (Transl. from *Izvestiya Vsesoyuznogo Geograficheskogo Obshchestva*, no. 4, 1979.)

Clayton, Elizabeth. *Agricultural land evaluation in the Soviet Union and Eastern Europe* (Center for International Studies, University of Missouri-St. Louis. Occasional Paper 8021): December 1980. 13 (unnumbered) pp. (hectograph).

Dienes, Leslie. "Pasturalism in Turkestan: its decline and its persistence." *Soviet Studies* 27 (July 1975) 3: 343–65.

Durgin, Frank A. "The Virgin Lands programme." *Soviet Studies* 13 (Jan. 1962) 3: 255–80.

Dzhaoshvili, V. Sh. "Territory. A natural resource of Georgia and problems of its use for settlement." *Soviet Geography* 15 (1974) 2: 73–81. (Transl. from *Izvestiya Akademii Nauk SSSR. Seriya Geograficheskaya*, no. 4, 1973.)

Jackson, W. A. D., and Towber, R. "The continuing perplexities of Soviet agriculture: the performance of Northern Kazakhstan." *The Soviet Union. The seventies and beyond.* B. W. Eissenstat, ed. Lexington, Mass.: D. C. Heath; London: Lexington Books, 1975: 169–80.

Järvesoo, Elmar. "Progress despite collectivization. Agriculture in Estonia." *Problems of mininations. Baltic perspectives.* San José, Calif.: Association for the Advancement of Baltic Studies, 1973: 137–49.

Jensen, Robert G. "The Soviet concept of agricultural regionalization and its development." *Soviet and East European agriculture* (See I.1): 77–98.

Khan, Azizur Rahman, and Ghai, Dharam. *Collective agriculture and rural development in Soviet Central Asia.* A study prepared for the International Labour Office within the framework of the World Employment Programme. London, Basingstoke: Macmillan, 1979. XI+120 pp.

Koropeckyj, I. S. "National income of the Soviet union republics in 1970: revision and some applications." *Economic development in the Soviet Union and Eastern Europe. Vol. 1.* Zbigniew M. Fallenbuchl, ed. New York, Washington, London: Praeger, 1975: 287–331.

Kosov, B. F. (et al.). "The gullying hazard in the Midland region of the USSR in conjunction with economic development." *Soviet Geography* 18 (1977) 3: 172–78. (Transl. from *Geomorfologiya*, no. 2, 1976.)

Krylov, M. P. "A regionalization of the rural areas of Tambov Oblast based on development indicators." *Soviet Geography* 17 (1976) 6: 406–11. (Transl. from *Vestnik Moskovskogo Universiteta. Seriya Geografiya,* no. 2, 1975.)

Laykin, V. I. "Some methods for an agricultural typology of the South of Krasnoyarsk Kray." *Soviet Geography* 14 (1973) 9: 564–72. (Transl. from *Doklady Instituta Geografii Sibiri i Dal'nego Vostoka,* no. 36, 1972.)

Lola, A. M., and Savina, T. M. "Regularities and prospects of transformation of rural settlements in the Nonchernozem zone of the RSFSR." *Soviet Geography* 20 (1979) 3: 170–84. (Transl. from *Izvestiya Akademii Nauk SSSR. Seriya Geograficheskaya,* no. 1, 1978.)

Lonsdale, Richard E. "Regional inequity and Soviet concern for rural and small-town industrialization." *Soviet Geography* 18 (1977) 8: 590–602.

Lydolph, Paul E. "The agricultural potential of the Nonchernozem zone." *The future . . .* (see I.1): 49–77.

Lydolph, Paul E. *Geography of the U.S.S.R.* Elkhart Lake, Wisconsin: Misty Valley Publishing, 1979. VIII+522 pp.

Lydolph, Paul E. "Schemes for the amelioration of soil and climate in the USSR." *Soviet agricultural and peasant affairs* (see II.1.b): 204–14.

Maciuika, Benedict V. "Contemporary social problems in the collectivized Lithuanian countryside." *Lituanus* 22 (1976) 3: 5–27.

Maurel, M. C. "The spatial organization of the Soviet farmland." *Annales de géographie* 88 (1979): 549–80.

McCauley, Martin. *Khrushchev and the development of Soviet agriculture. The Virgin Land Program 1953–1964.* London and Basingstoke: Macmillan, 1976. XIII+232 pp.

Mills, Richard M. "The formation of the Virgin Lands policy." *Slavic Review* 29 (1970) 1: 58–69.

Mills, Richard M. "The Virgin Lands since Khrushchev: Choices and decisions in Soviet policy making." *The dynamics of Soviet politics.* Paul Cocks, ed. Cambridge, Mass.: Harvard Univ. Press, 1976: 179–92.

Nachkebia, N. V., and Gudjabidze, V. V. "Some aspects of the transformation of rural settlements in Kolkheti [Georgian SSR. Soviet Union]." *Urbanization in Europe* (European Regional Conference of the International Geographical Union). Budapest: 1975: 251–54.

Rauner, Yu. L. "The periodicity of droughts in the grain-growing areas of the USSR. *Soviet Geography* 18 (1977) 9: 625–45. (Transl. from *Izvestiya Akademii Nauk SSSR. Seriya Geograficheskaya,* no. 6, 1976.)

Rostankowski, Peter. "The Nonchernozem Development Program and prospective spatial shifts in grain production in the agricultural triangle of the Soviet Union." *Soviet Geography: Review and Translation* 21 (1980): 409–19.

Schroeder, Gertrude E. "Regional differences in incomes and levels of living in the USSR." *The Soviet economy in regional perspective.* V. N. Bandera, Z. L. Melnik, eds. New York, Washington, London: Praeger, 1973: 167–95.

Shotsky, V. P. *Agro-industrial complexes and types of agriculture in Eastern Siberia.* Transl. by Bela Kecskés. Budapest: Akadémiai Kiadó, 1979. 130 pp.

Stroyev, K. F. "Agriculture in the Nonchernozem zone of the RSFSR." *Soviet Geography* 16 (1975) 3: 186–97. (Transl. from *Geografiya v shkole,* no. 5, 1974.)

Turushina, L. A. "Problems in the use of agricultural land in the zones of hydropower complexes in East Siberia." *Soviet Geography* 17 (1976) 3: 172–79. (Transl. from *Doklady Instituta Geografii Sibiri i Dal'nego Vostoka,* no. 44, 1974.)

Wädekin, Karl-Eugen. "Soviet agriculture and agricultural policy." *Soviet agriculture. The permanent crisis* (see II.1.d.): 52–67 and 71–98.

Zoerb, Carl. "The Virgin Land territory: plans, performance, prospects." *Soviet agriculture. The permanent crisis* (see II.1.d.): 29–44.

II.5. Irrigation and other land improvement measures (see also on
       Nonchernozem zone in the preceding section)

Gangardt, G. G. "On the question of diverting part of the unused runoff of
    northern and Siberian rivers into regions suffering from a shortage of water
    resources." *Soviet Geography* 13 (1972) 9: 622–28. (Transl. from *Gidrotekhniches-
    koye Stroitel'stvo*, no. 8, 1971.)
Gustafson, Thane. "Transforming Soviet agriculture. Brezhenv's gamble on land
    improvement." *Public Policy* 25 (1977) 3: 293–312.
Kornilov, B. A. "The impact of the Kara Kum canal on the environment." *Soviet
    Geography* 16 (1975) 5: 308–14. (Transl. from *Vodnye Resursy*, no. 3, 1974.)
L'vovich, M. I. (et al.). "The water aspect of the geography of agriculture in the
    USSR." *Soviet Geography* 20 (1979) 9: 515–33. (Transl. from *Izvestiya Akademii
    Nauk. Seriya Geograficheskaya*, no. 4 1978.)
Matley, Ian M. "The Murgab oasis." *Canadian Slavonic Papers* 17 (1975) 2/3:
    417–35.
Micklin, Philip P. "Dimensions of the Caspian Sea problem." *Soviet Geography* 13
    (1972) 9: 589–603.
Micklin, Philip P. "Disciplinary plans for USSR rivers." *The Geographical Magazine*
    51 (1979) 10: 701–6.
Micklin, Philip P. "Environmental factors in Soviet interbasin water transfer
    policy." *Environmental Management* 2 (1978) 6: 567–80.
Micklin, Philip P. "The falling level of the Caspian Sea." *Environmental deterioration
    in the Soviet Union and Eastern Europe*. Ivan Volgyes, ed. New York: Praeger
    1974: 67–79.
Micklin, Philip P. "Irrigation development in the USSR during the 10th five-year
    plan (1976–80)." *Soviet Geography* 19 (1978) 1: 1–24.
Micklin, Philip P. "Large-scale interbasin diversions in the USSR: Implications for
    the future." *Soviet resource management and the environment*. W. A. Douglas
    Jackson, ed. Columbus, Ohio: AAASS, 1978; 63–82.
Micklin, Philip P. "Nawapa and two Siberian water-diversion proposals: A
    geographical comparison and appraisal." *Soviet Geography* 18 (1977) 2: 81–99.
Micklin, Philip P. "Soviet plans to reverse the flow of rivers: The Kama-
    Vychegda-Pechora project." *Canadian Geographer* 13 (1969) 3: 589–603.

II.6. Incomes and living standards

Anderson, I. A. "The rise in level of living and the rural development of the
    intellectual needs of the rural population." *Soviet Sociology* 18 (1979) 1: 3–15.
    (Transl. from *Sotsiologicheskie Issledovaniia*, no. 1, 1978.)
Bronson, David W., and Krueger, Constance B. "The revolution in Soviet farm
    household income, 1953–1967." *The Soviet rural community . . .* (see II.7.):
    214–58.
Crook, Frederick W. and Crook, Elizabeth F. "Payment systems used in collective
    farms in the Soviet Union and China." *Studies in Comparative Communism* 9 (1976)
    3: 257–69.
Jasny, N. "Peasant-worker income relations: a neglected subject." *Soviet Studies* 12
    (July 1960) 1: 14–22.
Ladenkov, V. N., and Gorina, N. D. "A study of the living standard of the rural
    population." *Problems of Economics* 19 (1976) 2: 32–47. (Transl. from *Izvestiya
    Sibiriskogo Otdeleniia Akademii Nauk SSSR. Seriia Obshchestvennykh Nauk*, no. 11,
    1974.)
Lantsev, M. "Progress in social security for agricultural workers in the USSR."
    International Labour Review 107 (1973) 3: 239–52.

Mashenkov, V. and Nikitin, M. "Remuneration and productivity in Soviet agriculture." *International Labour Review* 117 (1978) 1: 69–79.

Nove, Alec. "The incomes of Soviet peasants." *The Slavonic and East European Review* 37 (1960) 91 (2): 314–33.

Nove, Alec. "Rural taxation in the USSR." *Soviet Studies* 5 (Oct. 1953) 2:159–66.

Paskhaver, B. "Indicators and analysis of differentiation of collective farm income." *Problems of Economics* 19 (1977) 9: 74–88. (Transl. from *Vestnik Statistiki*, no. 4, 1976.)

Taichinova, K. "Raising the living standard of the rural population." *The Soviet Review* 16 (1975/76) 4: 33–52. (Transl. from *Voprosy Ekonomiki*, no. 6, 1974.)

Wädekin, Karl-Eugen. "Income distribution in Soviet agriculture." *Soviet Studies* 27 (1975) 1: 3–26.

Wädekin, Karl-Eugen. "Payment in kind in Soviet agriculture." *Bulletin* (Institute for the Study of the USSR) 18 (Sept. 1971) 9: 5–18, and 18 (Oct. 1971) 11: 5–24.

Wronski, Henri. "Consumer cooperatives in rural areas in the U.S.S.R." *Agrarian policies . . .* (see I.1): 159–73.

Wronski, Henri. "Peasant incomes." *Soviet agriculture. The permanent crisis* (see II. 1.d.): 123–36.

II.7. Social and settlement patterns

Arutyunyan, Yurij V. "A comparative study of rural youth in the national regions of the USSR: general and specific features." *The integrated development of human and natural resources: the contribution of rural sociology.* Wrocław: Institute of Philosophy and Sociology, PAS, 1979: 231–37.

Dodge, Norton T., and Feshbach, Murray. "The role of women in Soviet agriculture." *Soviet and East European agriculture.* (see I.1): 265–302.

Dunn, Ethel "Factors affecting social mobility for women in the Soviet countryside." *Agricultural policies . . .* (see I.1): 71–93.

Dunn, Ethel. "The importance of religion in the Soviet rural community. "*The Soviet rural community . . .* (see below): 346–75.

Dunn, Stephen P. "Structure and functions of the Soviet rural family." *The Soviet rural community . . .* (see below): 325–45.

Hough, Jerry F. "The changing nature of the kolkhoz chairman." *The Soviet rural community* (see below): 103–20.

Kovalev, S. A. "Regularities in the formation of territorial systems of rural settlements in the European part of the Soviet Union." *Urbanization in Europe* (European Regional Conference of the International Geographical Union). Budapest: 1975: 237–42.

Liely, Helmut. "Shepherds and reindeer nomads in the Soviet Union." *Soviet Studies* 31 (1979) 3: 401–16.

Maurel, M. C. "Women in the Soviet rural society." *Cahiers du monde russe et sovietique* 20 (1979) 3–4: 323–44.

Mill, Ian H. "The end of the Russian peasantry?" *Soviet Studies* 27 (1975) 1: 109–27.

Monich, Zinaida I. "The professional and paraprofessional component in the structure of the rural population [based on data from the Belorussian SSR]." *Soviet Sociology* 12 (1973) 1: 56–76 and 2: 3–26. (Transl. from *Intelligentsiia v strukture sel'skogo naseleniia [na materialakh BSSR]*. Minsk: 1971.)

Newth, J. A. "The kolkhoz household: Ukraine, 1950–1955." *Soviet Studies* 11 (Jan. 1960): 307–16.

Pallot, Judith. "Rural settlement planning in the USSR." *Soviet Studies* 31 (1979) 2: 214–30.

Prudnik, I. V. "Agroindustrial workers." *Soviet Sociology* 15 (1977) 4: 3–37. (Transl. from Monich, Z. I.; Izokh, V. G.; and Prudnik, I. V. *Rabochii klass v strukture sel'skogo naseleniia.* Minsk: 1975.)

Richter, Luba. "Plans to urbanize the countryside, 1950–1962." *Soviet planning– Essays in honour of Naum Jasny.* J. Degras, A. Nove, eds. Oxford: B. Blackwell, 1964: 32–45.

Sakoff, A. N. "Rural and urban society in the U.S.S.R.—comparative structure, income, level of living." *Monthly Bulletin of Agricultural Economics and Statistics* 21 (1972) 10: 1–13.

Shimkin, Demitri B. "Current characteristics and problems of the Soviet rural population." *Soviet agricultural and peasant affairs.* (see II.1.b.): 79–127.

Shinn, William T. "The law of the Russian peasant household." *Slavic Review* 20 (1961) 4: 601–21.

Simush, P. I. "Differences between town and country in light of the development of the socialist way of life." *The Soviet Review* 17 (1976) 1: 26–48. (Transl. from *Voprosy filosofii,* no. 3, 1975.)

Simush, P. "Social changes in the countryside." *Soviet Law and Government* 16 (1978) 4: 52–73. (Transl. from *Kommunist,* no. 16, 1976.)

The Soviet rural community. James Millar, ed. Urbana, Chicago, London: Univ. of Illinois Press, 1971. XV+420 pp.

Town, country and people. G. V. Osipov, ed. With an introduction by Maurice Hookham. London: Tavistock Publ., 1969. VII+260 pp.

Vucinich, Alexander. "The peasants as a social class." *The Soviet rural community . . .* (see above): 307–24.

Wädekin, Karl-Eugen. "Housing in the U.S.S.R.—The countryside." *Problems of Communism* 18 (1969) 3: 12–20.

Wädekin, Karl-Eugen. "The nonagricultural rural sector." *The Soviet rural community* (see above): 159–79.

Wädekin, Karl-Eugen. "Soviet rural society. A descriptive stratification analysis." *Soviet Studies* 22 (April 1974) 4: 512–38.

Women in Russia. G. Atkinson, A. Dallin, G. W. Lapidus, eds. Hassocks, Sussex: Harvester Press, 1978. XIII+410 pp.

Zahn, Michael D. *Rural settlement in the USSR and its relation to farm size* (Univ. of Illinois, Urbana. Department of Agricultural Economics. AERR 148). Urbana, Ill.: 1977. 25 pp.

Žekulin, Gleb. "The contemporary countryside in Soviet literature: A search for new values." *The Soviet rural community . . .* (see above): 376–404.

II.8. Agricultural labor, migration

DePauw, John W. *Measures of agricultural employment in the U.S.S.R.: 1950–1966* (Foreign Demographic Analysis Division. International Population Reports Series P-95, No. 65). Washington, D. C.: U.S. Department of Commerce, 1968. 77 pp.

Dodge, Norton T. "Recruitment and the quality of the Soviet agricultural labor force." *The Soviet rural community . . .* (see II.7.): 180–213.

Feshbach, Murray, and Rapawy, Stephen. "Labor and wages." *Economic performance and the military burden in the Soviet Union.* A compendium of papers submitted to the Joint Economic Committee, Congress of the United States. Washington, D.C.: U.S. Government Printing Office, 1970: 71–84.

Ladenkov, V. N. "Studies of migration of skilled personal in agriculture." *Problems of Economics* 15 (1973) 10: 62–80. (Transl. from *Izvestiia Sibirskogo Otdeleniia Akademii Nauk SSSR. Seriia Obshchestvennykh Nauk,* no. 1, 1972.)

Nimitz, Nancy. *Farm employment in the Soviet Union, 1928–1963* (Memorandum RM-4623-PR, November 1965). Santa Monica, Calif.: The RAND Corporation. 155 pp.

Nimitz, Nancy. "Farm employment in the Soviet Union, 1928–63." *Soviet and East European agriculture* (see I.1): 175–205.

Potichnyj, Peter J. *Soviet agricultural trade unions, 1917–1970.* Toronto and Buffalo: Univ. of Toronto Press, 1972. XIX+258 pp.

Rapawy, Stephen. *Estimates and projections of the labor force and civilian employment in the U.S.S.R. 1950 to 1990* (Foreign Demographic Report no. 10) Washington, D. C.: U.S. Department of Commerce, 1976. V+76 pp.

Stuart, Robert C. "Aspects of rural-urban migration in the USSR." *The ACES Bulletin* 18 (1976) 4: 35–48.

Stuart, Robert C., and Gregory, Paul R. "A model of Soviet rural-urban migration." *Economic development and cultural change* 26 (Oct. 1977) 1: 81–92.

Wädekin, Karl-Eugen. "Internal migration and the flight from the land in the USSR, 1939–1959." *Soviet Studies* 17 (Oct. 1966) 2: 131–52.

Wädekin, Karl-Eugen. "Manpower in Soviet agriculture. Some post-Khrushchev developments and problems." *Soviet Studies* 20 (Jan. 1969) 3: 281–305.

II.9. Capital and technology

Batra, Raveend N. "Technological change in the Soviet collective farm." *The American Economic Review* 64 (Sept. 1974) 4: 594–603.

Conklin, D. W. "Barriers to technological change in the USSR. A study of chemical fertilizers." *Soviet Studies* 20 (Jan. 1969) 3: 353–65.

Dovring, Folke. "Capital intensity in Soviet agriculture." *Agricultural policies . . .* (see I.1): 5–26.

Dovring, Folke. "Progress on mechanization in Soviet agriculture." *The Soviet rural community . . .* (see II.7.): 259–75.

Dovring, Folke. "Soviet farm mechanization in perspective." *Slavic Review* 25 (1966) 2: 287–302.

Durgin, Frank A. "Toward the abolition of the RTS." *Soviet Studies* 12 (July 1960) 1: 83–86.

Emel'ianov. A. "Technological progress and structural change in agriculture." *Problems of Economics* 14 (1971) 4: 3–24. (Transl. from *Voprosy Ekonomiki*, no. 4, 1971.)

Humphrey, Arthur E. "Soviet technology: the case of single cell protein." *Survey* 23 (winter 1977–78) 1 (102): 81–86.

The impact of fertilizer on Soviet grain output, 1960–1980 (National Foreign Assessment Center, ER 77-1055). Washington, D. C.: 1977. IV+22 pp.

Kahan, Arcadius. "Shifts to off-farm agricultural inputs in the tenth economic plan: The economic and institutional implications." *The future . . .* (see I.1.): 9–25.

Poletaev. P. "Fixed capital in agriculture and indices of the effectiveness of its use." *Problems of Economics* 17 (1974) 1: 22–33. (Transl. from *Planovoye khozyaistvo*, no. 8, 1973.)

Rubenking, Earl M. "The Soviet tractor industry: Progress and problems." *Soviet economy in a new perspective . . .* (see II.1.d.): 600–619.

Semenov, V. N. "Technical progress and the problem of dual prices for agricultural machinery." *Problems of Economics* 19 (1976) 3: 63–79. (Transl. from *Finansy SSSR*, no. 4, 1975.)

Structure and accounting of working capital in the U.S.S.R. (International Population Reports Series P-95, no. 70). Washington, D. C.: U.S. Department of Commerce, 1972. V+82 pp.

Walker, Frederick A. "Industrialization of Soviet agriculture." *Studies on the Soviet Union* 11 (1971) 3: 46–66.

II.10. The place of agriculture in the economy at large, agro-industrial integration

Ambartsumov, A. "On the question of equivalence in exchange between industry and agriculture." *Problems of Economics* 21 (1979) 12: 39–45. (Transl. from *Ekonomicheskiye Nauki*, no. 1, 1978.)
Bobylev, S. "Agricultural production and the development of infrastructure." *Problems of Economics* 21 (1979) 12: 23–38. (Transl. from *Voprosy Ekonomiki*, no. 6, 1978.)
Bronshtein, M. "Economic and social problems in the industrialization of agriculture." *Problems of Economics* 19 (1976) 7: 77–96. (Transl. from *Voprosy Ekonomiki*, no. 8, 1975.)
Dementsev, V. V. "Associations in agriculture and finances." *Problems of Economics* 19 (1977) 11: 24–49. (Transl. from *Finansy SSSR*, no. 6, 1976.)
Driuchin, V. "Agriculture in the national economic balance." *Problems of Economics* 22 (1979) 7: 25–36. (Transl. from *Vestnik Statistiki*, no. 6, 1979.)
Eidel'man, M. "Methodological problems in defining a national economic agroindustrial complex." *Problems of Economics* 19 (1976) 3: 24–44. (Transl. from *Voprosy Ekonomiki*, no. 4, 1975.)
Kahan, Arcadius. "The problems of the 'agrarian-industrial complexes' in the Soviet Union." *Economic development . . .* Fallenbuchl, ed. (see I.1): 205–22.
Kassirov, L. "The effectiveness of the national industrial complex." *Problems of Economics* 22 (1979) 8: 48–68. (Transl. from *Voprosy Ekonomiki*, no. 2, 1979.)
Miller, Robert F. "The politics of policy implementation in the USSR: Soviet policies on agricultural integration under Brezhnev." *Soviet Studies* 33 (1980) 2: 171–94.
Nikiforov, L. "Socioeconomic problems in the industrialization of agricultural production." *Problems of Economics* 19 (1976) 3: 3–23. (Transl. from *Ekonomicheskie Nauki*, no. 3, 1975.) Also in *The Soviet Review* 17 (1976) 3: 77–97.
Pospielovsky, Dimitry. "The 'link system' in Soviet agriculture." *Soviet Studies* 21 (1970) 4: 411–35.
Sakoff, Alexander N. "Production brigades: organizational basis of farm work in the U.S.S.R." *Monthly Bulletin of Agricultural Economics and Statistics* 17 (1968) 1: 1–8.
Smetanin, N. "Agroindustrial integration in the USSR." *Problems of Economics* 17 (1974) 8: 46–64. (Transl. from *Voprosy Ekonomiki*, no. 3, 1974.)
Trei, B. A., and Frolova, L. A. "Long-range planning in the development of the agro-industrial complex of a Union Republic." *Matekon* 14 (1977/78) 3: 22–38. (Transl. from *Ekonomika i Matematicheskie Metody*, no. 2, 1977.)

II.11. Socialist farms

Ballard, Allen B., Jr., "Problems of state farm administration." *Soviet Studies* 17 (Jan. 1966) 3: 339–52.
Ballard, Allen. "Sovkhoz Kuban." *Survey* 48 (July 1963): 66–78.
Belov, Fedor. *The history of a Soviet collective farm.* New York: Praeger, 1955; and London: Routledge and Kegan, 1956. 237 pp.
Bernstein, Thomas P. "The state and collective farming in the Soviet Union and China." *Food, policies, and agricultural development. Case studies in the public policy of rural modernization.* Raymond F. Hopkins, et al., eds. Boulder, Colorado: Westview Press, 1979: 73–105.
Bonin, John P. "Work incentives and uncertainty on a collective farm." *Journal of*

Comparative Economics 1 (1977) 1: 77–97.
Bradley, Michael E. "Incentives and labour supply on Soviet collective farms." *The Canadian Journal of Economics* 4 (1971) 3: 342–52.
Bush, Keith. "The third All-Union Congress of Kolkhozniks." *Bulletin* (Institute for the Study of the USSR) 17 (1970) 1: 16–23.
Durgin, Frank A. "The growth of inter-kolkhoz cooperatives." *Soviet Studies* 12 (Oct. 1960) 2: 183–89.
Jasny, Naum. "Kolkhozy, the Achilles' heel of the Soviet régime." *Soviet Studies* 3 (Oct. 1951) 2: 150–63.
Kahan, A. "The collective farm system in Russia: some aspects of its contribution to Soviet economic development." *Agriculture in economic development*. C. Eicher, L. Witt, eds. New York: MacGraw-Hill, 1964: 251–71.
Kucherov, Alexander. "The peasant." *Problems of Communism* 14 (1965) 2: 98–104.
Laird, Roy D. *Collective farming in Russia. A political study of the Soviet kolkhoz.* Lawrence, Kansas: Univ. of Kansas Publications, 1958. 176 pp.
Laird, Roy D. "The new zveno controversy." *Osteuropa Wirtschaft* 11 (1966) 4: 254–61.
Laird, Roy D. "The plusses and minusses of state agriculture in the USSR." *The political economy . . .* (see I.1.): 3–20.
Laird, Roy D., and Beasley, Kenneth. "Soviet tractor stations. Policy control by auxiliary services." *Public Administration Review* 20 (1960) 4: 213–18.
Laird, Roy D., Sharp, Darwin E., and Sturtevant, Ruth. *The rise and fall of the MTS as an instrument of Soviet rule.* Lawrence, Kansas: Governmental Research Center, 1960. 97 pp.
Laptev, O. "Rigorously enforce the laws on collective farm democracy." *Soviet Law and Government* 13 (1974/75) 3: 79–85. (Transl. from *Sotsialisticheskaia Zakonnost'*, no. 12, 1973.)
Maggs, Peter B. "The law of farm-farmer relations." *The Soviet rural community . . .* (see II.7.): 139–56.
McAuley, Alastair N. D. "Kolkhoz problems in recent literary magazines." *Soviet Studies* 15 (1963/64) 3: 308–30.
Miller, Robert F. "The future of the Soviet kolkhoz." *Problems of Communism* 25 (1976) 2: 34–50.
Miller, Robert F. "A good kolkhoz." *Survey* 51 (April 1964): 32–43.
Miyamoto, Katsuhiro. "A mathematical model of kolkhoz." *Bulletin of University of Osaka Prefecture.* Series D: Economics 19 (1975): 27–32.
Oi, Walter Y., and Clayton, Elizabeth M. "A peasant's view of a Soviet collective farm." *The American Economic Review* 58 (1968) 1: 37–59.
Pervushin, A. "The length of service of collective-farm members and its legal meaning." *Soviet Law and Government* 12 (1973) 2: 85–93. (Transl. from *Sovetskaia Iustitsiia*, no. 14, 1972.)
Reston, R. *Aftermath to revolution: the Soviet collective farm.* London: Collier Macmillan, 1975. IX+87 pp.
Sakoff, A. N. "Soviet agriculture and the New Model Constitution of the kolkhoz." *Monthly Bulletin of Agricultural Economics and Statistics* 19 (1970) 9: 1–9.
Stuart, Robert C. "Structural change and the quality of Soviet collective farm management, 1952–1966." *The Soviet rural community . . .* (see II.7.): 121–38.
Stuart, Robert C. "The changing role of the collective farm in Soviet agriculture." *Canadian Slavonic Papers* 16 (1974) 2: 145–59.
Stuart, Robert C. *The collective farm in Soviet agriculture.* Lexington, Mass., Toronto, London: D. C. Heath & Co., 1972. XVI+254 pp.
Wädekin, Karl-Eugen. "Kolkhoz, sovkhoz, and private production." *Agrarian policies . . .* (see I.1): 106–37.
Wädekin, Karl-Eugen. "The Soviet kolkhoz: vehicle of cooperative farming or of

control and transfer of resources?" *Cooperative and commune.* Peter Dorner, ed. Madison, Wisconsin: Univ. of Wisconsin Press, 1977: 95–116.

Zoerb, Carl R. "From the promise of land and bread to the reality of the state farm." *Studies on the Soviet Union* (New Series) 6 (1967) 4: 89–107.

II.12. The private sector

Clayton, Elizabeth. "Notes on the productivity of Soviet private agriculture." *The ACES Bulletin* 21 (1979) 2: 85–91.

DePauw, John W. "The private sector in Soviet agriculture." *Slavic Review* 28 (1969) 1: 63–71.

Järvesoo, Elmar. "Private enterprise in Soviet agriculture in the 1970's." *Journal of the Northeastern Agricultural Economics Council* 3 (1974) 1: 48–63.

Karcz, Jerzy F. *Appendix to quantitative analysis of the collective farm market.* Santa Barbara, Calif.: Univ. of California, 1963. 46 pp. (hectographed.)

Karcz, Jerzy F. "Quantitative analysis of the collective farm market." *The American Economic Review* 54 (1964) 4, part I: 315–34.

Lovell, C. A. Knox. "The role of private subsidiary farming during the Soviet seven-year plan, 1959–65." *Soviet Studies* 20 (1968) 1: 46–66.

Newth, J. A. "Soviet agriculture: The private sector 1950–1959: animal husbandry." *Soviet Studies* 13 (1961) 2: 160–71 and 4: 414–32.

Sakoff, A. N. "The private sector in Soviet agriculture." *Monthly Bulletin of Agricultural Economics and Statistics* 11 (1961) 9: 1–12.

Severin, Barbara S. "USSR. The all-Union and Moscow collective farm market price indexes." *The ACES Bulletin* 21 (1979) 1: 23–36.

Wädekin, Karl-Eugen. *The private sector in Soviet agriculture* (Second, enlarged, revised edition of *Privatproduzenten in der sowjetischen Landwirtschaft.* Cologne: 1967). Translated by K. Bush. Edited by J. F. Karcz. Berkeley, Los Angeles, London: Univ. of California Press, 1973. XVIII+407 pp.

Whitman, John T. "The kolkhoz market." *Soviet Studies* 7 (1955) 4: 384–408.

II.13. Directing agriculture (planning, prices, procurements, administration; see also II.11: Socialist farms)

Bernstein, Thomas P. "Leadership and mass mobilization in the Soviet and Chinese collectivization campaigns of 1929–30 and 1955–56: a comparison." *China Quarterly* 31 (1966/67) 2: 98–101

Bornstein, Morris. "The administration of the Soviet price system." *Soviet Studies* 30 (1978) 4: 466–90.

Bornstein, Morris. "The Soviet debate on agricultural price and procurement reforms." *Soviet Studies* 21 (1969) 1: 1–20.

Bornstein, Morris. "Soviet price policy in the 1970's." *Soviet economy in a new perspective . . .* (see II.1.d.): 17–66.

Boyev, V. "Pricing as a tool for the stimulation and regional distribution of farm production." *The future of agriculture* (Papers and reports. Fifteenth International Conference of Agricultural Economists). Oxford: Agricultural Economics Institute, 1974: 202–7.

Braginskii, L. "On the interaction of economic levers in collective-farm management." *Problems of Economics* 14 (1971) 5: 60–73. (Transl. from *Voprosy Ekonomiki,* no. 3, 1971.)

Conklin, David W. *"An evaluation of the Soviet profit reforms: With special reference to agriculture.* New York, London: Praeger, 1970. XIII+192 pp.

Durgin, Frank A. "Monetization and policy in Soviet agriculture since 1952." *Soviet Studies* 16 (1963) 4: 375–404.

Em, V. "Transportation and rental relations in agriculture." *Problems of Economics* 17 (1975) 9: 74–80. (Transl. from *Ekonomicheskie Nauki*, no. 5, 1974.)

Gofman, K. G., Khrabrov, I. M., Gusev, A. F., and Pronin, G. V. "Optimal land management." *Matekon* 9 (1973) 3: 71–88. (Transl. from *Ekonomika i Matematicheskie Metody*, no. 6, 1971.)

Gray, Kenneth R. *The efficient location and specialization of Soviet agricultural procurement.* Univ. of Wisconsin: Ph.D. dissertation, 1976. 420 pp.

Gray, Kenneth R. "Soviet agricultural specialization and efficiency." *Soviet Studies* 31 (1979) 4: 542–58.

Grushetskii, L. "The stimulating function of purchase prices." *Problems of Economics* 22 (1979) 7: 70–79. (Transl. from *Ekonomicheskie Nauki*, no. 5, 1979.)

Gumerov, R. "Procurement prices and the stimulation of agricultural production." *Problems of Economics* 22 (1979) 6: 23–40. (Transl. from *Planovoe Khozyaistvo*, no. 3, 1979.)

Iur'ev, V. "Stimulating the growth of procurement of agricultural output." *Problems of Economics* 22 (1979) 7: 80–95. (Transl. from *Voprosy Ekonomiki*, no. 5, 1979.)

Jacobs, Everett M. "Recent developments in Soviet agricultural planning." *Jahrbuch der Wirtschaft Osteuropas—Yearbook of East European Economics* 3 (1972): 303–25.

Jasny, Naum. "Production costs and prices in Soviet agriculture." *Soviet and East European agriculture* (see I.1): 212–57.

Jensen, Robert G. "Regional pricing and the economic evaluation of land in Soviet agriculture." *The Soviet economy in regional perspective.* V. N. Bandera, Y. L. Melnik, eds. New York, Washington, London: Praeger, 1973: 305–27.

Karcz, Jerzy F. "Farm marketings and state procurements: definitions and interpretations." *Soviet Studies* 15 (1962) 2: 152–66.

Karcz, Jerzy F. "Soviet inspectorates for agricultural procurements." *California Slavic Studies* 3 (1964): 149–72.

Karnaukhova, E. "Some problems in measuring differential rent under socialism." *Problems of Economics* 19 (1976) 7: 48–64. (Transl. from *Ekonomicheskie Nauki*, no. 9, 1975.)

Kassirov, L. "The profitability of socialist agriculture [Methodological issues]." *Problems of Economics* 20 (1977) 6: 55–74. (Transl. from *Voprosy Ekonomiki*, no. 4, 1977.)

Kochkarev, V. "The role of credit in strengthening the collective farm economy." *Problems of Economics* 22 (1979) 8: 86–104. (Transl. from *Voprosy Ekonomiki*, no. 7, 1979.)

Krueger, Constance B. "A note on the size of subsidies on Soviet government purchases of agricultural products." *The ACES Bulletin* 16 (1974) 2: 63–69.

Laird, Roy D. "The dilemma of Soviet agricultural administration: the short unhappy life of the TPA." *Agricultural History* 40 (1966) 1: 11–18.

Laird, Roy D. "Khrushchev's administrative reforms in agriculture: an appraisal." *Soviet and East European agriculture.* (see I.1): 29–50.

Lukinov, I. "The methodology of forming prices of farm produce, history of price formation in the USSR." *Policies, planning and management . . .* (see I.1.): 239–52.

Lysov, E. "An important factor in the formation of differential rent." *Problems of Economics* 19 (1976) 7: 65–76. (Transl. from *Ekonomicheskie Nauki*, no. 9, 1975.)

Mikhasiuk, I. "The economic evaluation of land and the improvement of rental relationships in the USSR. *Problems of Economics* 21 (1979) 10: 65–78. (Transl. from *Ekonomicheskie Nauki*, no. 7, 1978.)

Millar, James R. "Financial innovation in contemporary Soviet agricultural policy." *Slavic Review* 32 (1973) 1: 91–114.

Millar, James R. "Financing the modernization of kolkhozy." *The Soviet rural community* . . . (see II.7.): 276–303.

Miller, Robert F. "Continuity and change in the administration of Soviet agriculture since Stalin." *The Soviet rural community* . . . (see II.7.): 73–102.

Miller, Robert F. *One hundred thousand tractors. The MTS and the development of controls in Soviet agriculture.* Cambridge, Mass.: Harvard Univ. Press, 1970. XV+423 pp.

Nimitz, Nancy. *Soviet government grain procurements, dispositions, and stocks, 1940, 1945–1963* (Memorandum RM-4127-Pr, November 1964). Santa Monica, Calif.: The RAND Corporation. XI+113 pp.

Nove, Alec. "Incentives for peasants and administrators." *Soviet agricultural and peasant affairs* (see II.1.b): 69–76.

Nove, Alec. "Peasants and officials." *Soviet and East European agriculture* (see I.1): 57–72.

Obolenski, K. P. "Agricultural planning in the U.S.S.R." *Economic problems* . . . (see I.1.): 43–60.

Orlikovskaia, Iu. "Differential rent and its realization in collective farm production." *Problems of Economics* 18 (1975) 7: 43–60. (Transl. from *Ekonomicheskie Nauki*, no. 1, 1975.)

Rumyantsev, A. M. "Planning Soviet agriculture: current problems." *Policies, planning and management* . . . (see I.1.): 40–55.

Rusinov, I. "Prospects in the development of planning of agricultural production." *Problems of Economics* 10 (1967) 5: 25–32. (Transl. from *Planovoe khozyaistvo*, no. 3, 1967.)

Schinke, Eberhad. "The organization and planning of Soviet agriculture." *World Agricultural Economics and Rural Sociology Abstracts* 12 (1970) 1: 1–18.

Semenov, V. N. "The financing of agriculture." *Problems of Economics* 17 (1975) 10: 69–88. (Transl. from *Finansy SSSR*, no. 5, 1974.)

Stuart, Robert C. "Aspects of Soviet rural development." *Agricultural Administration* 2 (1975): 165–78.

Stuart, Robert C. "Managerial incentives in Soviet collective agriculture during the Khrushchev era." *Soviet Studies* 22 (1971) 4: 539–55.

Swearer, Howard R. "Agricultural administration under Khrushchev." *Soviet agricultural and peasant affairs* (see II.1.b.): 9–40; cf. "Commentary" by J. F. Karcz, *ibid.*: 41–50.

Treml, Vladimir G. *Agricultural subsidies in the Soviet Union* (Foreign Economic Report No. 15). Washington, D.C.: U.S. Department of Commerce, Bureau of the Census, 1978. IV+46 pp.

Yaney, George L. "Agricultural administration in Russia from the Stolypin land reform to forced collectivization: an interpretative study." *The Soviet rural community* . . . (see II.7.): 3–35.

II.14. Food supply and consumption

Checinski, Michael. *Agriculture and nutrition in the USSR 1940–45* (Radio Liberty Research Supplement, April 18, 1975). Munich: 1975. 25 pp.

Kahan, Arcadius. "Natural calamities and their effect upon the food supply in Russia." *Jahrbücher für Geschichte Osteuropas.* New series 16 (1968) 3: 353–77.

Levin, A. "The study of the demand of the population and problems of information." *Problems of Economics* 18 (1975) 1: 3–23. (Transl. from *Voprosy Ekonomiki*, no. 8, 1974.)

Schroeder, Gertrude E., and Severin, Barbara S. "Soviet consumption and income policies in perspective." *Soviet economy in a new perspective* . . . (see II.1.d.): 620–60.

Severin, Keith. "Soviet policies on agricultural trade and the consumer." *Agricultural policies* . . . (see I.1): 37–48.
Stuart, Robert C. *The Soviet food question.* Washington, D.C.: U.S. Information Agency, 1976. 111 pp.
Symons, Leslie. "Feeding the Soviet Union." *Contemporary Review* 228 (1976) 1321: 57–63.

III. GDR (East Germany)

Bajaja, Vladislav. "Concentration and specialization in Czechoslovak and East German agriculture." *Agricultural policies* . . . (see I.1): 263–95.
Francisco, Ronald A. "Agricultural collectivization in the German Democratic Republic." *The political economy* . . . (see I.1): 63–85.
Francisco, Ronald A. "The future of East German agriculture: The feasability of the 1976–80 plan." *The future* . . . (see I.1): 185–203.
Jacobeit, Wolfgang. "Research into peasant work and economy in the German Democratic Republic." *Technology and Culture* 5 (1964) 3: 379–85.
Koenig, Ernest. *The agricultural situation in East Germany* (The Agricultural situation in Eastern Europe. Foreign Agricultural Service. M-31). Washington, D.C.: USDA,1958. 16 pp.
Marcinko, David J. "Total factor productivity and collective agriculture in East Germany." *The ACES Bulletin* 18 (Winter 1976) 4: 49–58.
Merkel, Konrad. "The agrarian problem in divided Germany." *Agrarian policies* . . . (see I.1): 210–26.
Nadkarni, M. V.: *Socialist agricultural price policy. A case study of GDR.* New Delhi: People's Publ. House, 1979. XVI+215 pp.
Roubitschek, Walter. "Development and regional pattern of ownership forms and size of farms in the German Democratic Republic." *Geographia Polonica* 16 (1969): 5–16.
Schlicht, H. "Rural nutrition and living conditions in relation to GDR agricultural and economic policies." *Decision making* . . . (see I.1): 257–62.
Schmidt, Walter. "Agrarian reform with special reference to its effects on employment and social questions in the German Democratic Republic." *Wissenschaftliche Beiträge des Instituts Ökonomik der Entwicklungsländer an der Hochschule für Ökonomie, Berlin* 5 (1970); special issue no. 1: 41–53.
Thoms, A. "The development of socialist agriculture in the German Democratic Republic." *Wissenschaftliche Beiträge* . . . (as above): 159–70.
Weber, Adolf. "Agricultural modernization in market and planned economies: the German experience." *Studies in Comparative Communism* VI (autumn 1973) 3:280–300.
Wirsig, H. "Agricultural policy and development of agriculture in the GDR." *Economies et Sociétés* 10 (1976) 7/8: 1493–1514.
Wunderlich, Frieda. *Farmer and farm labor in the Soviet zone of Germany.* New York: Twayne, 1958. 162 pp.

IV. Poland

IV.1. Polish agriculture in general

IV.1.a. Up to 1944/45

Cieplak, Tadeusz N. "Polish agrarian movement, 1875–1939." *Polish Review* 22 (1977), 2: 3–28.
Horak, Stephen M. "Belorussian and Ukrainian peasants in Poland, 1919–1939. A

case study in peasantry under foreign rule." *The peasantry,* . . . vol. 1 (see I.1): 133–56.

Kagan, George. "Agrarian regime of pre-war Poland." *Journal of Central European Affairs* 3 (1943/44): 241–69.

Kieniewicz, Stefan. *The emancipation of the Polish peasantry.* Chicago, London: Univ. of Chicago Press, 1969. XIX+285 pp.

Orczyk, Józef. "The main features of the agricultural crisis in Poland in the years 1929–1935." *Studia Historiae Oeconomicae* (Uniwersytet w Poznaniu) 3 (1968): 221–41.

Orczyk, Józef. "Profitability of agricultural farms in the years 1929–1937." *Studia Historiae Oeconomicae* (Uniwersytet w Poznaniu) 7 (1972): 147–158.

IV.1.b. Since 1944/45

Biegajlo, W. "Polish land utilization survey in the years 1960–64." *Land utilization* . . . (see I.1): 28–34.

Brzeski, Andrzej. "Poland's uncertain five-year plan." *The future* . . . (see I.1): 139–47.

Brzoza, Anatol. "Changing agricultural policy in Poland: Individual agriculture in a planned economy." *Illinois Agricultural Economics* 5 (1965) 1: 1–10.

Cieplak, Tadeusz N. "Private farming and the status of the Polish peasantry since World War II." *The peasantry,* vol. 2 (see I.1): 25–38.

Collins, Christine. *The feed-livestock economy of Poland* (Economic Research Service, Foreign Agricultural Economic Report, 99). Washington, D. C.: USDA, 1975. VI+93 pp.

Dabrowski, Przemysław. "The shaping of the socio-economic and technical infrastructure of agriculture expressed in spatial terms, as exemplified on Poland." *Integration of the agriculture* . . . (see I.1): 217–43.

Duymovic, Andrew A. "Poland's agricultural policies in the 1970's: impact on agricultural trade with U.S." *Agricultural policies* . . . (see I.1): 185–98.

Dziewoński, Kazimierz. "Transformations of the rural landscape in Poland during the last two centuries." *Geographia Polonica* 38 (1977): 83–88.

Ellinger, K. R. "Polish agriculture and the dairy industry." *Monthly Bulletin of Agricultural Economics and Statistics* 14 (1965) 3:1–11.

Feiwel, George R. "The mismanagement of agriculture in Poland." *Rivista Internazionale di Scienze Economiche e Commerciali* XXI (July 1974), 7: 773–90.

Gorzelak, Eugeniusz, and Tomczak, Franciszek. *Agrarian policy and agricultural planning.* Warsaw: Research Institute for Developing Countries, 1977. 128 pp.

Heneghan, Thomas E. *The summer storm in Poland* (Radio Free Europe Research, Background Report/176). Munich: 1976. 24 pp. (On the unrest in connection with the abortive food price increases of 1976.)

Hunek, T., Tomczak F. "Agrarian policy and the future of agriculture in Poland." *Economies et Sociétés* 10 (1976) 7/8: 1553–74.

Koenig, Ernest. *The agricultural situation in Poland* (Foreign Agricultural Service, M-54). Washington, D.C.: USDA 1959. 32 pp.

Koenig, Ernest. "Collectivization in Czechoslovakia and Poland." *Collectivization* (see I.1.): 103–125.

Korbonski, Andrzej. *Polish agricultural production, output, expenses, gross and net product, and productivity, 1934–38, 1937, and 1946–1970* (Occasional Papers of the Research Project on National Income in East Central Europe. OP-37). New York: Economic Studies, Riverside Research Institute, 1972. IV+60 pp.

Korbonski, Andrzej. *Politics of socialist agriculture in Poland: 1945–1960.* New York, London: Columbia Univ. Press, 1965. 330 pp.

Kos, Czesław and Woś, Augustyn. "The food sector as a factor in Poland's strategy of social and economic development." *Eastern European Economics* 14 (winter 1975–76) 2: 37–59. (Translated from *Ekonomista,* no. 1, 1975.)

Kostrowicka, Irena. "Stages of the development of agriculture in People's Poland." *Studia Historiae Oeconomicae* (Uniwersytet w Poznaniu) 9 (1974): 329–41.

Kostrowicki, Jerzy. "Agricultural typology, agricultural regionalization, agricultural development." *Geographia Polonica* 14 (1968): 265–74.

Kostrowicki, Jerzy. "An attempt at the determination of transformation trends in the spatial organization of agriculture in Poland between 1960 and 1990." *Geographia Polonica* 32 (1975): 27–41.

Kostrowicki, Jerzy. "Polish land utilization survey." *Problems of applied geography* 25 (1961): 45–56.

Kostrowicki, Jerzy. "Types of agriculture in Poland. A preliminary attempt at a typological classification." *Geographia Polonica* 19 (1970): 99–110.

Kostrowicki, Jerzy, and Szczesny, R. "A new approach to the typology of Polish agriculture." *Agricultural typology and land utilisation.* Verona: Center of Agricultural Geography, 1972: 213–21.

Krzyzaniak, Marian. *Transformation of Polish agriculture from 1920 on: a historical perspective* (Rice University. Program for Development Studies. Paper no. 13). Houston, Texas: Rice University, 1971. 51 pp.

Lipski, Witold. *Agriculture in Poland.* Warsaw: Interpress Publishers, 1969. 93 pp.

Lipski, Witold. "Changes in agriculture." *Canadian Slavonic Papers* 16 (spring and summer 1973) 1–2: 101–2.

"Long-term changes in Polish agriculture." *Growth and adjustment in national agricultures.* M. P. O'Hagen, ed. Montclair and New York: Allanheld, Osmun & Co. and Universe Books, 1978 (first published in England in 1978 by Macmillan Press), pp. 67–145.

Lotarski, Susanne S. "Reform of rural administration." *Canadian Slavonic Papers* 15 (1973) 1/2: 108–21.

Makarczyk, Wacław. "Innovation in agriculture and the use of information sources." *The Polish Sociological Bulletin* (1962) 2: 48–60.

Pienkos, Donald. "Regional differences and rural change in Poland." *East Central Europe* 3 (1976) part 2: 177–94.

Radkiewicz, Wacław. "Agriculture in Poland's Western Territories: under Prussian rule and in People's Poland." *Polish Western Affairs* 4 (1963) 2: 295–329.

Rajtar, Jan, and Wiśniewski, L. "Effectiveness of technological changes in Polish agriculture." *Bulletin,* 40 (Research Institute for Agricultural Economics, Budapest): 35–58.

Raszeja-Tobjasz, Ewa. "Land policy in Poland." *Toward modern land policies.* D. McEntire, D. Agostini, eds. Padua: Institute of Agricultural Economics and Policy, University of Padua [1970?]: 275–331.

Report of the FAO mission for Poland. Washington, D. C.: Food and Agriculture Organization, May 1948. 159 pp.

Romanowski, Jacek I. "Prospects for the future of Polish agriculture." *The future . . .* (see I.1): 97–137.

Skrzypek, Stanislaw. "Agricultural policies in Poland." *Journal of Central European Affairs* 16 (1956) 1: 45–70.

Social and political transformations in Poland. Stanisław Ehrlich, et al., eds. Warsaw: PWN—Polish Scientific Publishers, 1964. 329 pp. (see especially Tepicht, Jerzy. "Problems of the development of Polish agriculture," pp. 71–98.

Szulc, Halina. "The development of the agricultural landscape in Poland." *Geographia Polonica* 22 (1972): 85–103.

Tepicht, J. "Problems of the re-structuring of agriculture in the light of the Polish experience." *Economic problems* (see I.1): 534–54.

Tyszkiewicz, Wiesława. "The transformations in the agrarian structure in Poland 1945–1970." *Transformations of rural areas.* Warsaw: Institute of Geographical and Spatial Organization, PAN, 1978: 123–40.

Wierzbicki, Z. T. "A case study in planned social change: the collective farm experiment in Poland." *Sociologia Internationalis* 6 (1968) 1: 29–39.

Woś, Augustyn. "The transformations of property relations in Polish agriculture." *Oeconomica Polona* 4 (1977) 2: 147–64.

Woś, Augustyn, and Grochowski, Zdzisław. *Recent developments in Polish agriculture.* Warsaw: Interpress Publishers, 1979. 155 pp.

Zawadski, Michal I. "Ten years of Communist planning in Polish agriculture." *Journal of Farm Economics* 38 (1956) 3: 792–98.

IV.2. Social and settlement patterns, labor

Banasiak, Stefan. "Settlement of the Polish Western Territories in 1945–1947." *Polish Western Affairs* 6 (1965) 1: 121–49.

Dobrowolska, M. "Demographic and social changes in post-war Polish villages." *Recent population movements in the East European countries.* Budapest: 1970; 35–41.

Dziewicka, M. "Peasant-workers: A new social group in Poland." *International Social Science Bulletin* 9 (1957) 2: 174–80.

Frenkel, Izasław. "Employment problems in Polish agriculture." *International Labour Review* 83 (1961) 2: 156–77.

Galaj, Dyzma. "Part-time farmers in Poland." *Part-time farming. Problem or resource in rural development.* A. M. Fuller, J. A. Mage, eds. (Proceedings of the first rural geography symposium, University of Guelph, 1975). Guelph, Ontario, and Norwich (University of East Anglia): 258–66.

Gałęski, Bogusław. "Determinants of rural social change: sociological problems of the contemporary Polish village." *The social structure . . .* (see I.1): 229–58.

Gałęski, Bogusław. "Farmers' attitudes to their occupation." *The Polish Sociological Bulletin* (1963) 1: 57–68.

Gałęski, Bogusław. "From peasant to farmer." *The Polish Sociological Bulletin* (1964) 2: 84–96.

Grabowska, Urszula. "The effect of technical innovations on peasant family and farm in Poland." *Bulletin,* 40 (Research Institute for Agricultural Economics, Budapest): 23–33.

Kolankiewicz, G. "The new 'awkward class': the peasant worker in Poland." *Sociologia Ruralis* 20 (1980) 1–2: 28–43.

Pawełczyńska, Anna. "The dynamics of cultural changes in the rural areas." *The Polish Sociological Bulletin* (1966) 1: 67–76.

Pienkos, Donald E. "Changes in peasant political and religious attitudes and behavior in Poland." *The Polish Review* 23 (1978) 1: 58–68.

Pohorille, Maksymilian. "The productivity of labour in agriculture and the problem of the parity of incomes in agriculture and industry." *Problems of economic dynamics and planning. Essays in Honour of Michał Kalecki.* Warsaw: 1966; 317–25.

Rural social change in Poland. J. Turowski, L. M. Szwengrub, eds. Wrocław: Ossolineum, 1976. 336 pp.

Turowski, Jan. "Changes in the rural areas under the impact of industrialization." *The Polish Sociological Bulletin* (1966) 1: 123–30.

Vielrose, Egon. "Changes in the percentage of agricultural population in Poland in 1950–70." *Studia Demograficzne* 34 (1973): 49–60.

IV.3. The place of agriculture and its integration in the economy at large

Hunek, Tadeusz. "Concept of food economy in Poland as a way of modernizing agriculture." *Modernization* (see I.1): 257–65.

Smoleński, Zygmunt. "Links between agriculture and the food industry." *Integration of the agriculture . . .* (see I.1): 101–22.

Woś, Augustyn. "Basic problems of the integration of agriculture with other branches of the national economy in Poland." *Integration of the agriculture . . .* (see I.1): 57–70.

Woś, Augustyn, and Grochowski, Zdzisław. "Relationship between agricultural policies and general economic policy." *Integration of the agriculture . . .* (see I.1): 3–22. Also in *Decision-making . . .* (see I.1): 245–56.

IV.4. Farming and cooperatives

Bergmann, Theodor. "Agricultural co-operation in Poland." *Year Book of Agricultural Co-operation*. Oxford: The Plunkett Foundation for Co-operative Studies, 1967: 150–161. (Cf. the same author's earlier report in the 1960 volume of the same *Year Book*: 142–153.)

Galaj, Dyzma. "The Polish peasant movement in politics, 1895–1969." *Rural protest. Peasant movement and social change*. Henry A. Landsberger, ed. New York, London: Macmillan, 1974: 316–47.

Korbonski, Andrzej. "Peasant agriculture in socialist Poland since 1956: An alternative to collectivization." *Soviet and East European agriculture*. (see I.1): 411–31.

Kozlowski, Zygmunt. "Socialism and family farming. The Polish experience." *Osteuropa Wirtschaft* 22 (1977) 1: 21–36.

Kulikowski, Roman, and Szyrmer, Jacek. "Changes in the production orientations of the individual agriculture in Poland in the years 1960, 1965 and 1970." *Transformation of rural areas*. Warsaw: Institute of Geography and Spatial Organization, PAN, 1978: 141–54.

Lipowski, Adam. "Interdependence between goals and incentives in an experimental system of management." *Eastern European Economics* 8 (1969) 1: 20–71 (Transl. from *Ekonomista*, no. 3, 1968.)

Manteuffel, Ryszard. "Comparisons of state, cooperative, and family-owned farms in Poland from the technical and economic point of view." *Eastern European Economics* 8 (1970) 3: 253–77. (Transl. from *Zagadnienia Ekonomiki Rolnej*, no. 2, 1969.)

Marcinko, David John. *Centralized versus decentralized production. A comparative study of production functions for Polish state and private agriculture*. Boston College: Diss. 1973. II+98 pp. (Photocopy).

Marek, Jadwiga. "The significance of the 'good farmer'." *Sociologia ruralis* 6 (1966) 2: 144–55.

"The peasant in Poland today. Reactions to land reform and collectivization." *The World Today* 8 (1952) 2: 71–79.

Piekalkiewicz, Jaroslaw A. "Kulakization of Polish agriculture." *The political economy . . .* (see I.1): 86–107.

Przybyla, Jan. "Gomulka and the peasants." *Problems of Communism* 7 (1958) 3: 23–30.

Socio-educational activity of the supply and marketing co-operatives in Poland. Warsaw: Publ. House of the Central Agricultural Union of Co-operatives, 1967. 62 pp.

Szczesny, Roman. "Changes and trends in the spatial pattern of types of individual agriculture in Poland 1960–1970. *Transformation of rural areas*. Warsaw: Institute of Geography and Spatial Organization, PAN, 1978: 155–64.

Tepicht, Jerzy. "Agricultural circles in the light of the general problems of agriculture." *Eastern European Economics* 4 (1966) 4: 29–49. (Transl. from *Ekonomista,* no. 5, 1964.)

Zembrzuska, Helena. *Agricultural circles as a social and economic organization of peasants in Poland.* Warsaw: Publ. House of the Central Agricultural Union of Co-operatives, 1967. 63 pp.

IV.5. Directing agriculture (planning, prices, procurements, administration)

Herer, W. "Transmission of decisions of the central planning authority to peasant farms in a centrally planned economy." *Contributed papers read at the 16th International Conference of Agricultural Economists.* Oxford: published by the University of Oxford Institute of Agricultural Economics for the International Association of Agricultural Economists, 1977: 211–20.

Małkowski, Jan. "The influence of prices on the development of production and supply of cattle and milk in Poland." *Integration of the agriculture . . .* (see I.1): 141–60.

Mieszczankowski, Mieczysław. "Indirect planning in Polish agriculture." *Modernization of agriculture . . .* (see I.1): 25–44.

Pohorille, Maksymilian. "Plan and market in agriculture." *Planning and markets.* New York, London: Publications of the Wertheim Committee, 1969: 128–42.

Pohorille, M. "Purchasing contracts and price policy as means of planning agricultural production." *Economic problems of agriculture . . .* (see I.1): 441–48.

Tepicht, Jerzy. *Essays on agriculture in a planned economy.* (Teaching materials. Szkoła Główna Planowania i Statystyki, the Advanced Course in National Economic Planning. Vol. 22) Warsaw: 1966. 86 pp.

V. Czechoslovakia

V.1. Agriculture in general and various of its aspects

Anderson, Philip E. "New directions in Czechoslovak agricultural policy under Novotny." *Slavic Review* 21 (1962) 3: 471–86.

Baca, Jan. "The development of large-scale production in agriculture in the Czechoslovak Socialist Republic." *Czechoslovak Economic Papers* (Czechoslovak Academy of Sciences) 3 (1964): 67–87.

Bača, Ján, and Šilar, Jiři. "A thesis concerning the concept of economic reform in the management of agriculture." *Eastern European Economics* 8 (1970) 3: 211–33; cf. *ibid.,* pp 234–52, Olmová, Gabriela. "Comments on 'A thesis . . .' ". (Transl. from *Zemědělská Ekonomika,* nos. 7/8 and 11/12, 1968.)

Bajaja, Vladislav. "Concentration and specialization in Czechoslovak and East German farming." *Agricultural policies . . .* (see I.1.): 263–93.

Basch, Antonin. "Land reform in Czechoslovakia." *Family farm policy.* J. Ackerman, M. Harris, eds. Chicago: 1947: 309–28.

Bernitz, Alexander. *A survey of Czechoslovak agriculture* (Economic Research Service, ERS-Foreign-38). Washington, D.C.: USDA, 1962. IV+48 pp.

Brainard, Lawrence J. "A model of cyclical fluctuations under socialism." *Journal of Economic Issues* (Michigan State University, East Lansing) 8 (1974) 1: 67–81.

Brainard, Lawrence J. *Policy cycles in a planned economy: the case of Czechoslovak agricultural policy, 1948–1967.* Chicago: Univ. of Chicago (dissertation), 1971. 214 pp. (hectographed).

Brainard, Lawrence J. "Policy cycles in socialist economies: examples from Czechoslovak agriculture." *Reorientation and commercial relations . . .* (see I.1): 214–28.

Divila, Emil, and Goulli, Rochdi. "New stage of development in the Czechoslovak agriculture and food industry." *Czechoslovak Economic Papers* (Czechoslovak Academy of Sciences) 16 (1976): 115–33.

Divila, Emil, and Jeníček, Vladimír. "Development problems of agriculture in Czechoslovakia." *Czechoslovak Economic Papers* (Czechoslovak Academy of Sciences) 6 (1966): 97–111.

Economic policy of the state in Czechoslovak agriculture during the period 1945–1969. Prague: Economic Research Institute of Agriculture and Food, 1970. 45 pp.

Fabry, Valer. *Agricultural laws of the Czechoslovak Republic, May 1945–March 1949.* Prague: Brazda, 1949. 37 pp.

Feierabend, Ladislav. *Agricultural cooperatives in Czechoslovakia.* New York: Mid-European Studies Center, 1952. 125 pp. (On pre-collectivization cooperatives of Raiffeisen Type.)

Hajda, Joseph. "The politics of agricultural collectivization and modernization in Czechoslovakia." *The political economy* . . . (see I.1): 130–54.

Hajda, Joseph. "Principal characteristics of agricultural policy trends in Czechoslovakia." *The future* . . . (see I.1): 149–69.

Jeníček, Vladimír. "Approaches and criteria for studying the macro-structure of Czechoslovak agriculture." *Czechoslovak Economic Papers* (Czechoslovak Academy of Sciences) 15 (1975): 71–90.

Jeníček, Vladimír. "Czechoslovak agriculture and economic equilibrium." *Czechoslovak Economic Papers* (Czechoslovak Academy of Sciences) 12 (1970): 93–105.

Kalvoda, Josef. "Soviet-style agricultural reform in Czechoslovakia." *Journal of Central European Affairs* 22 (1962) 2: 200–219.

Kocvara, Stefan. *The sovietization of Czechoslovak farming. Standard charter of a unified agricultural cooperative in Czechoslovakia of Feb. 17, 1953. Analysis and translation* (Mid-European Law Project, Law Library, Library of Congress). Washington, D.C.: National Committee for a Free Europe, 1953. 42 pp.

Koenig, Ernest. "Agriculture." *Czechoslovakia.* V. Busek, N. Spulber, eds. New York, London: Praeger, 1957: 245–67.

Koenig, Ernest. "Collectivization in Czechoslovakia and Poland." *Collectivization* . . . (see I.1): 103–25.

Kolařik, Josef. "Czechoslovak land reforms." *Za Sotsialisticheskuiu Sel'skokhoziaistvennuiu Nauku (Prague)* 7 (1959) 3: 223–46.

Kot'átko, J. *Land reform in Czechoslovakia.* Transl. from the Czech by R. Rohan, Fr. Stein. Prague: Orbis, 1948. 38 pp.

Krblich, Jan. *Survey of Czechoslovak agriculture.* Prague: Institute for international collaboration in agriculture and forestry, 1947. 35 pp.

Lazarcik, Gregor. *Czechoslovak agricultural and non-agricultural incomes: 1948–1965* (Occasional Paper no. 20 of the Research Project on National Income in East Central Europe). New York: Columbia University, 1968. 76 pp.

Lazarcik, Gregor. *Czechoslovak agriculture and output, expenses, gross and net product, and productivity: 1934–1938 and 1946–1962* (Occasional Paper no. 7 of the Research Project on National Income in East Central Europe). New York: Columbia University, 1965. 33 pp.

Lazarcik, Gregor. "Factors affecting production and productivity in Czechoslovak agriculture, 1934–38 and 1946–60." *Journal of Farm Economics* 45 (1963) 1: 205–18.

Lazarcik, Gregor. "The performance of Czechoslovak agriculture since World War II." *Soviet and East European agriculture* (see I.1): 385–406.

Lazarcik, Gregor. *The performance of socialist agriculture. A case study of production and productivity in Czechoslovakia, 1934–38 and 1946–61.* New York: L. W. International Financial Research, Inc., 1963. V+121 pp.

Meissner, Frank. "Economies of scale in relation to agrarian reforms in Czechoslovakia." *The American Slavic and East European Review* 14 (1955) 1: 67–83.

Meissner, Frank. "Mandatory delivery quotas in Czechoslovak agriculture." *Journal of Central European Affairs* 15 (1955) 1: 30–48.

Meissner, Frank. "The socialization progress in Czechoslovak agriculture." *Journal of Farm Economics* 35 (1953) 1: 88–98.

Menclová, Jarmila, and Stočes, Ferdinand. *Land reforms in Czechoslovakia.* Transl. H. Pštrossová. Prague: Agricultural Publ. House, 1963. 96 pp.

Moulis, Miloslav, and Větvička, Miloš. *The co-operative village.* Prague: 1963. 81 pp.

Nikl, Josef. "Czechoslovak agriculture in the period of socialist reconstruction in rural areas." *Czechoslovak Economic Papers* (Czechoslovak Academy of Sciences) 2 (1962): 73–179.

Nikl, Josef. "The share of agriculture in the formation and in the primary and secondary distribution of national income." *Eastern European Economics* 6 (1967) 1: 3–14. (Transl. from *Politická Ekonomie,* no. 7, 1966.)

Pospichal, J. "Cooperative-state relationships in a centrally planned economy." *The role of group action in the industrialization of rural areas.* (International Research Centre on Rural Cooperative Communities). New York, London: Praeger, 1971: 319–27.

Samal, Mary Hrabik. "The Czechoslovak Republican Party of Smallholder Peasants and the German minority, 1918–1919." *The peasantry . . .* (see I.1), vol. 1: 157–79.

Šedivý, Bohuslav. "A brief survey of the development and chief principles of activity of agricultural producers' co-operatives in Czechoslovakia." *Za Sotsialisticheskuiu sel'skokhoziaistvennuiu Nauku* (Prague) 9 (1960) 2: 101–34.

Skotta, Ladislav. "Supplier-customer relationships of cooperative farms." *Eastern European Economics* 4 (1966) 4: 58–62. (Transl. from *Planované Hospodářství,* no. 3, 1965.)

Šprincová, Stanislava. "Functional changes in agricultural settlements in backward regions." *Regional studies, methods and analyses.* Budapest: Hungarian Academy of Sciences, Research Institute of Geography, 1974: 133–138.

Tauber, Jan. "Economic and sociological aspects in raising Czechoslovak agriculture to the level of industry. A summary of articles by K. Svoboda et al., publ. in *Zemědělská Ekonomika* and other magazines in 1962 and 1963. *Za Sotsialisticheskuiu Sel'skokhoziaistvennuiu Nauku* (Prague) 12 (1963) 4: 325–54.

Testing and technical state control in agricultural production in the ČSSR. Prague: 1965. 26 pp.

Thomasová, Marie. *Agricultural and forestry education in the Czechoslovak Socialist Republic.* Engl. transl. from Czech by H. Pštrossová. Prague: Agricultural Publ. House, 1966. 151 pp.

Václavů, Antonín. "The postwar problems of Czechoslovak agricultural production." *The Czechoslovak economy 1945–1948* (Acta Oeconomica Pragensia. 28). Prague: Prague School of Economics, 1968: 155–97.

Vraný, Jan. "Development of production and labour productivity of Czechoslovak agriculture." *Czechoslovak Economic Papers* (Czechoslovak Academy of Sciences) 2 (1962): 111–27.

V.2. Social aspects, income

Erben, Bohumil. "Social insurance of co-operative farmers in Czechoslovakia." *Za Sotsialisticheskuiu Sel'skokhoziaistvennuiu Nauku* (Prague) 12 (1963) 1: 1–12.

Falťan, Michal. "Some thoughts on class-relations in our village." *Czechoslovak Economic Papers* (Czechoslovak Academy of Sciences). 1959: 53–84. (Transl. from *Ekonomický časopis* 4 [1956] 1.)

Flek, Josef. "Life and income of Czechoslovak co-operatives." *Economic problems* . . . (see I.1.): 304–15.

Glaserova, Jaroslava. "Labor in Czechoslovak agriculture." *Demosta* (Prague) 3 (1970) 4: 293–300

Karlik, Jiři, and Flek, Josef. "Current problems of manpower in Czechoslovak agriculture." *Eastern European Economics* 4 (1965) 1: 34–41 (From *Czechoslovak Economic Papers*, no. 4, 1964.)

Klinko, Ladislav. "Material incentives in agriculture." *Eastern European Economics* 4 (1966) 4: 50–57. (Transl. from *Plánované Hospodářství*, no. 3, 1966.)

Kunc, Jaroslav. "Regional social planning in agriculture." *Decision-making* . . . (see I.1.): 439–42.

Tauber, Jan. "Co-operation between rural sociology and agricultural economics in solving agriculture questions. With special regard to conditions in C.S.S.R." *Proceedings of the 12th International Conference of Agricultural Economists, held at Lyon, France.* London: 1966: 350–360 (plus discussion: 361–378).

Tauber, Jan. "Problems, tasks, and methods in the study of social development of the Czechoslovak village and agriculture." *Za Sotsialisticheskuiu Sel'skokhoziaistvennuiu Nauku* (Prague) 11 (1962) 2: 119–48.

VI. Hungary

VI.1. Agriculture in general

Agriculture in the new economic system (Research Institute for Agricultural Economics. Bulletin 23). Budapest: 1968. 52 pp.

Asztalos, I. "A geographical study of stockbreeding in Hungary." *Research problems in Hungarian applied geography.* Béla Sarfálvi, ed. Budapest: 1969: 183–203.

Balogh, András, and Kulcsár, Viktor. *Agricultural policy and development of agriculture in Hungary (1945–1975).* Budapest: Research Institute for Agricultural Economics, 1975. 43 pp. plus 18 pp. of tables.

Balogh, A., and Kulcsar, V. "Development of agriculture and its effects on rural society in Hungary." *Economies et sociétés* (1976) 7/8: 1515–36.

Benet, Iván. "The role of capital in Hungarian agriculture." *Economic studies* . . . (see below): 85–106.

Bernát, T. "The delimitation and characterization of agricultural areas deficient in physical resources." *Regional development and planning.* Budapest: Geographical Research Institute, Hungarian Academy of Sciences, 1976: 71–81.

Csendes, Béla. "Development trends of Hungarian agriculture." *Economic development and planning.* Budapest: 1973: 117–34.

Csendes, Béla. "The principal questions on the progress and further development of Hungarian agriculture." *Essays on economic policy and planning in Hungary.* Budapest: 1978: 171–23.

Csikòs-Nagy, Béla. "Productivity in Hungarian agriculture." *Modernization of agriculture* . . . (see I.1.): 45–49.

Csizmadia, E. "New features in Hungarian agriculture in the 1970s." *Acta Oeconomica* 12 (1974) 3–4: 349–65.

Csizmadia, Ernö. "The relationships between intensive economic development and agriculture." *Eastern European Economics* 8 (1970) 3: 278–98. (Transl. from *Közgazdasági Szemle*, no. 5, 1969.)

Csizmadia, Ernö. *Socialist agriculture in Hungary.* Budapest: Akadémiai Kiadó, 1977. 178 pp.

Czekner, John, Jr. "A comparison of the agricultural systems of Austria and Hungary in the two decades before World War I." *East European Quarterly* 12 (1978) 4: 461–73.

Dohrs, Fred E. "Nature versus ideology in Hungarian agriculture. Problems of intensification." *Eastern Europe. Essays in geographical problems.* New York,Washington: 1971: 271–95.

Donáth, Ferenc. *Ownership and efficiency in the industrialising Hungarian agriculture* (Institute of Economics, Hungarian Academy of Sciences: Studies, No. 18). Budapest: 1980. 86 pp.

Donáth, Ferenc. *Reform and revolution. Transformation of Hungary's agriculture 1945–1970.* Budapest: Corvina Kiadó, 1980. 489 pp.

Economical and integration problems of the large-scale farming in Hungary (Studies) (Research Institute for Agricultural Economics. Bulletin. 38). Budapest: 1976. 132 pp.

Economic studies on Hungary's agriculture. I. Benet, J. Gyenis, ed. Budapest: Akadémiai Kiadó, 1977. 193 pp.

The effect of the development and efficiency on the planning and forecasting of agricultural production. A. Csepeli-Knorr, A. Bisztray, eds. (Research Institute for Agricultural Economics. Bulletin 40). Budapest: 1977. 142 pp.

Elek, Peter S. "Hungary's new agricultural revolution and its promise for the fifth five-year plan." *The future . . .* (see I.1): 171–84.

Éliás, András. "The application of Western technology in the Hungarian agriculture and food industries." *The ACES Bulletin* 22 (1980) 1: 61–82.

Éliás, András, "Development of Hungarian foreign trade in agricultural and food products." *The effect . . .* (see above): 99–105.

Éliás, A., Marillai, V. F., Sebestyén, K. L., and Varga, G. *Development of Hungarian agriculture from the liberation down to date* (Research Institute for Agricultural Economics. Bulletin 44). Budapest: 1979. 89 pp.

Enyedi, Gy. "A brief characterization of the agricultural land utilization in Hungary." *Land utilization in Eastern Europe.* Budapest: 1967: 74–88.

Enyedi, Gy. "Geographical types of agriculture in Hungary." *Applied geography in Hungary.* Budapest: 1964: 58–105.

Enyedi, Gy. "A regional subdivision of the agriculture in Hungary." *Regional development and planning.* Budapest: Geographical Research Institute, Hungarian Academy of Sciences, 1976: 93–100.

Erdei, Ferenc. *Agriculture and co-operation.* (Institute for Farm Economics of the Hungarian Academy of Sciences. Bulletin 5). Budapest: 1963. IV+52 pp.

Erdei, Ferenc. "The socialist transformation of Hungarian agriculture." *The New Hungarian Quarterly* 5 (1964) 15: 3–28.

Fehér, L. "Current problems of our co-operative policy." *Year-book 1971 . . .* (see below): 9–48.

Fekete, F. "The major social and economic features of co-operative farming in Hungary." *Acta Oeconomica* 11 (1973) 1: 19–32.

Fekete, Ferenc, Heady, E.O., and Holdren, B. R. *Economics of cooperative farming. Objectives and optima in Hungary.* Leyden: Sijthoff; Budapest: Akadémiai Kiadó, 1976. 183 pp.

Fekete, F., and Sebestyén, K. "Organization and recent development in Hungarian agriculture." *Acta Oeconomica* 21 (1978) 1/2: 91–105.

Fischer, Lewis A. "Efficiency in Hungarian agriculture after six years of economic reform." *Economic development.* Fallenbuchl, ed. (see I.1): 306–18.

Fischer, Lewis A., and Uren, Philip E. *The new Hungarian agriculture.* Montreal, London: McGill-Queen's Univ. Press, 1973. XXI+138 pp.

Friss, I. "Ten years of economic reform in Hungary." *Acta Oeconomica* 20 (1978) 1–2: 1–19.

Held, Joseph. "The interwar years and agrarian change." *The modernization . . .* (see below): 197–292.

Held, Josef. "Some aspects of the transformation of the Hungarian peasantry in the 20th century." *The peasantry* . . . (see I.1): 141–59.

Hidas, Peter. "The dance of death: World War I and the Hungarian peasants." *The modernization* . . . (see below): 169–95.

Hidas, Peter. "The emancipation and its impact on the Hungarian peasantry in 1848–1850." *The modernization* . . . (see below): 7–20.

Kiraly, Bela K. "Organized peasant mass movements since 1867." *The moderniza- tion* . . . (see below): 131–67.

Kiraly, Bela K. "Peasant movements in the twentieth century." *The modernization* . . . (see below): 319–50.

History of the Research Institute for Agricultural Economics, 1954–1969 (Research Institute for Agricultural Economics. Bulletin 26). Budapest: 1970. XVI+39 pp.

Hungary's problem and her agriculture. Ch. Ihrig, ed. (Compiled from the research work of the Agrarian Research Institute and the National Institute of Farm Management and Calculations). Budapest: 1946. VII+59 pp.

Kádár, B. "Major specialization tendencies of Hungarian exports to the West." *Acta Oeconomica* 20 (1978) 1–2: 147–69.

Komlos, John. "The emancipation of the Hungarian peasantry and agricultural development." *The peasantry* . . . (see I.1): 109–18.

Kovács, Kálmán. "Long term projection for some social and economic factors in the development of agricultural enterprises." Research Institute for Agricul- tural Economics. *Abstracts* 41 (1976–77): 23–31.

Kovacs, Martin L. "Aspects of Hungarian peasant emigration from pre-1914 Hungary." *The peasantry* . . . (see I.1): 119–32.

The modernization of agriculture: rural transformation in Hungary, 1848–1975. Joseph Held (et al.), eds. Boulder, Colorado: East European Monographs, 1980 (distributed by Columbia Univ. Pres). 508 pp.

Murray, Kenneth L. *The agricultural situation in Eastern Europe. 4: Hungary* (FAS-M-79). Washington, D.C.: USDA, 1960. 29 pp.

O'Relley, Z. Edward. "Hungarian agricultural performance and policy during the NEM." *East European economies post-Helsinki* (see I.1): 356–78.

Orolin, Zsusza. "Hungarian agriculture and problems with the supply of labor." *Economic Studies* . . . (see above): 64–84.

Pálovics, Béla. "Perspective development of Hungarian agriculture. *Acta Oeconomica* 6 (1971) 4: 333–46.

Papers on some problems of the Hungarian village (Research Institute for Agricultural Economics. Bulletin 9 and 15). Budapest: 1964 and 1967. 107 and 86 pp.

Pclva, Ágoston. *Representative survey methods in Hungarian agricultural statistics* (Some questions of sampling for current statistics. [Hungarian co-referates for the 1963 UN Regional Seminar]). Budapest: 1965. 20 pp.

Pounds, Norman J. G. "Land use on the Hungarian plain." *Geographical essays on Eastern Europe.* N. J. G. Pounds, ed. Bloomington, Indiana: Indiana Univ. Publications, 1961: 54–74.

Romány, Pál. "Agricultural economics and agrarian development in Hungary." Research Institute for Agricultural Economics. *Bulletin* 45 (1979): 5–13.

Romány, Pál. "Hungarian agriculture in the seventies. The period of intensive progress." *The New Hungarian Quarterly* 19 (1978) 71: 74–81.

Rott, Nándor, and Szakonyi, László. "Trends of agricultural investments and fixed assets efficiency." Research Institute for Agricultural Economics. *Abstracts* 41 (1976–77): 33–37.

Rural transformation in Hungary. Gy. Enyedi, ed. Budapest: Akadémiai Kiadó, 1976. 116. pp.

Studies on the foreign economic factors affecting Hungarian agriculture (Research Institute for Agricultural Economics. Bulletin 36). Budapest: 1975. 80 pp.

Sutherland, P. H. "Country reports: Hungary." *Agriculture Abroad* 34 (1979) 6: 11–17.

Szalai, B. "Agricultural exports and the equilibrium of the balance of foreign trade in Hungary." *Acta Oeconomica* 18 (1977) 3–4: 343–54.

Third National Conference on Agricultural Economics (Kecskemet, 2–3 September 1977). (Research Institute for Agricultural Economics. Bulletin 42). Budapest: 1978. 167 pp.

Tóth, Antal Ernö. "Production factors in Hungarian agriculture." *Economic studies* . . . (see above): 21–45.

Ujhelyi, T. "Foreign trade in agricultural and food products—the Hungarian national economy and the world market." *Acta Oeconomica* 11 (1973) 1: 3–18.

Vankai, Thomas A. *The agricultural economy and trade of Hungary.* (ERS-Foreign 269). Washington, D. C: USDA, 1969. 18 pp.

Volgyes, Ivan. "Dynamic change: rural transformation, 1945–1975." *The modernization . . .* (see above): 351–508.

Volgyes, Ivan. "Modernization, collectivization, production, and legitimacy: agricultural development in rural Hungary." *The political economy . . .* (see I.1.): 108–29.

Voros, Antal. "The age of preparation: Hungarian agrarian conditions between 1848–1914." *The modernization . . .* (see above): 21–129.

Year-book 1971. Compiled by A. Gyenes (Co-operative Research Institute). Budapest: Akadémiai Kiadó, 1971. 183 pp.

VI.2. Land reform and collectivization

"Agrarian reform in Hungary." *International Labour Review* 2 (1921): 261–71.

Balassa, Bela A. "Collectivization in Hungarian agriculture." *Journal of Farm Economics* 42 (1960) 1: 35–51.

Dovring, Folke. "Land reform in Hungary." *A. I. D.* [Agency for International Development] *Spring review of land reform, June 1970.* Country papers. 2nd. ed. Vol. 10. Washington, D. C.: Department of State, 1970. 59 pp.

Erdei, Ferenc. "Land reform. 1945." *The New Hungarian Quarterly* 16 (1975) 57: 44–50.

Erdei, Ferenc. "The peasants and the 1945 land reform." *The New Hungarian Quarterly* 12 (1971) 43: 166–72.

Földes, Iván. "Land policy and land ownership in Hungary." *Land Reform. Land settlement and cooperatives.* Rome: FAO, 1977; 47–54.

"Land reform in Hungary." *The World Today* 5 (1949) 1: 22–28.

Petri, E. "The collectivization of agriculture and the 'tanya' system." *Research problems in Hungarian applied geography.* Bela Sarfálvi, ed. Budapest: 1969: 169–81.

Seres, Imre. "Land tenure in Hungary after the liberation." *Development of the political and legal system of the Hungarian People's Republic in the past 30 years.* Budapest: Hungarian Lawyers' Association, 1975: 203–12.

VI.3. Social and settlement patterns

Bango, Eugen. "The new Hungarian village." *UKI Reports* (1970) 3: 21–73.

Coulter, Harris L. "The Hungarian peasantry, 1948–1956." *The American Slavic and East European Review* 18 (1959) 4: 539–54.

Enyedi, György. "The transformation of the Hungarian village." *The New Hungarian Quarterly* 18 (1977) 67: 67–86.

Erdei, Ferenc. "The changing Hungarian village." *The New Hungarian Quarterly* 11 (1970) 38: 3–16.

Erdei, Ferenc. "The development of the economic-social relations of the agricultural population after the liberation." *Papers on some problems of the Hungarian village.* Budapest: 1964: 67–102

Erdei, Ferenc. "The Hungarian village in the process of transformation." *Sociological aspects . . .* (see below): 5–27.

Erdei, Ferenc. *Social problems of co-operative farms.* (Research Institute for Agricultural Economics. Bulletin 25). Budapest: 1969. 38 pp.

Fél, Edit, and Hofer, Tamás. *Proper peasants. Traditional life in a Hungarian village.* Chicago: Aldine Publ. Co., 1969. XIII+440 pp.

Gyenes, A. "Re-stratification of the agricultural population in Hungary." *Acta Oeconomica* 11 (1973) 1: 33–49.

Gyenes, A. "Some aspects of stratification in Hungarian co-operative farms." *Sociologia ruralis* 16 (1976) 3: 161–76.

Gyenes, Antal. " 'Workers' and 'peasants' in the Hungarian co-operative farms." *Year Book of Agricultural Co-operation.* Oxford: 1975: 135–51.

Hegedüs, András. *The structure of socialist society.* London: Constable, 1977. VIII+230 pp.

Holács, Ibolya. *Change of the way of living in six Transdanubian co-operative villages.* Keszthely: Agricultural University Keszthely, 1974. 91 pp.

Holács, Ibolya. *Sociological survey on the situation of country women in the South Western Transdanubian region.* Keszthely: Agricultural University Keszthely, 1971. 47 pp.

Illyés, Gyula. *People of the Puszta.* Transl. by G. F. Cushing. Budapest: Corvina Press, 1967. 307 pp.

Kardos, László. "Past and present of a village." *The New Hungarian Quarterly* 11 (1970) 37: 56–82.

Kenéz, Enikö. *Rural households, rural families. Social and economic changes in rural area, peasant families and in the work and living conditions of women* (Research Institute for Agricultural Economics. Bulletin 30). Budapest: 1971. 119 pp.

Kovács, Kálmán. "Demographical questions of the agrarian settlements." *Studies on the settlement . . .* (see below): 27–53.

Kunszabó, Ferenc. "Social strata and interest relations." *Sociological aspects . . .* (see below): 51–78.

Lettrich, E. "The Hungarian 'tanya' system. History and present-day problems." *Research problems in Hungarian applied geography.* Béla Sarfálvi, ed. Budapest: 1969: 151–68.

Márkus, István. *The agricultural population of Hungary planned changes in their way of life.* Budapest: Institute of Sociology, Hungarian Academy of Sciences, 1970. 21 pp.

Márkus, István. "Post-peasants and pre-citizens." *The New Hungarian Quarterly* 13 (1972) 46: 79–90.

Molnár, Lászlóné. *The major trends of social change in Hungarian villages.* Budapest: Institute of Sociology, Hungarian Academy of Sciences, 1970. 17 pp.

Móricz, Miklós. "Landless agricultural workers in Hungary." *International Labour Review.* 28 (1933): 518–30.

Pénzes, I. "The goods and structure of the markets of Szeged." *Agricultural typology and agricultural settlements* (Papers of symposium held in Szeged and Pécs, August 15–18, 1971). Szeged: 1972: 266–85.

Petri, E. "Settlement system of scattered farmsteads and problem of the new communities with scattered farmsteads on the Great Plain." *Agricultural typology . . .* (as above, for Pénzes): 303–14.

Robinson, William F. "Paying the Hungarian cooperative farmer." *Studies in Comparative Communism* 9 (1976) 3: 270–74.

Sárfalvi, Béla. "Internal migration and decrease of agricultural population in Hungary." *Applied geography in Hungary* (Studies in Geography. 2) Budapest: 1964; 150–63.

Sas, Judit H. "Expectations from and demands made upon children in a rural community." *Hungarian Sociological studies* (The Sociological Review. Monograph 17) Newcastle, 1972: 247–68.

Sociological aspects of rural life in Hungary (Research Institute for Agricultural Economics. Bulletin 28). Budapest: 1970. 93 pp.

Sozan, Michael. "Sociocultural transformation in East Central Europe: the case of the Hungarian peasant-worker in Burgenland." *East Central Europe* 3 (1976) Part 2: 195–209.

Studies on the settlement relations in Hungary (Research Institute for Agricultural Economics. Bulletin 14). Budapest: 1967. 133 pp.

Szijjártó, András. "Professions chosen by rural youth and rural labor force relations." *Sociological aspects . . .* (see above): 29–49.

Sztáray, Zoltán. "From agricultural labourer to small-holder, 1936–1956." *The Review* (Imre Nagy Institute for Political Research) (1959) 1: 47–69.

Tanner, József. "A cooperative village." *The New Hungarian Quarterly* 1 (1960) 1: 89–107.

Vágvölgyi, András, and Taganyi, Zoltán. "Trends and opinions on the Hungarian village and peasantry." *Problems of the development of agriculture and information on the state of rural sociology in various countries*. Warsaw: Institute of Philosophy and Sociology, Polish Academy of Sciences, 1971: 101–8.

Varga, Gyula. "Changes in the peasant living standard." *The New Hungarian Quarterly*. 7: (1966) 21: 86–101.

Varga, Gyula. "Changes in the social and economic status of Hungary's peasantry." *The New Hungarian Quarterly* 6 (1965) 20: 28–46.

Varga, Gyula. "A cooperative village." *The New Hungarian Quarterly* 6 (1965) 19: 16–34. Also in: *Studies on the settlement . . .* (see above) 109–33.

Volgyes, Nancy. "The Hungarian tanyas: persistence of an anachronistic settlement and production form." *The process* (see I.1.): 175–90.

VI.4. The place of agriculture and integration in the economy at large

Balogh, K., and Engszter, B. *Organization and operation of the vertical system in the food economy (after the example of the "Petőfi" co-operative farm at Gyermeli)* (Research Institute for Agricultural Economics. Bulletin 32). Budapest: 1973. 49 pp.

Benet, Iván. *The agro-industrial complex in Hungary*. Transl. by Jenő Rácz. Budapest: Institute of Economics, Hungarian Academy of Sciences, 1979. 134 pp.

"The contribution of agriculture and food industry to national income." *Agriculture and national economic planning* (Research Institute for Agricultural Economics. Bulletin 27). Budapest: 1970: 3–74.

Csaki, C. *First version of the Hungarian agricultural model ⟨HAM-1⟩* (International Institute for Applied Systems Analysis, IIASA. Research Memorandum 1978, 38). Laxenburg/Austria: 1978. VII+88 pp.

Csaki, Csaba. "National policy model for the Hungarian food and agriculture sector." *European Review of Agricultural Economics* 5 (1978) 3/4: 325–47.

Csizmadia, Magda. "Cooperation in the Hungarian food economy." *Economic studies . . .* (see VI.1): 123–36

Donáth, Ferenc. "Economic growth and socialist agriculture." *The New Hungarian Quarterly* 18 (1977) 65: 33–42, and 18 (1977) 66: 107–23.

Donáth, F. "Some phenomena of the industrialization of collectivized Hungarian agriculture." *Acta Oeconomica* 17 (1976) 1: 31–50.
Economic integration concepts, theories and problems. M. Simai, K. Garam, eds. Budapest: Akadémiai Kiadó, 1977. 423 pp.
Hegedüs, Miklós. *Relationships between industry and agriculture.* Budapest: Institute of Economics, Hungarian Academy of Sciences, 1972. 66 pp.
Hegedüs, Miklós. "Transformations of the structure of the food economy, and the main tendencies in the evolution of its production relations." *Economic studies . . .* (see VI.1): 174–93.
Komló, L. "The industrialization and integration of agriculture in a socialist country." *Acta Oeconomica* 10 (1973) 1: 67–79.
Komló, László. "The problems of vertical integration in agriculture: the Hungarian case." *Economic problems . . .* (see I.1.): 365–78.
Magyarváry, László. "Production systems in the Hungarian agriculture." *Integration of the agriculture . . .* (see I.1.): 71–100.
Marillai, Vilmos. "Current economic issues of agro-industrial complexes." Research Institute for Agricultural Economics. *Abstracts* 41 (1976–77): 95–101.
Marillai, Vilmos. "Management of economic associations of co-operatives in Hungary." *Integration of the agriculture . . .* (see I.1.): 161–81.
Márton, J. "The vertical integration of the Hungarian food economy." *Acta Oeconomica* 11 (1973) 1: 81–96.
Márton, János. *Vertical relations in food economy* (Research Institute for Agricultural Economics. Bulletin 31). Budapest: 1971. 64 pp.
Márton, János, and Németh, József. "Horizontal and vertical integration of the Hungarian agriculture." *Economical and integration problems . . .* (see VI.1.): 7–33. Also in *Integration of the agriculture . . .* (see I.1.): 23–56.
Sipos, A. "Agro-industrial integration." *Acta Oeconomica* 18 (1977) 2: 125–39.
Sövény, Sándor. "The joint commercial undertakings of the co-operative farms in Hungary." *Integration of the agriculture . . .* (see I.1.): 182–211.

VI.5. Socialist farms

Bango, J. F. "Some social-historical aspects in the development of Hungarian cooperative farming." *Journal of Rural Cooperation* 4 (1976) 2: 129–53.
Belák, Sándor. *Leading activities in large scale farms.* Keszthely: Agricultural University Keszthely, 1970. 24 pp.
"Changes in farming co-operatives." *The New Hungarian Quarterly* 12 (1971) 43: 181–91.
Csendes, Béla. "Characteristic features of cooperative farming." *Economic studies . . .* (see VI.1.): 109–22.
Csendes, Béla. "The expanding scope of economic activities in Hungarian agricultural co-operatives." *Acta Oeconomica* 9 (1972) 3/4: 331–42.
Csepely-Knorr, András. "Production costs and producers' income in large-scale farms." *The effect of the development and efficiency on the planning and forecasting of agricultural production.* (Research Institute for Agricultural Economics. Bulletin 40). Budapest: 1977: 69–82.
Csete, László. "Enterprisal system of agricultural production in Hungary." Research Institute for Agricultural Economics. *Bulletin* 45 (1979): 15–24.
Domé, Mária Gy. *Legal aspects of the associations of agricultural cooperatives.* Transl. by J. Decsényi. Budapest: Akadémiai Kiadó, 1973. 135 pp.
Erdei, Ferenc, and Enése, László. *Leading of socialist large farms* (Research Institute for Agricultural Economics. Bulletin 20). Budapest: 1967. 57 pp.

Gyenis, János. "Special farming cooperatives in the Hungarian cooperative movement." *Economic studies* . . . . (see VI.1): 154–73.

Kalocsay, Ferenc. *Farm units of socialist large-scale farms* (Research Institute for Agricultural Economics. Bulletin 17). Budapest: 1967. 50 pp.

Kalocsay, Ferenc. "Organizational activity, organization and its operation in large-scale farms." Research Institute for Agricultural Economics. *Abstracts* 41 (1976–77): 79–87.

Nagy, L. "Some timely problems of the development of co-operative law." *Year-Book 1971* (see VI.1): 121–51.

Nyers, R. "Current questions of co-operative theory as reflected in a debate." *Year-book 1971* (see VI.1): 49–66.

Simó, Tibor. *The career of an experiment. An experiment of independent enterprise, like rational farming in specialized vine and fruit growing co-operative societies in Hungary.* Budapest: Hungarian Co-operative Research Institute, 1971. 202 pp.

Some data concerning the Hungarian co-operative movement. Budapest: Hungarian Co-operative Research Institute, 1964. 85 pp.

Sövény, Sándor. "The joint commercial undertakings of the co-operative farms in Hungary." *Economical and integration problems* . . . (see VI.1.): 81–113.

Szabó, I. "Economic associations of farming co-operatives." *Acta Oeconomica* 11 (1973) 1: 51–64.

Szabó, I. "The federations of the co-operative farming societies." *Year-book 1971* (see VI.1.): 79–93.

VI.6. The private sector

Elek, Peter S. "Agro mass production and the private sector in Hungary." *The peasantry* . . . (see I.1) vol. 2: 61–76.

Fekete, F., and Varga, Gy. "Household plot farming of co-operative peasants in Hungary." *Acta Oeconomica* 2 (1967) 4: 345–61.

Lázár, István. "The collective farm and the private plot." *The New Hungarian Quarterly* 17 (1976) 63: 61–77.

Tóth, Antal Ernö. "The place and role of household plots and auxiliary farms in socialist agriculture." *Economic studies* . . . (see VI.1): 137–53.

Toth, A. E. "Small-scale agricultural production in Hungary and efficiency of the agro-industrial complex." *Acta Oeconomica* 21 (1978) 1/2: 107–19.

Tóth, Antal Ernö. *Some new economic and managerial features of small-scale agricultural production in Hungary.* Transl. by J. Rácz (Institute of Economics, Hungarian Academy of Sciences. Studies 12). Budapest: 1978. 149 pp.

Varga, Gyula. "The household plot." *The New Hungarian Quarterly* 7 (1966) 23: 7–23.

Varga Gyula. "Rural development and food production." *The New Hungarian Quarterly* 19 (1978) 71: 82–88.

VI.7. Directing agriculture (planning, prices, procurements, administration; see also VI. 5: socialist farms)

Csaki, S., Jonas, A. and Meszaros, S. *Modelling of centrally planned food and agricultural systems. A framework for a national policy model for the Hungarian food and agriculture sector* (International Institute for Applied Systems Analysis, IIASA. Research Memorandum 1978, 11). Laxenburg/Austria: 1978. IX+102 pp.

Csepely-Knorr, Andras. "Main characteristics of the economic regulation system for developing animal husbandry in Hungary." *Economical and integration problems* (see VI.1.): 63–79.

Dohrs, Fred E. "Incentives in Communist agriculture: the Hungarian models." *Slavic Review* 27 (1968) 1: 23–38.

Elek, Peter S. "The Hungarian experiment in search of profitability." *Agricultural policies* (see I.1): 165–84.

Erdei, F., and Fekete, F. "Economic problems of the strengthening of inefficiently farming co-operatives." *Papers on some problems of the Hungarian village.* Budapest: 1964; 5–16.

Fekete, F. "Accomplishment and challenges for agricultural economists working at the national level of centrally managed economies." Research Institute for Agricultural Economics. *Bulletin* 45 (1979): 25–36.

Kiss, Sandor. "Hungarian agriculture under the NEM." *East Europe* 17 (1968) 8: 10–19.

Kovács, Kálmán. "Development and some problems of planning in the framework of the management system of Hungarian agriculture. *"The effect of the development* . . . (see VI.1): 121–34.

László, J. "Some problems of agricultural price policy." *Economic development for Eastern Europe* (Proceedings of a conference held by the International Economic Association). London, New York: 1968; 265–77.

László, János. "The planned control of cooperative farm production and the reform of economic management." *Acta Oeconomica* 2 (1967) 3: 213–26.

Marillai, Vilmos. "Management of economic associations of co-operatives in Hungary." *Economical and integration problems* . . . (see VI.1.): 115–32.

Öri, J. "Pricing of agricultural products and foodstuffs in Hungary." *Acta Oeconomica* 17 (1976) 1: 83–91.

Papers on the economic regulation of food economy. I. Bukta, A. Csepely-Knorr (et al.), eds. (Research Institute for Agricultural Economics. Bulletin 34). Budapest: 1973. 63 pp.

Plan, market and interest in food economy (1st National Conference on Agricultural Economics, Budapest, March 24–26, 1969) (Research Institute for Agricultural Economics. Bulletin 24). Budapest: 1969. 109 pp.

Reform of the economic mechanism in Hungary. Nine studies. I. Friss, ed.; G. Hajdu, J. Rácz, transl. Budapest: Akadémiai Kiadó, 1969. 274 pp.

Robinson, William F. *The pattern of reform in Hungary.* New York: Praeger, 1973. 467 pp. (therein, pp. 108–20: "Positive trends—agriculture").

Sebestyén, Joseph. *Facts for planning* (Research Institute for Agricultural Economics. Bulletin 29). Budapest: 1970. 22 pp.

Sebestyén, Jozsef. "Problems of rational location in Hungarian agriculture." *Regional Science Association. Papers* (Regional Science Association. Philadelphia, Pa.) 22 (1969): 141–47.

Szabó, Gábor. "The problems of the economic valuation of land withdrawn from agricultural cultivation in the works of Hungarian economists." *Acta Oeconomica* 7 (1971) 3/4: 399–408.

Ujhelyi, Tamás. "The role of the world market forecasts in planning the development of Hungarian agriculture." *The effect of the development* . . . (see VI.1.): 135–42.

Zsuffa, Ildikó. "Agricultural price ratios in Hungary and in the European capitalist countries." *Studies on the foreign economic factors* . . . (see VI.1): 39–52.

VII. Rumania

VII.1. Agriculture in general

Bossy, George H. "Agriculture." *Romania.* Stephen Fischer-Galati, ed. New York, London: Praeger, 1956: 197–231.

Cernea, Mihail. "The cooperative farm as an organization: an attempt at conceptualization." *Revue Roumaine des Sciences Sociales. Série de Sociologie* 17 (1973): 109-30.

Cernea, Mihail. "The large scale formal organization and the family primary group." *Revue Roumaine des Sciences Sociales. Série de Sociologie* 18 (1974): 85–98. (Also in *Journal of Marriage and the Family,* November 1975: 927–36.)

Cernea, Mihail. "Organizational build-up and reintegrative regional development in planned agriculture." *Journal of the European Society for Rural Sociology* 14 (1974) 1/2: 30–44.

Cioranescu, George. "How efficient is Rumanian agriculture?" *East Europe* 16 (1967) 9: 9–13.

Collins, H. Christine. *The agricultural economy and trade of Romania* (ERS Foreign 320). Washington, D.C.: USDA, 1971. I+38 pp.

" 'Co-operativized' agriculture in Rumania in a blind alley?" (Radio Free Europe Research, 2 February 1970). Munich: 1970. 13 pp.

Dando, William A. "Wheat in Romania." *Annals of the Association of American Geographers* 64 (1974) 2: 241–57.

The development of agriculture in the Rumanian People's Republic. Compendium of statistics. Translation of texts. Bucharest: Central Statistical Board, Rumanian People's Republic, 1965. 94 pp.

Dumitriu, Dumitru. "Agricultural organization and planning in Rumania." *Planning and markets. Modern trends in various economic systems.* John T. Dunlop and Nikolay P. Fedorenko, eds. New York, London: McGraw-Hill, 1969: 143–54.

Fulea, Maria. "Material incentives in the co-operative farms. *Revue Roumaine des Sciences Sociales. Série de Sociologie* 18 (1974): 99–104.

Gilberg, Trond. "The costly experiment: collectivization of Romanian agriculture." *The political economy* . . . (see I.1): 23–32.

Gilberg, Trond. *Modernization in Romania since World War II.* New York, Washington, London: Praeger, 1975. XIII+261 pp. (therein, pp. 171–89: "The development of agriculture").

Gilberg, Trond. "Romanian agricultural policy in the quest of 'The Multilaterally Developed Society'." *Agricultural policies* . . . (see I.1): 137–164.

Iacob, Gh. "The rational use of sandy soils in Romania." *Revue Roumaine de Géologie, Géophysique et Géographie. Série de Géographie* 12 (1968) 1/2: 171–74.

Kideckel, David A. "The dialectic of rural development. Cooperative farm goals and family strategies in a Romanian commune." *Journal of Rural Cooperation* 5 (1977) 1: 43–61.

Matei, Ioan I. and Mionara, I. "Problems of the development of the rural environment in the Socialist Republic of Romania." *Revue Roumaine des Sciences Sociales. Série de Sociologie* 15 (1971): 47–56.

Mitrany, David. *The land and the peasant in Rumania. The war and agrarian reform, 1917–21.* New York: Greenwood, 1968; London: Milford, 1969. XXXIV+627 pp. (Reprint. Orig. 1930).

Moldovan, Petre. *The technical-material basis of Romanian agriculture.* Bucharest: Meridiane Publ. House, 1967. 63 pp.

Molnar, E. "Types of agriculture in the Transylvanian tableland." *Agricultural typology and land utilisation.* Verona: Center of Agricultural Geography, 1972: 419–26.

Montias, John Michael. *Economic development in Communist Rumania.* Cambridge, Mass., London: The M.I.T. Press, 1967. XIII+327 pp. (therein, pp. 87–134: "Agriculture").

Morariu, T., Mac, I., and Crişan, I. "The agricultural capacity of territories assessed by the components of the geographical landscape." *Revue Roumaine de Géologie, Géophysique et Géographie. Série de Géographie* 16 (1972) 1: 15–19.

Murray, Kenneth L. *The agricultural situation in Rumania* (ERS Foreign-4). Washington, D.C.: USDA; 1961. II+29 pp.

Negel, Dumitru. *Romanian agricultural taxation.* New York: Free Europe Committee, 1954. 20 pp.

Roberts, Henry L. *Rumania: political problems of an agrarian state.* New Haven, London, Oxford: Yale Univ. Press, 1951. XIV+414 pp. (Reprint: Hamden, Conn.: Archon Books, 1969.)

Rumanian agriculture since full collectivization (Radio Free Europe Research 30). Munich: 1965. 58 pp.

Staicu, Irimie. *Scientific research in Romanian agriculture.* Bucharest: Meridiane Publ. House, 1966. 93 pp.

Stefanescu, I. "Geographical types of agriculture and their evolution in the Romanian Sub-Carpathians between Şuşiţa-Zabrauţ and Buzau." *Agricultural typology and agricultural settlements* (Papers of Symposium held in Szeged and Pécs, August 15–19, 1971) Szeged: 1972; 359–74.

Thomas, C. "Regional fluctuations in Rumanian agriculture. A case study of grain production, 1956–1966." *Balkan Studies* 15 (1974) 1: 80–98.

Tsantis, Andreas C. *Romania. The industrialization of an agrarian economy under socialist planning* (Report of a mission sent to Romania by the World Bank). Coordination authors: A. C. Tsantis and R. Pepper. Washington, D.C.: 1979. XXXV+707 pp.

Tufescu, V., and Velcea, I. "Study and mapping of land use in Rumania." *Revue Roumaine de Géologie, Géophysique et Géographie* 8 (1964): 233–37.

Velcea, I., and Jacob, Gh. "Types of land use in the Danube delta." *Revue Roumaine de Géologie, Géophysique et Géographie* 8 (1964): 239–44.

Woolley, Douglas Charles, Jr. *The role of the agricultural sector in the development of the Romanian economy since 1950.* The University of Connecticut: Diss. 1976. VIII+183 pp.

VII.2. Social and settlement patterns

Barbat, Alexandru. " 'Urbanization' of rural areas?" *Revue Roumaine des Sciences Sociales. Série de Sociologie* 20 (1976): 45–51.

Cernea, Mihail. "Cooperative farming and family change in Romania." *The social structure of Eastern Europe* (See I.1.): 259–79.

Cernea, Michael. "Macrosociological change, feminization of agriculture and peasant women's threefold economic role." *Sociologia Ruralis* 18 (1978) 2/3: 107–24.

Cernea, Mihail. "The transformation of the peasantry's consciousness and the economic relations in the countryside." *The Rumanian Journal of Sociology* 1 (1962): 205–21.

Cobianu, Maria. "Aspects of modernization of the cultural life in the Romanian village of today." *Revue Roumaine des Sciences Sociales. Série de Sociologie* 19 (1975): 59–64.

Constantinescu, Virgil. "The contract payment system in terms of the individual-team relationship." *Revue Roumaine des Sciences Sociales. Série de Sociologie* 20 (1976): 85–93.

Dorel, Abraham. "The suburb: rural transition to urban condition ⟨with reference to the capital's subordinate localities⟩." *Revue Roumaine des Sciences Sociales. Série de Sociologie* 20 (1976): 131–37.

Drăgan, Ion. "Social incidences of the industrialization of rural areas." *Revue Roumaine des Sciences Sociales. Série de Sociologie* 20 (1976): 147–55.

Fulea, Maria. "Aspects of rural population mobility." *Revue Roumaine des Sciences Sociales. Série de Sociologie* 20 (1976): 75–83.

Fulea, Maria, and Alexandrescu, Petruş. "An analysis model of relationships between professional structure and labour force stability." *Revue Roumaine des Sciences Sociales. Série de Sociologie* 19 (1975): 39–48.

Gilberg, Trond. "Peasant workers in Romania." *Rural change* . . . (see I.1.): 179–94.

Gilberg, Trond. "Rural transformation in Rumania." *The peasants* . . . (see I.1.), vol. 2: 77–122.

Ionescu, Constantin. "Economic factors of change in the social structure." *Revue Roumaine des Sciences Sociales. Série de Sociologie* 16 (1972): 17–28.

Lăzărescu, Cezar. "Current problems of organizing the national territory and human settlements in Romania." *Revue Roumaine des Sciences Sociales. Série de Sociologie* 20 (1976): 37–44.

Malinschi, V. "Rural sociology and agrarian economy." *The Rumanian Journal of Sociology* 1 (1962): 67–77.

Velcea, Ion. "The urbanization process of the rural settlements in Romania." *Revue Roumaine de Géologie, Géophysique et Géographie. Série de Géographie* 16 (1972) 1: 93–101.

VIII. Bulgaria

VIII.1. Agriculture in general

The advance of agriculture and agricultural sciences in Bulgaria during the years of the people's power. B. Popov, ed. Sofia: Academy of Agricultural Sciences Bulgaria, Center for Scientific, Technical and Economic Information in Agriculture and Forestry, 1969. 118 pp.

Bazala, Razvigor. *The agricultural economy and trade of Bulgaria* (ERS-Foreign 256). Washington, D.C.: USDA, 1969. IV+16 pp.

Chary, Frederick B. "Agrarians, radicals, socialists, and the Bulgarian peasantry." *The peasantry* . . . (see I.1), vol. 1: 35–55.

Dimitrov, A., and Atanasova, G. "Food consumption in Bulgaria." *Eastern European Economics* 3 (1964/65) 1:60–67. (Transl. from *Ikonomicheska Misul*, no. 3, 1963.)

Dobrin, Bogoslav. *Bulgarian economic development since World War II.* New York, Washington, London: Praeger, 1973. XV+185 pp. (therein, pp. 43–87: "Agriculture").

Draganov, Mincho. "On the social-psychological typology of the patriarchal-traditional Bulgarian peasant." *Contemporary sociology in Bulgaria.* Sofia: Institute of Sociology, Bulgarian Academy of Sciences, 1978: 347–60.

Draganov, Rashko. "The auxiliary plot of the cooperative farmers." *East European Economics* 3 (1964) 1: 50–59. (Transl. from *Ikonomicheska Misul*, no. 4, 1963.)

Feiwel, George R. "Economic reform in Bulgaria." *Osteuropa Wirtschaft* 24 (1979) 2: 71–91.

Ganev, Atanas. *Implementation of Lenin's co-operative plan and the development of agricultural farming in Bulgaria during the years of people's power.* Sofia: Agricultural Academy "G. Dimitrov", 1974. 43 pp.

Georgeoff, John. "Rural education in Bulgaria: contemporary developments and policies." *The peasantry* . . . (see I.1), vol. 2: 123–39.

Georgiev, H. *The Bulgarian Agrarian Union. Seventy years since its foundation.* Sofia: Sofia-Press, 1970. 31 pp.

Hoffman, George W. "Transformation of rural settlement in Bulgaria." *Geographical Review* 54 (1964) 1: 43–64.

Ilieva, Nikolina. "Planning and specialization in Bulgarian agriculture." *Eastern*

European Economics 5 (1967) 3: 27–38. (Transl. from *Ikonomicheska Misul,* no. 6, 1965.)

Iordanov, T. "The application of agricultural profiles in land utilization maps." *Land utilization* . . . (see I.1.): 59–66.

Jankoff, Dimiter A. *Labor cooperative farms in Bulgaria.* New York: National Committee for a Free Europe, 1953. 17 pp.

Kunchev, Khristo. "Better utilization of the land." *Eastern European Economics* 1 (1962/63) 2: 38–45. (Transl. from *Planovo Stopanstvo i Statistika,* no. 4, 1962.)

Kyurkchiev, Mincho. *Bulgarian agriculture today and life in the villages.* Sofia: Sofia Press, 1974. 48 pp.

Kyurkchiev, Mincho. *The general and the specific in the reconstruction of agriculture.* Sofia: Sofia Press, 1971. 52 pp.

Kyurkchiev, Mincho. *A new road for Bulgarian agriculture.* Sofia: Foreign Languages Press, 1962. 75 pp.

Makhlebashiev, Iv. "Perspectives for purchase of the stock of machine-tractor stations by labor-cooperative farms." *Eastern European Economics* 1 (1962/63) 2: 45–49. (Transl. from *Finansi i Kredit,* no. 1, 1962.)

Mills, Theodora. *Bulgaria. Foreign agricultural trade* (ERS-Foreign 104). Washington, D.C.: USDA, 1964. 9 pp.

Mouzelis, Nicos. "Greek and Bulgarian peasants. Aspects of their sociopolitical situation during the interwar period." *Comparative Studies in Society and History* 18 (1976) 1: 85–105.

Nissan, Oren. *Revolution administered: agrarianism and Communism in Bulgaria.* Baltimore, London: John Hopkins Univ. Press, 1973. XV+204 pp.

Organization and management of co-operative farms in Bulgaria. Transl. by Z. Stankov, K. Noneva, and L. Dimitrova. Sofia: Sofia Press, 1968. 380 pp.

Rachinski, Todor, and Gotsov, Kosta. *Wheat breeding in Bulgaria.* Sofia: National Center for Scientific and Technical Information in Agriculture, Food Industry and Forestry, 1978. 111 pp.

Radoykov, Vladimir. "Economic regulation of the relations between industry and agriculture." *Eastern European Economics* 9 (1970/71) 2: 132–51. (Transl. from *Ikonomicheska Misul,* no. 2, 1970.)

Sanders, Irwin T. "Dragalevtsy household members then ⟨1935⟩ and now." *Population and migration trends in Eastern Europe.* Huey Louis Kostonick, ed. Boulder, Colo.: Westview Press, 1977: 125–33.

Severin, R. Keith. *Bulgaria's agricultural economy in brief* (ERS-Foreign 136). Washington, D.C.: USDA, 1965. 11 pp.

Stillman, Edmund O. "The collectivization of Bulgarian agriculture." *Collectivization* . . . (see 1.1): 67–98.

Syulemezov, Stoyan. *The cooperative movement in Bulgaria.* Transl. by Z. Stankov. Sofia: Sofia Press, 1976. 214 pp.

Tsvetkov, Georgi. "The proportion between industry and agriculture." *Eastern European Economics* 3 (1964/65) 1: 26–36. (Transl from *Ikonomicheska Misul,* no. 4, 1964.)

Zagorski, Nikolai. *Peasant life in Bulgaria.* Sofia: Foreign Languages Press, 1964. 86 pp.

Zakhariev, I. A. "Bulgarian industrialization and farm mechanization under socialism." *Economic development for Eastern Europe. Proceedings of a conference held by the International Economic Association.* London, New York: 1968: 33–45.

VIII.2. Agro-industrial integration

Bulgaria's agro-industrial complexes after seven years (Radio Free Europe Research. Background Report. Bulgaria 34). Munich: 1977. 16 pp.

Ganev, A., and Videv, V. *Agro-industrial complexes (AIC), a form of expanding the concentration and specialization of agriculture in the People's Republic of Bulgaria.* Sofia: Center for Scientific, Technical and Economic Information in Agriculture and Forestry, Agricultural Academy "G. Dimitrov", Bulgaria, 1973. 22 pp.

Industrialization of agriculture. Bulgaria's experience (International round-table ministerial meeting of the developing countries, held from May 20–24, 1976 in Varna, Bulgaria). Sofia: Sofia Press, 1977. 53 pp.

Kalchev, Kalcho K. "Some social and economic problems of agrarian-industrial complexes (AIC) and the developed socialist society in the People's Republic of Bulgaria." *Problems of the development of agriculture and information on the state of rural sociology in various countries.* Warsaw: Institute of Philosophy and Sociology, Polish Academy of Sciences, 1971: 55–58.

Kostadin, Ivanov. "Some problems of formation and distribution of income in the agroindustrial complexes." *Eastern European Economics* 11 (1972/73) 2: 64–79. (Transl. from *Planovo Stopanstvo*, no. 1, 1971.)

Prumov, Ivan. *Bulgarian agriculture today.* 2nd rev. ed. Sofia: Sofia Press, 1976. 107 pp. (1st ed., year not indicated, 92 pp.)

Wiedemann, Paul. "The origins and development of agro-industrial development in Bulgaria." *Agricultural policies . . .* (see I.1): 97–136.

Zhivkov, Todor. *Green light to complex mechanization, industrial technology and methods in agriculture.* Sofia: Sofia Press, 1969. 23 pp.

IX. Albania

Albania. Stavro Skendi, ed. New York: F. A. Praeger, 1956. XIII+389 pp. (therein, pp. 148–72: "Agriculture").

Bardhoshi, Besim, and Kareco, Theodhor. *The economic and social development of the People's Republic of Albania during thirty years of people's power.* Tirana: "8 Nëntori", 1974. 247 pp. (therein, pp. 98–147: "The development of the productive forces in agriculture along socialist lines").

The development of agriculture in the People's Republic of Albania. Tirana: Naim Frashëri, 1962. 55 pp.

Fusonie, Alan E. "An experiment in foreign agricultural education in the Balkans, 1920–1939." *East European Quarterly* 8 (1974) 4: 479–93. (On Charles T. Erickson.)

Hall, D. R. "Some development aspects of Albania's fifth 5-year plan." *Geography* 60 (1975) 2: 129–32.

Hoxha, Enver. "Forward towards new victories for the happiness and prosperity of our people and socialist fatherland (Report at the 6th Congress of the PLA, November 1–7, 1971)." *Albania Today* (Tirana) (1971) 1: 5–40.

Kaser, Michael. "Albania." *The new economic systems . . .* (see I.1): 251–73.

Kaser, Michael, and Schnytzer, Adi. "Albania—a uniquely socialist economy." *East European economies post-Helsinki . . .* (see I.1): 567–646.

Land's reform in the People's Republic of Albania. Tirana: Mihal Duri, 1961. 48 pp.

Marmullaku, Ramadan. *Albania and the Albanians.* Transl. from the Serbo-Croatian by M. and B. Milosavljević. London: Ch. Hurst, 1975. 178 pp. (therein, pp. 82–113: "The family and tribal tradition").

Osborne, Richard H. "Albania." *East Central Europe. An introductory geography.* New York: Praeger, 1967: 71–90.

Pali, Z. "Total collectivization of agriculture in Albania." *International Peasant Union Bulletin* 9 (1958): 25–27.

Shehu, Mehmet. "For the further and uninterrupted development of the revolution, for the successful construction of socialism in Albania. (Report at the 6th

Congress of the P.L.A., November 1–7, 1971)." *Albania Today* (Tirana) (1971) 1: 41–54.

Shehu, Mehmet. *Report on the directives of the 6th Congress of the Party of Labor of Albania for the 5th five-year plan (1971–1975) of economic and cultural development of the People's Republic of Albania.* Tirana: Naim Frashëri, 1971. 153 pp.

Shehu, Mehmet. *Report on the directives of the 7th Congress of the Party of Labour of Albania for the 6th five-year plan (1976–1980) of economic and cultural development of the People's Republic of Albania.* Tirana: "8 Nëntori", 1976. 120 pp.

Stillman, Edmund O. "A note on Albanian collectivization." *Collectivization . . .* (see I.1): 97–102.

Twenty years of socialism in Albania. Tirana: Naim Frashëri, 1974. 127 pp.

Vokopola, Kemal Aly. "Albania." *Government, law and courts . . .* (see I.1): 1725–39.

X. Yugoslavia

X.I. Agriculture in general

Aćimović, Viktor. "Consumption and production of fertilizers, 1961–1975." *Yugoslav Survey* 18 (1977) 1: 59–68.

"Agricultural land and its use." *Yugoslav Survey* 5 (1964) 16: 2293–2302.

Agricultural policy in Yugoslavia. Paris: Organisation for Economic Cooperation and Development, 1973. 47 pp. (New edition: 1980.)

"Agricultural production in Yugoslavia 1957–1959." *Yugoslav Survey* 1 (1960) 2: 193–206.

Allcock, John B. "The socialist transformation of the village: Yugoslav agricultural policy since 1945." *Agricultural policies . . .* (see I.1): 199–217.

Avdalović, Slavko. "The food industry, 1968–1973." *Yugoslav Survey* 16 (1975) 1: 97–110.

Brashich, Ranko M. *Land reform and ownership in Yugoslavia, 1919–1953.* New York: Mid-European Studies Center, 1954. VI+169 pp.

Brashich, Ranko M. *Taxation in Yugoslavia's agriculture.* New York: Mid-European Studies Center, 1953. 27 pp.

D'Andrea Tyson, Laura. "The Yugoslav economy in the 1970s: a survey of recent developments and future prospects." *East European economies post-Helsinki . . .* (see I.1): 941–96 (therein, pp. 967–70: "Agricultural performance and policies").

Dobos, Manuela. "The Nagodba and the peasantry in Croatia-Slavonia." *The peasantry . . .* (see I.1), vol. 1: 79–107.

Dovring, Folke. "Land reform in Yugoslavia." *A. I. D.* [Agency of International Development] *spring review of land reform, June 1970.* Country papers. 2nd. ed. vol. 10. Washington, D. C.: Department of State, 1970. 62 pp.

Drače, Džemal. "Development of agriculture, 1945–1970." *Yugoslav Survey* 11 (1970) 4: 13–40.

Drače, Džemal. "Development of agriculture." *Yugoslav Survey* 17 (1976) 1: 45–66.

Đurbabić, Marko. "Taxation of private farmers." *Yugoslav Survey* 9 (1968) 4: 109–22.

Hoffman, George W. "Agriculture and forestry." *Südosteuropa-Handbuch. vol. 1: Jugoslawien.* K. D. Grothusen, ed. Göttingen: Vandenhoeck and Ruprecht, 1975: 254–74.

Hoffman, George W. "Changes in the agricultural geography of Yugoslavia." *Geographical essays on Eastern Europe.* N. J. G. Pounds, ed. Bloomington: Indiana University Publications, 1961: 101–40.

Hoffman, George W. "Yugoslavia: changing character of rural life and rural economy." *The American Slavic and East European Review* 18 (1959) 4: 555–78.

Horvat, Branko. "The postwar evolution of Yugoslav agricultural organization. Interaction of ideology, practice, and results." *Eastern European Economics* 12 (1973/74) 2. 106 pp. (Transl. of *Projekt. Paritetne cene poljoprivredno-prehrambenih proizvoda i mehanizam njihovog odrzavanja.* Belgrade: 1973.)

Integrated food processing in Yugoslavia (Report of Seminar and digest of technical papers, Novi Sad, Yugoslavia, November 4–28, 1968). New York: United Nations Industrial Development Organization, 1970. 120 pp.

Kardelj, Edward. *Problems of socialist policy in the countryside* (From the Serbo-Croatian original). London: Lincolns-Prager, 1962. 303 pp. (Earlier Yugoslav edition in English: Belgrade 1960; contains only Kardelj's speech of May 5, 1959.)

Klein, George, and Patricia, V. "Land reform in Yugoslavia: two models." *The peasantry* . . . (see I.1), vol. 2: 39–59.

Klemenčič, Marijan. "Abandonment of agricultural land in the phase of transition from agrarian to industrial society. A case study of Slovenia." *Transformation of rural areas* (Proceedings of the 1st Polish-Yugoslav Geographical Seminar, Ohrid, May 24–29, 1975). Warsaw: Institute of Geography and Spatial Organization, PAN, 1978: 185–92.

Kuzmanovski, Tome. "Land irrigation and drainage." *Yugoslav Survey* 10 (1969) 2: 121–26.

Kuzmanovski, Tome. "Water management." *Yugoslav Survey* 18 (1977) 4: 79–98.

Lambert, Miles J. "Yugoslav agricultural goals for 1976–85." *The ACES Bulletin* 20 (1978) 3/4: 37–64.

The legal status of agricultural land (Collection of Yugoslav laws, vol. 1). B. T. Blagojević, ed. Belgrade: Institute of Comparative Law, 1962. 91 pp.

Marsenić, Dragutin V. "The price system—development and problems." *Yugoslav Survey* 14 (1973) 3: 141–54.

Mlinar, Zdravko. *Local government and rural development in Yugoslavia.* Ithaca, N. Y.: Center for International Studies, Cornell University, 1974. VI+125 pp.

Mlinar, Z. "Rural industrialization in Yugoslavia." *The role of group action* . . . (as for Pospichal, J., see V. 1., above): 166–76.

Neal, Fred Warner. *Titoism in action. The reforms in Yugoslavia after 1948.* Berkeley, Los Angeles: Univ. of California Press, 1958. XI+331 pp.

Pancer, Oton. "Land settlement in Yugoslavia." *International Journal of Agrarian Affairs* 2 (1955) 1: 76–90.

Rackov, Radovan. "The agricultural machinery industry." *Yugoslav Survey* 12 (1971) 1: 59–68.

Radenković, Budimir. "Agricultural land in Yugoslavia." *Yugoslav Survey* 12 (1971) 3: 41–52.

Radovanović, B. "Integration of agriculture and industry in Yugoslavia." *The role of group action* . . . (as for Pospichal, J., see V. 1., above): 83–90.

Radovanović, Boris. "Land policy in Yugoslavia." *Toward modern land policies.* D. McEntire, D. Agostini, eds. Institute of Agricultural Economics and Policy, University of Padua [after 1970; year of publication not indicated].

Radovanović, Milija. "Impact of the economic reform on agriculture." *Yugoslav Survey* 10 (1969) 3: 65–70.

Rašić, Petko. *Agricultural development in Yugoslavia.* Belgrade: 1955. 76 pp.

Report of the Study Team on the Working of the Cooperative Movement in Yugoslavia and Israel. Ministry of Community Development and Co-operation, Government of India. Delhi: Manager of Publ., 1960. III+118 pp.

Rosenblum-Cale, Karen. "After the revolution: women in Yugoslavia." *The peasantry* . . . (see I.1), vol. 2: 161–82.

"Rural Development in Yugoslavia." The Federal Board of the Socialist Alliance of the Working People of Yugoslavia. Commission for International Co-operation and Relations. *Information Bulletin* 1963. 1: 26–39.

Sirc, Ljubo. "Agriculture in Yugoslavia since 1945." *Pakistan Economic Journal* 10 (1960) 4: 73–95.

Stanojevič, A. K., and Obradovič, M. R. "Errors in Yugoslav censuses of agriculture." *Bulletin of the International Statistical Institute* 45 (1973) 2: 466–71.

Tadić, Dobrosav. "Changes in the countryside, 1961–1969." *Yugoslav Survey* 10 (1969) 4: 1–12.

Tomić, Dušan. "Agriculture and scientific approach to the price policy of agricultural products." *Contributed papers read at the 15th International Conference of Agricultural Economists.* (Papers 1–17). *International Journal of Agrarian Affairs,* double no. 1974–75. Oxford: 1975; 29–36.

Tomić, Dušan. "Decision-making in the self-management system in Yugoslav agriculture at the macro- and micro-level." *Decision-making and agriculture . . .* (see I.1): 267–75.

Tomić, Dušan. "Prices of agricultural products." *Yugoslav Survey* 14 (1973) 1: 43–54.

Translation of Yugoslav laws on agrarian reform and colonization. Prepared by Mid-European Law Project, Law Library, Library of Congress. [Washington, D.C.]: Mid-European Studies Center, 1952. 14 pp. (Contains the decrees and laws of Nov. 21, 1944; August 23, 1945; as amended by the laws of March 18, 1948 and Nov. 26, 1947.)

Trouton, Ruth. *Peasant renaissance in Yugoslavia 1900–1950. A study of the development of Yugoslav peasant society as affected by education.* London: 1952. Reprint, Westport, Conn.: Greenwood Press, 1973. XIII+344 pps.

Vandal, Katja. "Distribution of agricultural land by size of holding and sector of ownership, 1960–1970." *Yugoslav Survey* 13 (1972) 4: 93–104.

Zaninovich, M. G., and Brown, D. A. "Politics, modernization, and the Yugoslav peasantry." *The process . . .* (see I.1.): 289–321.

X.2 Collectivization and cooperation

The agricultural cooperative system in Yugoslavia. Selected articles. Transl. from the Serbo-Croatian by A. Vujović (et al.). Belgrade: General Federation of Yugoslav Agricultural Cooperatives, 1961. 117 pp.

"Agriculture in Yugoslavia. Problems of collectivization." *The World Today* 6 (1950) 11: 469–80.

The associated labour act. Transl. by M. Pavičić. Belgrade: Secretariat of Information of the SFR of Yugoslavia Assembly, 1977. 419 pp.

Bajalica, Dimitrije. *Agricultural cooperative in Yugoslavia.* Transl. from Serbo-Croatian by M. Pavićević. Belgrade: General Federation of Yugoslav Agricultural Cooperatives, 1961. 59 pp.

Digby, Margaret. "Agricultural co-operation in Yugoslavia.' *Yearbook of Agricultural Co-Operation.* Oxford: Blackwell, 1967: 162–75.

"General agricultural co-operatives. Conditions and development up to 1944." *Bulletin of Information and Documentation* (Belgrade) 1 (1952) 2: 3–8.

Lopandić, Dušan. "Productive co-operation in Yugoslav agriculture." *Bulletin of Information and Documentation* (Belgrade) 9 (1960) 2: 1–17.

Lopandić, D. "Yugoslav co-operation as the principal investor in agriculture." *Bulletin of Information and Documentation* (Belgrade) 7 (1958) 2: 9–25.

Miller, Robert F. "Group farming practices in Yugoslavia." *Cooperative and commune . . .* (see I.1): 163–97.

Šekularac, B. "Organization of agricultural co-operatives in Yugoslavia." *Bulletin of Information and Documentation* (Belgrade) 7 (1958) 2: 26–36.
Tomasevich, Jozo. "Collectivization of agriculture in Yugoslavia." *The collectivization* . . . (see I.1): 166–91.

X.3 Individual crops and animal production

Avdalović, Slavko. "Sugar beet and sugar production, 1970–1976." *Yugoslav Survey* 19 (1978) 3: 99–106.
Avramov, Lazar. "Viticulture and wine production." *Yugoslav Survey* 15 (1974) 4: 83–92.
Mastilović, Stana. "Livestock farming, 1969–73." *Yugoslav Survey* 15 (1974) 4: 59–70.
Širadović, B., and Avdalović, S. "Grape growing and processing." *Yugoslav Survey* 12 (1971) 2: 89–98.
Vujanić, Nenad. "Production and consumption of wheat." *Yugoslav Survey* 14 (1973) 1: 55–68.

X.4 Social patterns, income, living conditions

Baletić, Z., and Baučić, I. *Population, labour force and employment in Yugoslavia 1950–1990* (Wiener Institut für Internationale Wirtschaftsvergleiche. Forschungsberichte 54). Vienna: 1979. 91 pp.
Benc, M., and Cvjetićanin, V. "Part-time farmers' basic attitudes." *Yugoslav Survey* 20 (1979) 1: 9–24.
Current problems of the village in Yugoslavia (IVth Plenary Session of the Federal Board of the Socialist Alliance of the Working People of Yugoslavia). Belgrade: "Review of International Affairs," 1963. 41 pp.
Erlich, Vera S. *Family in transition. A study of 300 Yugoslav villages.* Princeton, N.J.: Princeton Univ. Press, 1966. XIX+469 pp.
First-Dilić, Ruza. "The life cycle of the Yugoslav peasant farm family." *The family life cycle in European societies.* (New Babylon Studies in the Social Sciences. 28.) The Hague, Paris: 1977: 77–91.
First-Dilić, R. "The productive roles of farm women in Yugoslavia." *Sociologia Ruralis* 18 (1978) 2/3: 125–39.
Gallagher, Robert P. "Yugoslav migrant labor." *The ACES Bulletin* 16 (1974) 3: 3–17.
Golob, Matija. "Some interest trends of Yugoslav rural sociology." *Problems of the development of agriculture* . . . (as for Kalchev, Kalcho K., see VIII.2., above): 195–202.
Halpern, Joel M. *The changing village community.* Englewood Cliffs, N. J.: Prentice-Hall, 1967. VIII+136 pp.
Halpern, Joel M. "Farming as a way of life: Yugoslav peasant attitudes." *Soviet and East European agriculture.* (see I.1): 356–81.
Halpern, Joel M. "Memoirs of recent change: some East European perspectives." *The process* . . . (see I.1.): 242–68.
Halpern, Joel M. "Peasant culture and urbanization in Yugoslavia." *Contributions to Mediterranean sociology* (Acts of the Mediterranian Sociological Conference, Athens, July 1963). J.-G. Peristiany, ed. Paris, The Hague: 1968; 289–311. (Also in *Human Organization* 24 (1965) 2: 162–74.)
Hoffman, George W. "Migration and social change." *Problems of Communism* 22 (1973) 6: 16–31.
Khaled, Abdallah. "Farm labour's use and its wage policies in Yugoslav and Dutch agriculture." *L'Egypte contemporaine* 55 (1964) 315: 15–53.

Kitaljević, Boško. "Receipts and expenditure of rural households." *Yugoslav Survey* 10 (1969) 2: 133–38.

Krašovec, Stane. "Peasant workers and the social structure." *The effect of current demographic change in Europe on social structure* (Proceedings of the 3rd European Population Seminar, Beograd, September 26–29, 1978). Belgrade: 1979; 195–203.

Lockwood, William G. "The peasant-worker in Yugoslavia." *The social structure* ... (see I.1.): 281–300.

Marković, Petar. "Buildings and housing conditions in the countryside." *Yugoslav Survey* 18 (1977) 4: 99–112.

Nikolić, Miloje. "Yugoslav agricultural labour temporarily employed abroad." *Yugoslav Survey* 14 (1973) 2: 15–38.

Smolić-Krković, N., Milinković, D., and Visinski, A. *Longitudinal study of the social-economical state of aged people in the rural areas of SR Croatia, 1972–1976. Final report.* Zagreb: Republic Institute of Social Work, 1976. 143 pp.

"Socio-economic status of farmers and independent personal work with privately-owned resources." *Yugoslav Survey* 15 (1974) 3: 32–35.

Šuvar, S., and Puljiz, V. "The role of rural sociology in Yugoslav agrarian policy." *Sociologia Ruralis* 11 (1971) 1: 66–74.

Todorović, Srbislav. "Income of agricultural households." *Yugoslav Survey* 14 (1973) 1: 69–84.

Vedriš, Mladen. "Modern migration movements from Yugoslavia to West European countries." *Yugoslav Survey* 20 (1979) 2: 87–102.

Wertheimer-Baletić, Alica. "Agricultural population." *Yugoslav Survey* 19 (1978) 2: 55–70.

The Yugoslav village (special issue of *Sociologia Sela*). V. Puljiz, ed. Zagreb: Department of Rural Sociology, 1972. 261 pp.

X.5 Regional aspects

The agriculture of the Socialist Republic of Serbia. Belgrade: Prosveta, 1971. 290 pp.

Bennet, Brian Carey. *Sutivan, a Dalmatian village in social and economic transition.* Southern Illinois University, Dissertation, 1971. XVII+262 pp.

Crkvenčić, Ivan. "Some important socio-geographical processes in Croatia." *Geographical Papers.* (Institute of Geography, University of Zagreb) 1 (1970): 55–68.

Economic management of state farms in the Kosmet, (Yugoslavia). (O.E.C.D. Publications, No. 20,651). Paris: 1966. 93 pp.

Hammel, E. A. "The 'Balkan' peasant. A view from Serbia." *Peasants in the modern world.* Philip K. Bock, ed. Albuquerque: U. of New Mexico Press, 1969: 75–98.

Hammel, E. A. "Economic change, social mobility, and kinship in Serbia." *Southwestern Journal of Anthropology* 25 (1969) 2: 188–97.

Johnston, W. B., and Crkvenčić, I. "Examples of changing peasant agriculture in Croatia, Yugoslavia." *Economic Geography* (Clark University, Worcester, Mass.) 33 (1957) 1: 50–71.

Klemenčić, Vl. "Indicators of the urbanization of agrarian settlements in Slovenia, Yugoslavia. *Urbanization in Europe* (European Regional Conference of the International Geographical Union). Budapest: 1975; 271–78.

Lockwood, William G. "Social status and cultural change in a Bosnian Moslem village." *East European Quarterly* 9 (1975) 2: 123–34.

Medved, J. "The types of the changes in the land use in the S.R. Slovenia." *Agricultural typology and agricultural settlements* (Papers of symposium held in Szeged and Pécs, August 15–19, 1971). Szeged: 1972: 253–64.

Panov, Mitko. "Social and economic transformation of rural areas in Macedonia." *Geographical Papers*. (Institute of Geography, University of Zagreb) 1 (1970): 155–64.

Puška, Aslan. "The distribution of rural population. Structural changes and the impact of such distribution on the general development of Kosovo." *Transformation of rural areas* (Proceedings of the 1st Polish-Yugoslav Geographical Seminar, Ohrid, May 24–29, 1975). Warsaw: Institute of Geography and Spatial Organization, PAN, 1978: 193–200.

Report to the Government of Yugoslavia on land reclamation and organization of agricultural production in the Mirna River Basin, based on the work of J. G. Gillow. Rome: Food and Agriculture Organization of the United Nations, 1973. V+37 pp.

Winner, Irene. *A Slovenian village: Žerovnica.* Providence, Rhode Island: Brown Univ. Press, 1971. 267 pp.

Index